Technoromanticism

LEONARDO

Roger Malina, series editor

Richard Coyne, *Designing Information Technology in the Postmodern Age: From Method to Metaphor,* 1995
Richard Coyne, *Technoromanticism: Digital Narrative, Holism, and the Romance of the Real*, 1999
Michele Emmer (ed.), *The Visual Mind,* 1993
Craig Harris (ed.), *Leonard Almanac: International Resources in Art, Science, and Technology,* 1993
Craig Harris (ed.), *Art and Innovation: The Xerox PARC Artist-in-Residence Program,* 1999
Peter Lunenfield (ed.), *The Digital Dialectic: New Essays on New Media,* 1999
Mary Anne Moser (ed.), *Immersed in Technology: Art and Virtual Environments,* 1996

Technoromanticism

Digital Narrative, Holism, and the Romance of the Real

Richard Coyne

The MIT Press Cambridge, Massachusetts London, England

This book was set in 10.5 pt. Garamond #3 by Crane Composition Services, Inc.

Printed and bound in the United States of America.

Library of Congress Cataloging-in-Publication Date
Coyne, Richard.
Technoromanticism : digital narrative, holism, and the romance of the real / Richard Coyne.
p. cm. — (Leonardo)
Includes bibliographical references and index.
ISBN 0-262-03260-0 (alk. paper)
1. Information technology—Philosophy. 2. Holism. 3. Romanticism. I. Title. II. Series: Leonardo (Series) (Cambridge, Mass.)
T58.5.C687 1999
303.48'33—dc21 99-21241
CIP

To Adrian Snodgrass

Contents

Series Foreword

We are living in a world in which the arts, sciences, and technology are becoming inextricably integrated strands in a new emerging cultural fabric. Our knowledge of ourselves expands with each discovery in molecular and neurobiology, psychology, and the other sciences of living organisms. Technologies not only provide us with new tools for communication and expression, but also provide a new social context for our daily existence. We now have tools and systems that allow us as species to modify both our external environment and our internal genetic blueprint. The new sciences and technologies of artificial life and robotics offer possibilities for societies that are a synthesis of human and artificial beings. Yet these advances are being carried out within a context of increasing inequity in the quality of life and in the face of a human population that is placing unsustainable burdens on the biosphere.

The Leonardo series, a collaboration between the MIT Press and Leonardo/International Society for the Arts, Sciences and Technology (ISAST), seeks to publish important texts by professional artists, researchers, and scholars involved in Leonard/ISAST and its sister society, Association Leonardo. Our publications discuss and document the promise and problems of the emerging culture.

Our goal is to help make visible the work of artists and others who integrate the arts, sciences, and technology. We do this through print and electronic publication, prizes and awards, and public events.

To find more information about Leonardo/ISAST and to order our publications, go to the Leonardo Online Web at <http://www.mitpress.mit.edu/e-journals/Leonardo/home.html> or send e-mail to <leo@mitpress.mit.edu>.

Preface

Digital narratives place the invention and refinement of the computer at the pinnacle of scientific and technological accomplishment. Therefore, it may seem strange that digital narratives should draw so heavily on eighteenth- and nineteenth-century romanticism. This book examines the spectrum of romantic narrative that pervades the digital age, from McLuhan's utopian vision of social reintegration by electronic communications to the claims of cyberspace to offer new realities. The characters that populate these technoromances are putative digital identities, cyborgs, computerized agents, and avatars, under the unitary gaze of the global brain, the transcendent intelligence emerging as the digital network grows.

It is easy to show how romanticism encourages inflated expectations, diminishes tangible concerns with equipment and embodiment, promotes the heroism of the digital entrepreneur, and dresses conservative thinking in the guise of radicalism. But addressing technoromanticism in a critical light not only lessens its hold but reveals valuable insights into the computer and the digital age. This book engages in an imaginative game of "what if." What if we adopt the hard-nosed, commonsense alternatives to technoromanticism of empiricism and scientific rationalism? What are the consequences for digital narrative if we adopt the insights of pragmatic or structuralist theories of language, the praxical focus of Heidegger and the phenomenologists, Foucault's concept of bodily discipline, Freud's Oedipal condition, or Lacan's concepts of the real and desire? The unifying theme of our inquiry is hermeneutics, and we find that the computer serves as an aid to interpretation by providing a space for application and exploration. As well as countering romanticism, these excursions reveal much about the nature of narrativity and its consequences in the digital age.

The discussion identifies how romanticism deals with the perennial theme of unity, its identification with concepts of the real, and how contemporary agonistic narratives that speak of friction, dislocation, and schizophrenia supplant romanticism. The book also serves as a useful introduction to the application of contemporary theory to information technology, raising issues of representation, space, time, interpretation, identity, and the real in the digital age. As such it provides a companion volume to my earlier *Designing Information Technology in the Postmodern Age: From Method to Metaphor*, further integrating the insights of Heidegger, Derrida, Ricoeur, and Foucault, and introducing the provocations of Freud, Lacan, and surrealism to digital narrative.

Technoromanticism

Introduction

McLuhan identified the era of preliterate culture as a golden age in which humankind was one with itself and with nature. Speaking and listening in the absence of writing involved highly interactive exchanges that come close to directly sharing thoughts. Aural culture was tribal, engaged, practical, and unitary. Then followed the age of literacy. When we write, we lay things out in order and divide the world. Society under literacy is urban, global, and fragmented, rather than local, integrated, and whole. For McLuhan, information technologies are implicated in this shift between the whole and its individuation, or more generally, between unity and multiplicity. Initially the introduction of writing brought about the proliferation of individuation. Now we are entering a third age in which the incessant buzz of electronic communications returns us to a tribal state, but now the whole world is the tribe.[1]

McLuhan presents one of many variants of the narrative of unity and multiplicity that pervade IT (information technology) discourse. Similar narratives cluster around the four great artifices of the digital age: virtual communities, virtual reality, artificial intelligence, and artificial life. In narratives of virtual communities, people who have never met face-to-face are drawn together to participate in the global tribe through the media of electronic mail, on-line chat, computer role games, and video conferencing in ways similar to how conventional communities form, but without depending on spatial proximity, and in ways that obscure the divisiveness of issues of appearance and status. Virtual reality (VR) invites us to experience immersion in cyberspace, to move about in an endless sea of data. VR supposes that we can be immersed in virtual landscapes and virtual architectures, meet one another there and carry out conversations, develop intimacies with one another and with data, assume virtual identities, and be who and what we want to be. The language of virtual reality involves the unitary concepts of immersion and engagement. Various forms of artificial intelligence (AI) present the case for a unity between human and machine. The mind is treated as a kind of mechanism, or software within the hardware of the brain, so we can replicate and simulate aspects of mind in a computer. Conversely, we can study the mind in computational terms. More lately, some AI has overtly adopted the language of unity, maintaining that intelligent behavior emerges from the complex interaction of many simple subsystems, taking to heart the adage of systems theory that the whole is greater than the sum of the parts. The study of artificial life (AL) develops from this axiom, renouncing "centralized thinking." The way biological organisms organize themselves is apparently more akin to the way a colony of ants behaves. Each insect operates locally with no apparent plan for the whole, and yet the whole colony is able to construct complex, air-conditioned termite mounds. These emergent behaviors apparently challenge the need for centralized, hierarchical, and autocratic control structures, and AL researchers devise computer systems to manifest evolution, growth, and holistic behavior in artificial organisms.

Clearly the unity theme presents in various forms. In some cases, it presents as a simple numerical description: one whole (unity) and many parts (multiplicity). In some cases, it presents as a matter of the unity between elements, as in the parts of a system working as one, people united as community, or distinct categories of things such as humankind, nature, and machines working in harmony. In some cases, the unity theme also suggests the dissolution of the boundaries between categories, as in the unity between organism and machine, or body and mind. Sometimes unity is revealed as something one participates in experientially as a place, a state, a time, or a condition that one enters, as in cyberspace. Sometimes unity implies a state of freedom, or freedom within bounds, belonging in the past or in the future, or residing in another place. Or the unity state may transcend what is normally accessible to the senses, as an idea, or an ideal.

The technologies that support virtual communities, virtual reality, artificial intelligence, and artificial life also imply a certain self exaltation or conceit on the part of humankind, a presumption that we can have total control or omnipotence, play God, by simulating, mastering, redefining, manipulating, and controlling space, time, community, thought, and life. This presumed omnipotence can imply hierarchy, control, suppression, manipulation of others, delusion, and other manifestations of fragmentation and multiplicity. But it also implies an innocence, a return to a primal unity, to a state when we were truly omnipotent, our will was suffused with that of nature, and there was no differentiation between our demands and the whole, which, according to some commentators, is the state of early childhood. So the audacity and presumption of digital narratives also invoke the unity theme.

Digital narratives do not present on the theme of unity in isolation, but unity contrasted with multiplicity, particularly as understood pejoratively as fragmentation or disintegration. The fragmentation is either outside the world created by the technology or within it. Virtual communities are posited against the fragmentation of current social forms, or the failure of conventional mass media to realize the goals of producing an informed and active citizenry, a truly civil society. Virtual reality presents a world where you can be yourself, against a duplicitous world in which you have to conform to the expectation of others: a fake and fragmented world of similarly disconnected individuals. In a virtual world, you have instant access to any coordinate in data space. You can be here, there, or everywhere, unlike the limited, spatially, and logically constrained world we usually experience with its discontinuities and fragmentations. The new AI and AL present a holistic order, against the fractured methods of formal systems and hierarchical control. But these technological narratives also acknowledge the potential for fragmentation within their worlds: uneven access to computer systems, alienation from normal interactions with people and things, an imbalance in priorities, a privileging of objects and issues that are amenable to computer representations, the reinforcement of the status quo, problems of surveillance, and delusion—all of which point to alienation among ourselves, from nature, and from our machines. The rhetoric of unity also embraces its own apparent contradictions, through the aphorism of "unity in diversity." So virtual communities are said to allow differences to flourish, cyberspace is said to be fragmented and polyvalent as well as whole, and in the new AI and AL a global unity emerges from diversity, and vice versa.

The theme of unity and multiplicity is an ancient one, but it finds full flower in the writing of the romantics in the eighteenth and nineteenth centuries. The themes developed by the romantic writer Schlegel should be familiar. He reacted against the progressive subjection of the world to the rationalization and reduction of the early Enlightenment, complaining that in the modern age, under the influence of Cartesian rationalism and the growth in science, "the intellect exhausts itself in the study of individualities," resulting in the complete loss of "all idea of perfection in unity."[2] In so arguing, he articulated the romantic quest for the soul's reunion with nature. He contrasted the perfection to be found in unity with the fragmentation of individuality. He also appealed to a golden age in which all things were united under "the perfect consistency of the ancients," to be contrasted with our own "dismemberment."[3] He also pointed to the power of "instinct," which is beyond reason, and which, "beginning and ending in nature" can "unite nature with mankind."[4] He recognized the longing to transcend the world of the individual toward a unity: "There is in the human breast a fearful unsatisfied desire to soar into infinity."[5] The unity people seek is ineffable, beyond language, and nature is "ever mute, incomprehensible, unsympathising, and unconsoling."[6]

The rationalists had said that we should eschew feeling, emotion, and other distortions to knowledge in favor of pure reason. In contrast, the romantics rehabilitated and promoted feeling and emotion. Whereas the rationalists and empiricists debated the nature of reality and how we can know it, the romantics elevated the intangible world of the imagination. Rationalism and romanticism converged on the theme of liberation from tradition and authority and elevated the concept of the individual, who for the romantics was the source of creativity. Subjectivity was the key to art. For Schlegel, creative writing "must indeed be entirely personal, subjective in design and intention, conveying indirectly, and almost symbolically, the deepest individual feelings and peculiarities of the author."[7] Whereas rationalism sought the division of the world into parts, the second step of Descartes' method,[8] the romantics brought the "union of senses and imagination"[9] to bear, through the "special qualities of the soul,"[10] on that about which the emerging world of science was becoming increasingly mute: the issues of love, beauty, and the unity of all things. Romanticism constructed its narratives around the theme of unity and multiplicity, the whole and the parts, the one and the many, in ways that implicated the emerging Enlightenment notions of the self, individuality, and subjectivity, the latest demonstrations of which are found in the narratives of information technology.

The romanticism of digital narratives represents one of two antagonistic strands of the Enlightenment: rationalism and romanticism. We can readily identify digital narratives with both. Virtual communities, virtual reality, artificial intelligence, and artificial life demonstrate the influence, and even the triumph, of reductive rationalism, of the kind that the romantics sought to redress. Much of the discourse on virtual communities seems to suggest that access to community resides in communication, which in turn relies on the passage of information from one person to another. According to this view, as paradigmatic conduits of bits and bytes, networked computers grant privileged access to the formation of communities but by the isolation and transmission of individual communicable units.[11] Much virtual reality discourse assumes that we can construct correspondences between the world we inhabit and geometrical worlds defined using spatial coordinates, and that we can immerse ourselves in such spaces so that we are there in ways that mean as much as, and possibly more than, being in a physical place. Artificial intelligence can be construed as a recognition that at a fundamental level all understanding is grounded in number, symbol, and rule. Similarly, artificial life seems to presume the basis of all life in information and simple rules, a further reduction. Although antagonistic, the rationalist and romantic legacies are not so far from each other. They both start with Descartes' notion of the autonomous subject. Their continual antagonism seems to impel much of the intrigue with information technology, and further support its participation in the myth of unity and multiplicity. Many people eschew rationalism, but in doing so simply move to a romantic orientation, reworking old ground.

Many commentators, including myself, have dealt at length with the rationalist aspects of information technology, but its romantic form has received less critical attention.[12] This book is an attempt to address this lack. The book is divided into three sections. In the first, I examine narratives that develop the claims that we can transcend the constraints of the embodied world toward *unity* through the power of information technology. In the second section, I show how rationalism and empiricism speak explicitly of *multiplicity*, and the process of categorization and individuation against which the romantics railed, with particular emphasis on the role of language. Third, I take on the theme that the human condition is caught in the antagonism between these two, between unity and fragmentation, transcendence and order, the *ineffable* and the presumption of language, or in Lacan's terms, between the real and the symbolic order.

This is the narrative of this book, a romance in its own right, the story of how the myths of unity and disintegration have been variously translated into the forms we see in the digital age. Ancient myths of a transcendent reality of the whole and the parts caught in a cosmic antagonism have been transformed through Neoplatonism, rationalism, romanticism, and the technologies of an age enamored with the primacy of information. I will also show that this grand romance, this story of transformation, implicates the adherents of "commonsense" empiricism as much as the extreme devotees of new age technoromanticism. No one is entirely immune from the romantic legacy.

The concept of narrative is important to the theme of this book.[13] Why do I present IT commentary as an issue of *narrative* rather than description or explanation? The concept of narrative need not presume something to be described, nor does it presume artifice or construction. Narratives are evaluated primarily on criteria of efficacy, and their ability to disclose, prior to a consideration of their correspondence with a state of affairs. Contemporary understandings of narrative do not presume some notion of *facts* in distinction to the elements of a story.[14] Narrative operates all the way to the determination of facts. In this area of study, there is no essential computer science in distinction to the stories we construct about computers.

Narratives present as open ended, fully implicated in the hermeneutical circle, the process by which we interpret a situation, a text, an image, a work of art, or a narrative.[15] Insofar as narratives follow a structure,[16] it is the indeterminate hermeneutical process of excursion and return. There is a position from which the subject of the story departs: Alice ventures from the comfort of the sitting room (*Through the Looking Glass*), Pilgrim leaves the city of destruction (*Pilgrims Progress*), humankind embarks from the world of the tribe (McLuhan), and the cybernaut leaves the body (in cyberspace narratives). The encounters in the new world, the looking-glass world, the land of temptation, the Gutenberg galaxy, and the virtually real bring one back to the start, which is a world transformed: a world in which one can entertain the possibility of being a part of someone else's dream (*Through the Looking Glass*), the Celestial City is the world redeemed, the electronic age is again the age of the tribe, the disembodied virtual world is informed and challenged by the embodied. The cyclical process applies to the details of narrative as well. In looking-glass world, Alice encounters and reencounters the familiar chess pieces, cats and the paraphernalia of afternoon teas, rendered unfamiliar through various inversions, Pilgrim encounters traits of his former self as virtuous and untrustworthy traveling companions, McLuhan's Gutenberg world re-presents the interaction of the senses in different measure in each epoch, and the world of the disembodied cybernaut is already invested with the language of the body (front, back, in, out, up, down, prosthesis). The process by which one interprets narratives, and by which one constructs narratives as interpretations, follows the same cyclical structure. One approaches a text from a position, a point of view, a particular set of prejudices, which are transformed through the encounter with the text.

The narrative of this book inevitably bears the same cyclical structure. A conception of the ancient theme of unity and multiplicity provides a point of departure, to which the narrative returns periodically, culminating in the world as presented through digital narratives. But this is not the end of the story. The endpoint is also a rediscovery of the unity theme, transformed through the provocative insights of Lacan, among others, which in turn informs the concept of the hermeneutical circle. Another account of the hermeneutical process implicates the relationship between the whole and the parts: to understand the whole of a narrative you need to construct an understanding from each of the parts, but the parts do not make sense until seen in the context of the whole. By this formulation there is a hermeneutical circle, which presents as a paradox, or a vicious circle, addressed by various hermeneutical theorists.[17] So my theoretical position is that of hermeneutics, which will resurface periodically as a means of resolving some of the disputes between the antagonists in the grand narrative, and in the end, the hermeneutical process falls subject to its own scrutiny as an unresolved aporia.

What is the role of information technology in this drama? Information technology is intimately bound to language, and hence interpretation. IT discloses the strengths and limitations of various views of language. It operationalizes, as far as it is able, the correspondence view of language. If words correspond to things, then the words, codes, and symbol strings in a computer can represent the world and construct new worlds. If, by a more contemporary account, language trades in endless chains of reference, within a vast system of self-reference, then the linking of texts in global communications networks (hypertext) speaks profoundly of language and human practice.[18] If narrativity is at the core of the information technology world, then IT is subject to the workings of the hermeneutical process but is also disclosive of it. IT brings issues of language into sharp relief in ways unlike other technologies and, in the process, discloses aspects of the themes of unity and fragmentation.

The rest of this introduction traces the main narrative in greater detail, albeit in a cursory way, the nuances of which will be elaborated in the chapters that follow. Of necessity, the discussion takes us into areas of inquiry that at times seem removed from the immediate issues of information technology, but this broad, eclectic, and interdisciplinary investigation should reward us with a robust, invigorated, and informed understanding of our narratives in the digital age, and their consequences.

Unity: How IT Narratives Attempt to Transcend the Material Realm

In keeping with the romanticism of its narratives, information technology implicates itself in people's attempts to progress from one sphere of existence to another. The new sphere includes both the digital utopia promised by much IT commentary and literal transcendence through immersion in the "consensual hallucination" of the digital matrix, digital ecstasis, and participation in an ideal unity.

Digital Utopias

The return to a transformed golden age and the rhetoric of progress implicate digital narratives in the concept of utopia. The global village and the electronic cottage invoke a return to the ideal of preindustrial arts and crafts. Our induction into an egalitarian social order through electronic communications retells the message of early socialism and anarchism. The romantics reinvented the aesthetic, guilds, crafts, and feudal harmony of the medieval age. The IT world, from computer games to supposed anarchy on the net, similarly celebrates romantic medievalism, its tangled aesthetic, its sense of carnival, and the chaos of the marketplace. The dominant IT culture looks back to a golden age but is always projecting forward. In fact, the narratives of virtual communities, virtual reality, artificial intelligence, and artificial life seem to depend more on what is soon to be accomplished than on what is now possible.[19]

Cybernetic Rapture

Digital narratives represent the latest transformation of the theme of unity as initiated by Plato and the Neoplatonists and appropriated by the romantics. Plato divided the world into the realms of the material and the real, that is, the sensible and the intelligible. The material world is the world of the senses, where we are readily deceived by appearances. It is the world of particular things. The intelligible realm is the world of forms, categories, and the Good, where things do not change. The sensible world bears the imprint of the Intelligible. The Intelligible is the realm of the real. So begins an early account of the conflict between unity and multiplicity.[20] Plotinus adapted and amplified Plato's schema of the material and the real into a doctrine in which the soul seeks release from the body to join the unity of the real.[21]

Romanticism was also idealist in orientation, and the romantics read Plotinus. They readily equated the soul with individual genius, and they attributed to the unity of the real the source of creativity and beauty. Certain digital narrative is idealist and has taken to heart the Neoplatonic concept of ecstasis—release of the soul from the body—though here the soul is replaced with the mind, the means of ecstasis is immersion in an electronic data stream, and the realm of the unity is cyberspace. Cyberculture invokes a romantic apocalyptic vision of a cybernetic rapture, a new

Transcendence + consciousness.

electronically induced return to the unity, an age in which the material world will be transcended by information.

Multiplicity: The Empiricist Tradition of Realism, and Its Critics

As we have seen, according to Schlegel and other romantics, rationalism and empiricism distract the intellect with the study of "individualities" and conceal the perfection of unity. For the romantics this was highly undesirable, but from other positions the proliferation of parts and multiplicity speak of richness, complexity, and an inevitable and indeterminate profusion. We deal with narratives of multiplicity through an examination of the influences of empiricism, structuralism, and phenomenology on digital narratives.[22]

The Empiricist Legacy

If romanticism strives for the unity of all things, then the language of empiricism is of division and multiplicity. Thanks to the currency of the term "cyberspace," space provides a useful focus for the discussion of empiricist concepts of reality (as independent, divided, and ordered). The presence of technoromanticism is irrefutable in the extreme narratives of cyberpunk and the hyperbole of IT commentary. But a close examination of empiricism shows that rather than countering technoromanticism, empiricism provides the conditions for it to flourish, particularly through its development of notions of *representation*. Empiricism sustains a spectrum of positions on space, starting from representation: space as represented, resisted, reduced, and divided. Empiricism's sober reflections on the ability of the computer to represent space take us close to the romantic vision of cybernetic rapture, evident as we examine the spectrum of spatial narratives, from concepts of objective, empirical, and propositional space to narratives that call on relativity and quantum theory. Commentators on modern physics present the unity myth in terms of the conflict between simplicity and complexity, concepts that require the computer for their articulation.

The Symbolic Order

Representation is, after all, a matter of language. To acknowledge the primacy of language can help break the chain connecting representation to technoromanticism, empirical realism to techno-idealism, and unsettle the security of the romantic position. The use of computer networks as communications media, the AI project toward natural language understanding by computer, and the global hypertext of the World Wide Web bring issues of language to the fore in understanding IT. In this chapter, we examine major schools of thought on language, each of which carries different implications for how we understand IT. The correspondence theory of language seeks to divide the world into objects and label them with words. Although correspondence theory had support from positivism, it is now largely supplanted by the tenets of pragmatism and structuralism.[23] Structuralism is important if we are to understand the implications of poststructuralist writing on information technology, which ostensibly breaks the hold of the romantic tradition. Structuralism is also articulate on the subjects of myth and narrative, which provide ways of talking about the themes of unity and multiplicity, particularly as developed by critical theorists. Structuralism also provides the background to the provocative reflections on the real expounded by Lacan and others, to be developed in a later chapter.

Pragmatics of Cyberspace

In contrast to rationalist, empiricist, and romantic conceptions of computing, one can begin with the presupposition that computers and their accompanying technological systems are elements within constellations of practice. There are practices of designing, configuring, coding, distributing, using, teaching, and even writing about computers, and language is a practice. The orientation that begins with the *praxis* of computing is a phenomenological one. I introduce phenomenology through the issues of space and time.[24] The phenomenological position indulges the ubiquity of metaphor, as is evident on the issue of space. There are metaphors of space, and space provides metaphors for understanding other things. The primacy of mathematical and geometrical schemas by which we presume to represent the fundamentals of space in computer systems can be replaced by arguments from metaphor.[25] Phenomenology also puts IT utopian narrative in its place. Such narratives are reminders of our propensity to project ahead of ourselves.[26]

Ineffability: How Contemporary Narratives of Fractured Identities Challenge Technoromanticism

Technoromanticism seeks a new world order of unity through information.[27] Ancient and contemporary narratives do not point simply to the residence of the real in the whole or the parts, the unitary or the fragmented, but in the antagonism between the two, invoking narratives of rupture, paradox, and nondeterminacy. Surrealism and Freudian psychoanalysis make clear the applicability of these ideas to digital narratives.

Oedipus in Cyberspace

Surrealism was a movement in art and literature whose disciples drew on structuralism and phenomenology to develop provocative and potent cultural critique. The progeny of surrealism also develop critiques of subjectivity, challenging rationalism's notion of a unitary self. They also construct provocative narratives of unity and multiplicity that usurp the literal and limited discourses of romanticism. Certain IT commentators have adopted surrealist themes, invoking cyborgs, vision machines, bodies without organs, and other "objects" that are confounding and illogical from an empiricist point of view. We examine how these narratives function, beginning with surrealism and how it consorts with the psychoanalytic theories of Freud.[28]

The Oedipus myth and its variants retell the unity theme. Ostensibly, Freud's version is about the incestuous desire in the child (male) for union with the mother against the prohibition of the father.[29] For Freud, the Oedipus myth accounts for guilt and anxiety in the individual, but also in the formation of culture and society. The Oedipus myth is not hard to find in stories of political struggle and in fantasy and science fiction stories of evil genius, rebellion, and the quest for unity. Certain versions of the Oedipus myth are also evident in narratives of artificial intelligence and artificial life. The presumption of creating life from inanimate matter or information retells a version of the unity myth that involved the conflict between being born from the earth and being born of human lineage. The Oedipus theme discloses aspects of digital narratives that focus on identity.

Schizophrenia and Suspicion

In other versions of the Oedipus myth, notably those of Lacan, Irigaray, and even Freud himself, the Oedipus myth tells of attempts to return to a state of childhood innocence, before we were introduced to language, when we were indistinguishable from our parents and when we were omnipotent participants in a whole. This primal unity is the realm of the real for Lacan, and it is disrupted by the introduction of language, the symbolic order, which essentially divides the world. Whereas for empiricism, reality is what we represent in language, for Lacan the real is what resists symbolization.

The computer provides a potent metaphor for this sense of disconnectedness in the real, particularly through its putative presentation of openings into worlds, windows, and hyperlinks, which can return to themselves and which suggest a matrix of mirrorlike interreflections. The illusive and paradoxical nature of cyberspace provides one of the latest illustrations of the character of the real, which is not to privilege computer formations but to disclose what has been there all along in the real.[30]

The computer is inextricably associated with the symbolic order and is its apotheosis. As such, the computer exposes the real in certain ways through their mutual resistance. The computer is heavily implicated in contemporary narratives of rupture, nondeterminacy, and resistance surrounding unity and multiplicity, all but concealed by the romantic rendering.

Technoromantic Narratives

The final chapter enumerates some of the major digital narratives alluded to so far and summarizes what they reveal. We revisit the narratives of total immersion environments, digital communities, and the world of the cyborg, in both their empiricist and their romantic forms, as they pertain to unity and multiplicity. I also show how some narratives grant information technology a causal role in the transition from modernity to postmodernity, how such claims miss the point of postmodernity and further reinforce the hold of technoromanticism. The arguments presented here should provide a vigorous context against which we can test our understanding of the real in the information age, and through which new understandings of the computer will emerge and new narratives will be told.

There are several motivations for this study. Much is claimed of IT as an unsettling force that is overturning convention. Yet much of what is said about IT follows well worn paths of inquiry. We need to discern what is old from what is new. As well as empiricist and romantic narratives, there is an ad hoc "postmodern" rhetoric in which all distinctions seem to be lost, as though empiricist studies are but one ironical manifestation of a diverse postmodern whole. There are incommensurabilities and disagreements in digital narratives, many of which have yet to surface, but their airing can only be productive. IT culture and IT technology are inextricably entwined, and so the narratives we construct are consequential in the developments that take place. Technicians, chip designers, systems developers, inventors, researchers, managers, educators, entrepreneurs, legislators, commentators, and users are all implicated in digital narratives, not just on their days off as they entertain empiricist or romantic speculations, but in the IT praxis of which they are a part. Digital narratives are influential in the kinds of products and systems we create and demand.

This book presents a further vindication of the philosophical pragmatism that informs some IT research and development, and which I have addressed elsewhere.[31] But here I unravel how and where this pragmatism is easily caught in romanticism. In the process of this inquiry, we are able to canvas thinking on realism, idealism, phenomenology, empiricism, representation, language, surrealism, and psychoanalysis and, in passing, we broach the great themes of space, time, and identity as they impinge on the world of computing. Finally, digital narratives tell of our current ontology. At the transition between millennia, we are pouring vast resources into constructing and working out elaborate narratives, wrestling with unity and multiplicity. We need to understand what is at stake in the enterprise.

Unity

How IT Narratives Attempt to Transcend the Material Realm

1

Digital Utopias

Digital narratives are ever expectant. The unity of which digital narratives speak (harmonious digital communities, immersion in cyberspace, holistic lifelike systems, the unity of the animate and the inanimate) reside in the future. Digital narratives commonly emphasize what will be accomplished while downplaying current achievements, which are inevitably more modest than the predictions. The grand narrative in this romantic teleology[1] is of time-dependent progress, a surplus of expectation. As I will show, an analysis of the nature of such narratives is even more revealing than deciding whether the predictions will turn out to be true. A study of the nature of digital utopias also serves as an introduction to the perennial theme of unity and multiplicity.

Digital narratives, as narratives of progress, commonly take the form of extravagant predictions of unlikely outcomes, such as that which appeared in an article identifying developments in virtual reality: "Within a decade people will be taking utterly realistic virtual vacations to other countries—or even other worlds."[2] But such narratives also assume more subtle forms. In a well-known commentary of virtual reality, rather than focus on the current limited performance of headsets and data gloves, Rheingold rapidly moves on to what he regards as the interesting aspects of the technology: "In the future, less intrusive technologies will be used to create the same experience, and the computers will be both more powerful and less expensive, which means the virtualities will be more realistic and more people will be able to afford to visit them."[3] His extensive survey of institutions and corporations who are developing virtual reality systems in the here and now (as it was in the early 1990s) does not so much describe current achievements as it does future ones. And the future promise is of egalitarian access. In the same way, in a commentary on how information technology influences the built environment, Mitchell effects a ready transition to the future: "With higher band widths, ever-greater processing power, and more sophisticated input/output devices designed to take advantage of these capabilities, the boundary that has traditionally been drawn by the edge of the computer screen will be eroded. Through head-mounted stereo displays . . . or through holographic television (it's coming), you will be able to immerse yourself in simulated environments instead of just looking at them through a small rectangular window."[4] The message is of a future rather than an "actuality," in which the boundaries between human and machine will vanish.[5]

Such digital narratives suggest that computer technology will usher in a better future. This is particularly the case in discussions of Internet communications, which are claimed to bring us closer as a community. According to Sullivan-Trainor: "Technology today is taking the form of the information superhighway, a concept with the goal of exchanging ideas, information and commerce. The vision of this technology is no less than easy access for anyone, anywhere. Nearly unlimited business opportunities will be opened for the average person."[6] These statements introduce the major characteristics of the IT future: it is unified, fair, egalitarian, and highly productive.

Some narratives present such confidence in the ability of computer development to realize its promise that they conflate the distinction between potentiality and actuality. The difference between description and prediction is ambiguous in Heim's presumably ironical account of the cyberspace user: "The cybernaut seated before us, strapped into sensory input devices, appears to be, and is indeed, lost to this world. Suspended in computer space, the cybernaut leaves the prison of the body and emerges in a world of digital sensation."[7] Again, the specter is of an electronic unity (independent of the body), presented as a common occurrence but clearly not yet.[8]

As the media theorist James Carey cogently points out, dominant technologies feature in our narratives to reinforce the aspirations and the obsessions of an age. So steam railway transport was to unite the world, electricity was to be a force for peace, and now we have networked computers.[9] Some digital narratives present the expectation of a new leisure-filled democratic age. Unlike steam trains and electricity, however, IT has communication at its core. The technology is both the subject and the perpetrator of its narratives. IT's mouthpiece of the Internet, and the World Wide Web, presents IT as a technology of the Enlightenment, engaging the ideals of a literate, informed, and free-thinking citizenry.

Such digital narratives present themselves as progressive and radical, dealing as they do with reputedly leading edge technology. But the leading edge in these narratives is severely blunted. In what follows, I will identify the romantic conventionality evident in such digital narratives, considering other positions in subsequent chapters. In chapter 5, I will align the pragmatics of these narratives with Heidegger's radical concepts of time and expectation. Romantic digital narratives present on the subject of the utopia, a nostalgia for arts and crafts, a reverence for genius, a return to early socialism, a flirtation with systems theory and positivism, and the appropriation of the irrational.

Enlightenment and the Digital Utopia

Utopia, which means "no place," is a term popularized by Thomas More (1478–1535),[10] and one that resonates with those who hold to the ambiguous and unsettling nature of cyberspace. Many digital narratives are unabashedly utopian. Digital narratives share certain features with classical utopias. Utopias are literary forms that construct and describe an alternative world, in the future or somewhere far away, in which things are better than they are at the time of the writer.[11] The literary genre was especially popular in the nineteenth century, when utopians would each present their own version of the better world, each attempting to counter the utopia of the other. For example, William Morris's *News From Nowhere* was written at a time of worker unrest and general discontent with the social effects of industrialization.[12] The narrative presents a future socialist Britain in which modes of production based in the arts and crafts have taken over from factory production, and no one owns property. This utopia was largely a response to the popular American machine-oriented utopia *Looking Backward 2000–1887* of Edward Bellamy, of which Morris disapproved.[13] The concept of utopia is never far from its converse, the dystopia, and the components of one writer's utopia may constitute the dystopia of another.

Plattel provides a phenomenological account of utopias, presenting four main characteristics of utopian narratives that distinguish them from other genres such as science fiction and fantasy.[14] First, utopian narratives are moralistic. A utopian text is not just a diversion or entertainment but demonstrates how a particular ideology is to be worked out. It is a commentary on contemporary society, and it is intended to persuade people that things could be better and to spur them to action. Many digital narratives have this character, encouraging us to embrace and preserve the freedoms and opportunities IT offers.

Second, a utopian narrative presents a one-dimensional view of its alternative world, rarely engaging the tensions and contradictions in its own prognosis. Utopian writing is rarely reflexive and does not dwell on the nature of its own genre, a common feature of much unreflexive IT commentary and IT journalism.

Third, utopian writing focuses on description at the expense of story line and character development. The central characters of More's, Bellamy's, and Morris's utopias simply engage in protracted conversations through which the nature of the utopian world is disclosed. Twentieth-century science fiction and fantasy writing is therefore not strictly of the utopian genre, though it clearly borrows from it. George Orwell's *1984* presents a dystopic world dominated by bureaucratic surveillance.[15] William Gibson's *Neuromancer* presents dystopian and utopian aspects of a future in which people can plug into an electronically mediated world.[16] In these twentieth-century fictions, the future world is complex, and the reader is enjoined to be ambivalent about it, as the characters cope with situations peculiar to the strange world. Much IT commentary of the nonfiction variety fails to capture the ambivalence and irony of fiction, and it depicts the heroes of the IT world battling to bring utopia into being.

Fourth, utopian narrative is not strictly mythical. Plattel distinguishes between traditional myth and utopian thinking.[17] Within traditional myth the worlds depicted are beyond human control. The Garden of Eden and the Celestial City of the religious traditions are imposed orders. The new heaven and earth of the Bible are not human creations. Furthermore, much of the Judeo/Christian mythic tradition is eschatological, concerned with the end of time and some catastrophic eternal resolution of the tensions and conflicts of earthly life. The unitary world of the real descends to take over the divided sensible world.

Plato's *Republic* provides a description of a utopian world that ostensibly breaks with the traditional myth. The ideal state is not accomplished through the unity of the real taking over the world of the senses. At all levels, the ideal state participates in the unity of the real but without being replaced by it. (We will return to the mythic conception of the utopia in subsequent chapters when we discuss the legacy of digital "ecstasis.")

According to Plattel, the major transition from the traditional myth to the utopian narrative occurred in the Renaissance. In More's *Utopia*, published in 1516, humankind is responsible for building its better world: a republic with no aristocratic class and no ownership of property. In spite of its promotion of democracy and communalism, it is a society in which everyone knows his or her place. Slavery and violent punishment still exist, and people work hard and behave themselves because they are in community: "Everyone has his eye on you."[18] For More's early humanism, humankind is basically untrustworthy and lives within a regime of divine judgment—narrative that resonates with dystopian worlds under the repression of digital surveillance. But in the Renaissance, and in keeping with the discovery of the Americas, utopian possibilities are also discovered, not made. Traditional myth, eschatology, and utopia still go hand in hand.

For Plattel, the true utopia is an Enlightenment phenomenon. In Enlightenment utopias, reason takes over from obedience to authority and tradition. A utopian world is a thoroughly reasonable world. The complete humanistic utopia finds full flower in the nineteenth century. From then on, utopias are of the kind in which human beings are clearly in control by the exercise of reason. We make possibilities rather than discover them or have them imposed on us.[19] The residence of utopias in the Enlightenment concurs with the character of certain digital narratives, which present as transformations of the mythic utopia. Digital utopias are permeated by the hegemony of reason.

Enlightenment utopias differ from contemporary fantasies. For example, in Tolkien's *Lord of the Rings*, and derivative role games and computer games, control is out of the hands of the power of reason but resides with wizards and magicians,[20] a world dominated by unreason. The battles fought are no less the battles of the Enlightenment, but the world described is not yet liberated by it.[21]

At the core of utopian writing is the triumph of reason. Bellamy describes his future world as one in which people enjoy "the blessings of a social order at once so simple and logical that it seems but the triumph of common sense."[22] With the triumph of common sense comes freedom. For Morris, reason is also tied to freedom, authentic sensibility, and beauty, even human beauty: "there are some who think it not too fantastic to connect this increase of beauty directly with our freedom and good sense."[23] Technology also features prominently in the utopias of Bellamy and Morris. For Bellamy, a well-ordered system of industry and compulsory national service exists. For Morris, the future lies with simple Arts and Crafts technologies.

To what extent are contemporary digital narratives that speak of a better, more egalitarian world utopian? Aspects of digital narrative engage the Enlightenment preoccupation with reason in conjuring up an alternative world. A future is invoked where reason holds sway over unreason and disorder, in this case abetted by information technology, as illustrated in H. G. Wells's utopian vision of a "world brain." The world brain was to be a vast interconnected communications network with distributed knowledge bases that functioned as an aid to "human progress towards unity."[24] Without such a network, we are "inadequately informed": "We are not being told enough, we are not being told properly, and that is one of the main reasons why we are all at sixes and sevens in our collective life."[25] Information technology is also sometimes presented as a palliative to the disordering effects of other technologies. For example, Benedikt indicates that IT will rescue us from the chaos of the current technological age: "Bombarded everywhere by images of opportunity and escape, the very circumstances of a free and meaningful human life have become kaleidoscopic, vertiginous. Under these conditions, the definition of reality itself has become uncertain. New forms of literacy and new means of orientation are called for."[26] Technology has created the distress for which more and different technologies are offered as the palliative. (Certain digital narratives also trade in the advocacy of unreason, a theme we will pursue subsequently.)

Reason provides access to the authentic self. In their early book on the implications of computer-mediated communications, Hiltz and Turoff assert that electronic communications provide one with the "freedom to be oneself."[27] For Rheingold, information technology assists toward the Enlightenment aim of diminishing inequality and prejudice: "Because we cannot see one another in cyberspace, gender, age, national origin, and physical appearance are not apparent unless a person wants to make such characteristics public. People whose physical handicaps make it difficult to form new friendships find that virtual communities treat them as they always wanted to be treated—as thinkers and transmitters of ideas and feeling beings, not carnal vessels with a certain appearance and way of walking and talking (or not walking and not talking)."[28] This passage also reflects many Enlightenment themes, including the distancing of mind from body, to be developed in the next chapter. For these narratives, the essential person is the being who thinks, not the mere body with all its imperfections that houses the mind. Information technology is a technology of the mind that allows the transmission of aspects of the essential self. Mitchell indicates a similar transcendence beyond appearance that enhances our autonomy and our ability to create a more equal world: "My representation on the Net is not an inevitability of biology, birth, and social circumstance, but a highly manipulable, completely disembodied intellectual fabrication."[29] The body is also transcended: "Unlike Leonardo's Vitruvian Man, we telemanipulating cyborgs cannot be encircled by neat arcs swept through our outstretched limbs. Our grasp has no limits—upper or lower. We have no fixed scale."[30]

These claims that IT will bring about a free, better, and enlightened world are also echoed by commentators such as Schneiderman, writing from a social science perspective and for whom multimedia technologies in education usher in an improved sense of community: "These technologies can support teachers in fostering student engagement with peers and outsiders, and construction of projects that contribute to a better world. These approaches also promote each student's self-worth while learning the subject material. I believe that as teacher effectiveness increases and learning becomes interactive, creation generates satisfaction, process and product become entwined, and cooperation builds community."[31] These sentiments echo the Enlightenment formula of reason, exercised through education leading to freedom and cooperation.

As well as their positive program, digital narratives present in terms of what they are against. For utopian digital narratives, the enemies are those who seek to control information and diminish the power of the medium to set people free. The electronic socialist network is in danger of being hijacked by big business, the large companies that manufacture chips and hardware, large software companies, cable television, and entertainment companies working together in various combinations: "The Net these players are building doesn't seem to be the same Net the grassroots pioneers predicted back in the 'good old days' on the electronic frontier."[32]

There are differences between nineteenth-century utopian and contemporary digital narratives. The latter are commonly accompanied by the prospect of dystopia. In their more sophisticated forms, they bring out the ambiguities of the IT phenomenon, presenting the prospect of unmediated and free interaction between people and the creation of new modes of community. On the other hand, they also present the undesirable prospect of electronic surveillance and attempts to control people's lives. Less sophisticated treatment presents the dystopic possibility as a warning, as though we are under some moral imperative to ensure that the technology is used for good rather than evil.

Many digital narratives are utopian in the sense that they give credence to information technology as a means of realizing the Enlightenment project of a world where reason holds sway over unreason, and as a consequence people are free, equal, and in harmony. Clearly, this project is not yet realized but is a projection into another world, a burden for the future.

The Arts and Crafts of Computer Programming

Later Enlightenment thinking also entailed a reaction against industrialization and the dominance of the machine. William Morris invoked a return to craft as the technology of his utopia. The Arts and Crafts socialists of the nineteenth century fought against large-scale centralized industry. In Morris's utopia people make what they need and want using only simple machines. The artifacts people produce are exquisite, and there is an unaffected sense of good taste throughout the population. Road mending is on a par with fine wood carving and writing novels, and many people exercise skills in several areas, without privileging intellectual work over manual labor. There is no ownership, so people make things for the pleasure of it and give freely. There are no factory estates, and the slums of Victorian London's East End have been cleared.

Even though electronic communications formed no part of Morris's utopia, digital narratives resonate with the sentiments of the Arts and Crafts movement. For digital narrative, the "return" is to the new technologies of electronic communications and the skills associated with them, which are treated much as Morris regarded manual crafts. For Toffler, the electronic age enables a return to styles of work analogous to cottage industries: "a genuinely new way of life based on diversified, renewable energy sources; on methods of production that make most factory assembly lines obsolete; . . . on a novel institution that might be called the 'electronic cottage.' "[33] Marshall McLuhan also inverts the expectation of Morris and others that advanced technologies can lead to a dehumanizing, bureaucratic world order and proclaims a return to a kind of preliterate "tribal culture" in which people will enjoy an immediate engagement with one another and with the world around them—the global village.[34] According to these prognostications, communications networks and computerization inevitably reduce centralization. Computer skills are analogous to crafts. They can be practiced as local, home-scale operations.

According to some variants of the arts and crafts narrative, the computer also revives creativity and a new appreciation of nonindustrial production. According to Benedict, cyberspace technologies provide "the maximum number of individuals with the means of creativity, productivity, and control over the shapes of their lives within the new information and media environment."[35] Paradoxically, it also engenders a new appreciation of the nontechnological: "isolating and clarifying, by sheer contrast, the value of *un*mediated realities—such as the natural and built environment, and such as the human body—as the source of older truths, silence of a sort, and perhaps sanity."[36]

Certain digital narratives trace the arts and crafts lineage through to the computer entrepreneur. The successful politicians, academics, and corporate executives of the 1990s are of the baby boomer generation who were part of, or at least influenced by, the counterculture movements of the 1960s and their lineage.[37] Rheingold begins his explanation of life on the electronic network by considering the WELL system (Whole Earth 'Lectronic Link), which is an international electronic conferencing and email system for personal computers that began in the early 1980s in San Francisco. Its founders emerged from the 1960s counterculture, dabbling in "Eastern mysticism" and meditation: "Personal computers and the PC industry were created by young iconoclasts who had seen the LSD revolution fizzle, the political revolution fail. Computers for the people was the latest battle in the same campaign."[38] The 1960s counterculture movements were components in a web of cultural developments and variations dating back to the nineteenth century: art communes, the valorization of craft, "bohemian" lifestyles, and a romance with transience are part of a continuing tradition that began as a reaction against industrialization. These countercultures and their variants represent a countermovement to the ordered and disciplined bourgeois class—a class that sought to realize the Enlightenment ideal buoyed along by industrialization and capitalism.

Digital narratives also promote various computer crafts and those who practice them, much as nineteenth-century artists and intellectuals valorized the crafts of weaving and carving. There are esoteric knowledges and communities of craftspeople who support one another, particularly through electronic communications. The heroes in digital narratives are the practitioners of the art, some of whom have become directors of large software companies. But the origins of many of them as "hackers" is an important part of the digital narrative.[39]

Other simple analogies between the structure of the software industry and craft industries that sustain aspects of digital narratives exist. Even though the computer software industry is as advanced as any industry can be, it is not as heavily automated, bureaucratized, or streamlined as, say, automobile design and manufacture or the design and manufacture of microchips. The software industry is analogous to writing and publishing. It can depend substantially on the initiative of individuals (authors). Even the mass production and distribution of software does not have to depend on heavy industrial infrastructures. At the less commercial end of the software spectrum, the means of mass production is file copying, and the distribution mechanism is the ubiquitous Internet and the World Wide Web. Both processes are readily accessible to practitioners of computing crafts. Even aspects of the computer hardware industry present a craft orientation. The early days of computer hardware manufacture involved assembly from components. Even though the manufacture of components relies on sophisticated technologies and mass production, the industry is so structured that many personal computer manufacturers could begin simply with home-based assembly plants.[40]

Irrespective of the degree of industrialization of computer hardware and software today, and the threat of software monopolies, digital narrative trades in the theme of craft origins. The application of computer programs also demonstrates a craft orientation, especially in the area of computer graphics and modeling since artists and designers embraced the medium. There is an interest in fine detail, intricate layering in the manner of an illuminated manuscript or complex etching, and even patterning analogous to the arts and crafts aesthetic. Computer games such as *Riven* are admired for the craft skills of their creators. During the production process, the *Riven* team published images of objects that appear in the game on the World Wide Web as they were rendered, invoking admiration for their artistry but also a sense that one was privy to the workings of a team of skilled craft workers. The craft orientation of the IT world is in the company of various other art forms dependent on advanced technology such as electronic music.[41]

Romantic Imagination and Computer Genius

Digital narratives present the invention and refinement of the computer as at the pinnacle of scientific, technological, and Enlightenment accomplishment. As the purveyors of this remarkable achievement, we may also expect digital narratives to promote the particular Enlightenment legacy that appeared to bring them about, namely, scientific rationalism. The early inventors of computer systems, such as Church, Turing, and von Neumann, demonstrated allegiance to rationalism, and this legacy survives today in the advocacy of systems theory, design methods, and much artificial intelligence research. But the romantic rather than the rationalistic orientation seems to hold sway in many quarters.[42] Digital narratives borrow much of their romanticism from science fiction, which Gibson, the author of *Neuromancer*, remarks is largely founded on ignorance: "Most of the time I don't know what I'm talking about when it comes to the scientific or logical rationales that supposedly underpin my books."[43] Such inattention to the practices of inventing, producing, and using information technology provides ample scope for the romantic imagination.[44]

As we have seen, romanticism is the Enlightenment movement that reacted against the rationalism of the day. In attempting to subjugate all thinking to reason, understood in terms of mathematics and logic, rationalism clearly created a void in its account of human experience. As pointed out by Bernstein and Rorty, rationalism did not subjugate the world to mathematics and logic but rather divided it.[45] This division was in the terms laid out by Descartes. The division was not between rationalism and empiricism, which are predicated on similar notions (objectivity, the importance of reason, and the validation of knowledge), but between rationalism and romanticism. If rationalism explicated the world of objects, certainties, and the reducible, then romanticism disclosed the culture of the subject, human feeling, emotion, and wholeness. Romanticism represents the converse of the rationalist and the empiricist project without breaking from the tenets of the Enlightenment. Under romanticism, the individual is not only the source of reason but also the source of creativity, the major human faculty is not abstract reason but the imagination, and science is not the archetypal discipline but art. Romanticism was not against science but rather science as reduction, and some claim that romanticism gave science much of its impetus,[46] promoting the concept of nature and the reverence for it that features prominently in popular science writing. The "new sciences" of complexity theory, quantum physics as a holistic account of an intricately interconnected universe, and the development of "environmentalism" owe much to the early romanticism of Goethe and others.[47]

Romanticism had its origins in the philosophical movement of German idealism, to which such prominent thinkers as Kant, Fischte, Schlegel,

Schelling, and Hegel belonged, but it found full flower with artists, poets, and musicians of the eighteenth and nineteenth centuries. Morris, the pre-Raphaelites, and the adherents of the arts and crafts movement were sustained by the romantic legacy. Morris was a poet, novelist, and designer by practice, and the social reforms of which he spoke were bound to aesthetic concerns. The growing utilitarian industrialized world was ugly as well as unjust. There is also a romantic strand to the twentieth-century countercultural movements referred to above—the rejection of bourgeois values and the return to a craft aesthetic.[48]

Romanticism is also idealistic in that it elevates the existence and importance of ideas over matter, in the manner of philosophical idealism but also in the everyday sense. An idealist has a conception of how things are in their essence and discards what is incidental to them. So Morris was an idealist in that he had a clear view of what constituted aesthetic quality—craft, lack of pretension, art by the people—which did not admit an aesthetic appreciation of variegated industrial nineteenth-century English cities. He was also an idealist in that he thought that under the right conditions the essential goodness of people would prevail and they would happily eschew the trappings of property ownership to participate in a harmonious world order. Morris's utopia is uncomplicated: socialism reigns, government and formal education are unnecessary, and communities resolve their problems at the local level.

To the Enlightenment preoccupation with the freedom of the individual, the romantics added the freedom of the human spirit, or the natural spirit, well expressed in Rousseau's formulations of the process of education—ideas taken up by liberal educational reformers such as Dewey in the twentieth century. Education should provide the means for the child to give expression to the creative spirit within. Morris shared this view. It is the child within us that produces imaginative works. In his utopia, people were at last back to childhood.[49]

Romanticism promotes the importance of feeling and the imagination, but also that other central aspect of Enlightenment thinking—genius. As a culture of the subject, romanticism locates the source of creativity in the free-thinking individual who is able to cast aside the prejudices imposed by tradition and culture and give full scope to the creative spirit. An individual who so stands out from the crowd may be labeled a genius or a hero. Genius featured prominently as a tenet of art, but nineteenth-century socialism was also a party to the notion of the revolutionary hero who would lead people to a better future, though the notion never sat comfortably with the idea of an empowered working class.[50] In Morris's utopia, the charismatic socialist leader is conspicuously absent, though there is scope for everyone to be a hero.[51]

Romanticism and what it stood for never really came to an end. The Enlightenment legacy, and romanticism with it, flourished in the New World. According to Commager, Europe invented but America realized the Enlightenment project.[52] The twentieth century is every bit a romantic age re-presented through those other inventions of Enlightenment, the entertainment and leisure industries that pervade the mass media and inform digital narratives. In *Travels in Hyperreality*, Umberto Eco traces some of the exuberances of simulation and mimicry, theme park reconstructions of European folk and aristocratic culture, exhibits of prominent people molded into lifelike effigies in wax and using "animatronics"—spectacles centered on genius.[53]

What are the indications of genius in digital narratives? Contemporary technoromantic culture removes the mask from public office or the anonymity of the corporate world to expose the individual. It also removes the anonymity of technology to expose its promoters and heroes. The Internet, the World Wide Web, and IT publishing are extensions to the mass media and circulate narratives featuring the heroes of information technology. IT magazines such as *Wired* probe behind the anonymity of hardware and software providers to focus on talented individuals. These forums have created their own brand of heroes, their philosophers and seers of information technology. Leary labels such heroes "cyberpunks," who are "the inventors, innovative writers, techno-frontier artists, risk-taking film directors, icon-shifting composers, expressionist artists, free-agent scientists, innovative show-biz entrepreneurs, techno-creatives, computer visionaries, elegant hackers, bit-blipping *Prolog* adepts, special-effectives, video wizards, neurological test pilots, media explorers—all of those who boldly package and steer ideas out there where no thoughts have gone before."[54]

Socialism and Electronic Communities

The Internet and the World Wide Web provide a medium for self-defined countercultural movements. Leary affirms the emerging importance of the cyberpunk, who he associates with the individual who thinks for him or herself against the dictates of the "governing system" in a manner that is "unauthorized,"[55] a theme developed in *Neuromancer* and other science fiction.[56] Cyberculture is also against censorship, control, and the use of computer networks to promote surveillance, as in the debate about the "clipper chip" in the early 1990s—a microprocessor chip proposed for all computers that would allow safe and confidential encoding and decoding of messages but would be transparent to certain government agencies.

Aspects of digital narrative inherit the Enlightenment preoccupation with social change by evolution or revolution. The term *socialism* first appeared in the 1830s when the socialist movement presented itself as heir to the Enlightenment, reacting against industrialization and the social consequences it entailed. There is a strong link between romanticism and socialism, with its lineage passing through Rousseau (1712–1778),[57] who exerted an influence on Morris, as well as on Karl Marx. Rousseau's romantic thinking comes through most strongly in his book on education, *Emile*, in which parents and tutors are enjoined to allow a child's natural spirit to emerge.[58] By way of contrast, the empiricists such as John Locke saw the mind of the child as a blank slate onto which ideas and principles were to be written through life's experiences. For Rousseau, society rather writes over or smothers the child's innate abilities. Rousseau introduces the notion of the free individual to the socialist theme, preparing the way for the revolutionary hero.[59]

Reformers such as Robert Owen attempted to build socialist utopias as worker communes,[60] and Morris incorporated the socialist theme of a world governed by the people into his fictional utopia, where the new order comes about by violence—strikes, mass demonstrations, and a (prophesied) massacre of the socialists in Trafalgar Square in 1952. This was no less than the revolution by violence proposed by Marx, who advocated that the proletariat "seize power as the ruling class, and vest all property from the middle class to the state."[61] In Morris's case, the revolution was to come through a consensus among the population who were exhausted by the distress of resisting the inevitable good sense of socialism.[62]

Beginning in the 1920s, the intellectual credibility of socialism and Marxism was rehabilitated through the writings of the Frankfurt School, which sought to free Marxism from its ideological base. The Frankfurt School is the philosophical legacy of the intellectuals of various twentieth-century art and counterculture movements, including those caught up in the mass demonstrations in universities during the 1960s. Insofar as they are influenced by socialism, Marxism, and the Frankfurt School, the leaders in IT culture are heirs to various forms of socialist and revolutionary rhetoric and practice, albeit in diminished form. Information technology provides one of the few remaining areas where the Enlightenment concept of free and communal property has not yet been totally discarded.[63] Digital narratives project much of their political attention on the concept of free information. In media such as the Internet, information is readily available and can be appropriated with ease, whether legally or not. The relatively easy commerce in information reinforces aspects of the socialist ideal in digital narratives. With its free commerce in information, the Internet realizes aspects of the utopian socialist state of the kind put forward by Morris, and much digital narrative preserves this utopian illusion. In this light, it should come as no surprise that much digital narrative invokes a utopian rhetoric of revolution and social change. We will return to some of the insights on information technology provided by the Frankfurt School and critical theory in subsequent chapters.

The Romance with Systematization

Utopian narrative appears to be the preserve of speculative thought and runs counter to the sober reflections of positivists, empiricists, rationalists, and systems theorists. According to Popper (writing as an empiricist), utopian thinking is totalitarian, unenlightened, and politically dangerous.[64] Utopian speculations are ostensibly marginal to systems theory narratives but never far from its concerns. As I have already indicated, narratives of progress appear throughout artificial intelligence literature and pervade the reflections of cognitive scientists and analytical philosophers who make substantial use of imaginative scenarios as thought experiments to explore some theoretical construct in the here and now.[65] Such narratives are evident in the celebrated book *Society of Mind* by the AI researcher Marvin Minsky, which presents as a highly speculative work: "Since most of the statements in this book are speculations, it would have been too tedious to mention this on every page. Instead, I did the opposite—by taking out all words like 'possibly' and deleting every reference to scientific evidence."[66] The book is consequently an "adventure story for the imagination."

In the early 1970s, the Club of Rome saw systems theory as a way of solving major global problems.[67] There was also the Japanese Fifth Generation computing project in the early 1980s, a substantial artificial intelligence research effort aimed at developing computers that reason and process knowledge.[68] The project was promoted as a way of coping with society in the 1990s, in which economic survival would depend on knowledge rather than industrial production. The new technology would provide solutions for the problems of an aging and decreasingly productive population, offer instant translation of spoken and written language to and from Japanese, and permeate every area of society. This utopian vision was taken up by many United States researchers and politicians.

The rigorous core of systems theory does not endorse speculation, but systems theory is about prediction, the application of rational methods to project a current situation into a future situation and thereby control it. The components are parameters, theories linking those parameters, data, and some description of a future situation. Systems are often described in terms of inputs, outputs, and feedback loops. The method is also sometimes recast in terms of goals, a description of an initial situation, actions, transformation rules, or formulas to get from one state in the system to another, and some strategy of searching through a space of possible states to determine the best action sequence or plan to accomplish the goals. A further variation is to regard complex goals as decomposable into more easily managed subgoals. The process is to search through the hierarchy of goals and subgoals and their combinations to arrive at a solution to the complex goal. The whole systems theoretical approach has been softened somewhat since its inception in the 1950s. In design, for example, systems theoretical design methods have been supplanted by "second generation methods," and more lately "soft systems" theory.[69]

The various disciplines that embrace the term "planning"—policy, land use, economic, environmental, and urban planning—have wrestled with the issues of systems theory applied to prediction and control in complex domains. According to Inayatullah, several key assumptions lie behind the empirical approach to planning: that the future exists ontologically and is discernible, that time must be controlled, that information about the future is indispensable for economic success, that more comprehensive views of the future result in better policy decisions, that methods of prediction should be scientific and objective, and that the goal is to achieve a more rational world.[70] To the extent that these descriptions apply to systems theory, planning is party to the project of the Enlightenment utopia. Needless to say, most of the fields mentioned here are moving away from systems-theoretical approaches, which they have found to be inadequate to the complexities of the planning process that is recognized as involving a complex play of technical, social, and political factors. In spite of early attempts within systems theory to provide formal languages to enable people to communicate better and thereby facilitate community participation, attempts to bring concepts of community and consensus within the purview of systems theory have not so far been very successful.

Popular digital narratives invoke little of the methods and practices of systems theory, and yet they retain a romance with it. Systems are no longer understood in terms of inputs and outputs but in terms of vast, multilayered, and interconnected networks that defy understanding in terms of inputs and outputs. This fascination is supported not only by the residue of systems theory thinking but by engagement with communications technologies, particularly hypertext programming and the World Wide Web. It is also sustained by an interest in complex systems theory, theories of fractals, and even quantum mechanics, which, like systems theory, have been extended by many popular writers beyond their technical domains to provide "the big picture."[71] The fascination with interconnectivity is also sustained by a flirtation with the thinking of contemporary philosophers such as Deleuze and Guattari,[72] who trade in rhizomic epistemological metaphors and a cursory reading of Derrida, who for some is prima facie preoccupied with the endless interreferentiality of texts. (We will return to these thinkers in subsequent chapters.) Digital narratives are largely eclectic and are able to adopt an uncritical romance with an extended kind of systems theory.

Techno-Medievalism and Irrationalism

Early Enlightenment rationalism drew its inspiration from the classical period. The seeds of rational, democratic, and well-ordered society lay with the ancient Greeks. As an aesthetic movement, rationalism associated itself with the well-ordered simplicity of classical art and architecture. By way of contrast, the medieval period was despised as aesthetically tasteless, socially it harbored ignorance, and politically it represented feudal despotism.[73] But the romanticism of the later Enlightenment found greater affinity with the medieval period. The neo-medieval return to the country, later articulated by the pre-Raphaelites and the arts and crafts movement, provided justification for the reactionary position of the landed gentry as paternal lords in their castles offering protection to a productive village community, though this return did not uphold Enlightenment individualism so much as aristocratic paternalism—noblesse oblige.[74] However, later nineteenth-century romantics and socialists saw the preindustrial medieval crafts as a model for their anti-industrial aesthetic. The organization of medieval trades and guilds also provided a model for the autonomous mobilization of labor.

It is fair to say that the medieval period became, as it is now, anything that people wanted it to be. According to Baer, it "became whatever a critic perceived most lacking or imperfect in the present or most needed emphasis."[75] This is a large part of its appeal. As the mysterious other, it is wide open to prodigious interpretation, and the medieval readily becomes entangled with myths from other eras and places, of frontier worlds, the Wild West, piracy, Celtic and Norse legends, the Orient, primitivism, the future, and now the constructions of cyberspace. The romantic period of art and literature and its various revivals have witnessed the systematic construction of a potent medieval mythology. As indicated by Eco, it is difficult to discern a true medieval period independently of romantic conceptions of it.[76]

The romance with the medieval implicates individualism. Thomas Malory's retelling of the legend of King Arthur and the Knights of the Round Table[77] was a source of great interest for the romantics and was variously interpreted and retold by them. The myth of knightly chivalry entails prowess in battle, exemplary conduct, bravery, devotion to service, courtly love, and piety. These were aristocratic virtues[78] that clearly appealed to the Enlightenment mind and required little translation into the notion of the independent and free-spirited individual.[79]

The medieval appeals to the Enlightenment mind as the domain in which the rational and the irrational are still being fought out.[80] Romanticism interprets medieval narrative, such as the Arthurian legends, not only as tales of virtue battling against evil, or regret over the decline of a once noble kingdom, but of reason battling against unreason. Magic, mystery, and superstition hold sway with the hero able to lead people toward enlightenment.[81] Similar themes recur throughout science fiction and fantasy writing. They continue the romantic medieval theme, sometimes overtly, as in the invocation of the space age equivalents of magicians in *Star Wars*. In some science fiction, the medieval appears simply as a mood of menace and mistrust. The world depicted is unjust and hierarchical. There are outlaws, traders, and other opportunists who frequent space age taverns and market places. The dark underworld is hierarchical but unpredictable. Strange controlling technologies replace magic in some cases, but the world is no less the world of unreason than if magic were involved.[82]

A further aspect of romantic medievalism is the celebration of the carnival, a theme developed at length by Bakhtin. A kind of "grotesque realism" pervaded the festivities of the medieval period (and the Roman Saturnalias), and survived into the Renaissance: "The material bodily principle in grotesque realism is offered in its all-popular festive and utopian aspect. The cosmic, social, and bodily elements are given here as an indivisible whole. And this whole is gay and gracious."[83] In many ways, the carnival complemented church worship in medieval life, but not as mockery for its own sake. Debasement, clowning, and inversion of the sacred order appealed to the earth, its enveloping and regenerative character. In Enlightenment culture, the carnival became disintegrated from everyday affairs, representing unbridled revelry and unreason—life out of control. In the modern age, carnival parody (at least in literature) has a "negative character and is deprived of regenerating ambivalence."[84] By this reading, the carnival spirit is now the preserve of what we call "counterculture," which maintains a remnant of the unifying motivation of grotesque realism,[85] persisting in raves, festivals, social drug taking, body piercing, and "primitive" hairstyles, as expressions of solidarity within certain communities.[86] Unreason becomes the social glue, at least in public display, for groups who seek to define themselves against the prevailing culture. Such activities constitute a rebellion, a carnivalesque respite from the ordinariness, order, and reason of everyday life.

Apocalyptic visions of the future are also common medieval themes. Eco draws certain parallels between the medieval period and now. Anxiety about divine judgment at the end times is replaced by a general insecurity and anxiety about impending ecological disaster. Life was always perilous for the medieval traveler, but we are made to think it equally so now with the constant specter of terrorism, subversion by fundamentalists, and runaway technologies. The "Dark Ages" also epitomizes the pre-Enlightenment state into which we may fall if we reject Enlightenment precepts or abandon reason. So the medieval period itself is a fallen place.[87]

How do these themes of romantic medievalism impinge on digital narratives? Digital narratives often present computer skills as mysterious arts, analogous to romantic conceptions of medieval alchemy and black arts. Computer programs and computer network connections are labyrinthine. Computer systems involve the interconnection of software and hardware components, many of which were designed and manufactured by others and so are mysterious "black boxes," even unpredictable and irrational. These properties of complex computer systems are taken up as the central themes of certain computer games, file-naming conventions, and digital narrative. Dungeons and Dragons, which began as an elaborate role play, has various computer versions. The ingredients of romantic medievalism are there: labyrinthine progression, hierarchies of place and status, irrational interventions through the forces of magic, powerful and irrational forces, and the acquisition of power and victory by magic but appropriated through superior reason.

The romance with numbers and the privileged access to them provided by computers is well known and has been exploited by latter-day numerologists and astrologers, continuing the romantic medieval theme. IT medievalism also presents a revived romance with texts, words and wordplay, no more so than in computer games such as *Myst* and *Riven*, which are concerned with decoding access to "link books" through which one moves between worlds that are in turn created by virtue of being "written" in books—a kind of cosmological hypertext. The significance of the name, and its etymology, given to an algorithm or system in computer research and development indicates the mystique of the word in remnant form, as if the performance of a procedure is contingent on the name given to it.[88] The parallel between passing instructions to a computer by commands and incantation or invocation is obvious.[89] The power of words to invoke is also translated into the invention of new vocabularies. A browse through *Wired* magazine yields the following: electrosphere, encyclomedia, e-mapping, techno-thriller, bitnik, cybrarian, digerati, microserf, realies, and synthespian.[90] Some of the credibility of Gibson's fiction, as with much science fiction, lies in the melding of street talk with an emerging "techno-babble": flatline, icebreaker, holographix, implants, cranial jacks, dermatrodes, matrix, and geodesics. As we shall explore in the next chapter, such verbal nonsense can play a role in disrupting categories and hierarchical order, thereby supporting the sense of access to an ineffable unity beyond the world as understood by rational means.

Digital narratives also seem to trade in a fascination with the magical power of imagery.[91] Computer graphics images become icons, talismans, and magical objects. Photorealistic computer graphics provide a plastic medium for the exploration of the bizarre, the mysterious, and the surreal along romantic medieval pictorial themes. In fact, the current fetish with the depiction, collection, and display of vast catalogues of curious objects seems to continue the thread of irrationalism, as much as the rationalism of the Renaissance and Enlightenment encyclopedic tradition. The latter was ostensibly concerned with classifying, cataloging, and ordering objects, specimens, and knowledge. Prior to the development of the museum, the idea of the collection was already imbued with an understanding of order beyond that suggested by the necessities of modern science and language beyond mere representation. According to Foucault, the encyclopedic tradition was not just about classification and scientific examination but was motivated by the desire "to reconstitute the very order of the universe by the way in which words are linked together and arranged in space,"[92] a project that is reaching its zenith with the development of vast on-line repositories of interlinked pictorial data on the World Wide Web.[93]

According to Yates and Hooper-Greenhill, a thread exists within the notion of the collection that trades in the occult and the irrational. In *The Art of Memory*, Yates shows how the idea of the collection developed in relation to the art of rhetoric and memory that employed objects, spaces, and images as an aid to memory. Such collections pertained to the memorization of biblical narratives, prayers, psalms, and moral lessons, as in the use of images, objects, and ornamentation in church architecture. But such imagery was also a means of memorizing any presentation in rhetoric, by association, and such associations could be established through obtuse, ugly, and bizarre references.[94] For Yates, the importance of memory, and its techniques, waned with the advent of printing but was taken up in the renaissance as a "Hermetic or occult art,"[95] as evident in Giulio Camillo's influential memory theater. This was a model amphitheater containing many wooden figures and images structured and ordered according to the layers of the cosmos. According to Yates, an object in the memory theater functioned as a talisman, "an object imprinted with an image which has been supposed to have been rendered magical."[96] Such objects were not mere collections but aids to memory, and "the cosmically based memory would be supposed, not only to draw power from the cosmos into the memory, but to unify memory."[97] The memory theater developed into the baroque cabinet of curiosities, also known as "irrational cabinets," or "cabinets of the world," which were models of rooms full of curious objects. For Hooper-Greenhill, this is the other (nonrationalist) tradition on which the museum draws,[98] a theme taken up by the surrealists. We see the remnant of the occult interest not only in collections of objects in computer games that present curious objects with magical powers but also in the object fetish of attempts to provide vast digital image resources, which now seem to draw on anything but rational necessity.

As with any new medium, there is also the overt use of computer systems as a means of entertaining occult practices: pagan rituals on the net, horoscope services, and mediums. We should not assume that the translation of occult medieval and Renaissance culture into contemporary society, whether in mockery or seriousness, closely resembles fifth- to thirteenth-century medieval practice. The medieval world of IT is a flirtation with the irrational and the other, continuing the Enlightenment and romantic tradition. Having defined what is certain, sure, and reasonable, the Enlightenment manufactured and promoted the unreasonable, the vast array of human experience that it was forced to leave out of its reflections, appropriated, enhanced, redefined, and repackaged by the culture of the romantic.

Heim sums up the occult pretensions of cyberspace: "The transhuman aspects of VR [virtual reality] can approximate something that shamans, mystics, magicians, and alchemists sought to communicate. They invoke the transhuman."[99]

The underground cyberworld referred to above is also constructed as a medieval world of unreason and mystery. The carnivalesque is also evident in the prospect of cyberspace as a vehicle for exploring sensual pleasures, analogous to drug taking, and ecstatic experience. It also emerges in the countercultural rhetoric of its proponents, and in aspects of IT culture aesthetics—the way the human body is depicted in certain forms of computer graphics. It is also evident in the influence of surrealism on computing that we will examine in chapter 6.

Digital narratives of progress commonly feature a world in which these aspects of romantic medievalism hold sway. The future is commonly told as a return to a medieval utopia of community, a rediscovery of village life and craft values, or as a medieval dystopia of chaos and feudal domination. The future is also often foretold apocalyptically with the foreboding of the millennialist soothsayer, a descent into chaos and disintegration.

Narratives Counter to the Digital Utopia

To the litany of utopian narratives presented so far, there is a ready supply of counternarratives that speak of dislocation, pathology, social disunity, instrumentalism, bureaucratisation, disparity between rich and poor, "the chaotic effects of randomness and indeterminacy generating neither options nor choices,"[100] where "we fail to distinguish between experience and reality that has been manipulated by media images," a world in which "media images become substitutes for direct experience"[101] and where cyberspace is a breeding ground of a cyberpunk low life, suggesting anything but progress.[102] Following the "redundancy of man's muscular strength in favor of the 'machine tool,' " we have the prospect of "redundancy of his memory and his consciousness."[103] For Virilio: "At the end of the century, there will not be much left of the expanse of a planet that is not only polluted but also shrunk, reduced to nothing, by the teletechnologies of general interactivity,"[104] and "the world becomes meaningless now it is no longer so much whole as reduced."[105]

Narratives exist too that address the Enlightenment foundations on which the utopia/dystopia distinction is based. Insofar as digital narratives participate in the Enlightenment project and the centrality of reason, they are targets of poststructuralist and hermeneutical narrative. Contemporary critical narratives present the contingency of the distinction between reason and unreason. The Enlightenment project comes out of a particular period of history. There have been other times and there are other cultures in which the distinction between reason and unreason is not so presented. Contemporary hermeneutical scholarship seeks to redress the Enlightenment "prejudice against prejudice," accepting that we always announce our judgments on some matter from a position of prejudice. To think "reasonably" is not to free ourselves from prejudices but to acknowledge and examine them, and to understand the prejudices of others with whom we are speaking. Such examination is never from a position of impartiality but reflects the biases and preoccupations of some community or other, of which we are inevitably a part. For the hermeneutical narrative, we are far more involved in our judgments than the philosophers of the Enlightenment suggested we should be. According to hermeneutics, the terms *reason* and *unreason* are not obsolete, but their use is contingent. To "be reasonable" can mean different things: to conform to the norms of a community, to follow accepted methods and practices, to construct explanatory narratives that conform to a particular style of presentation, to take something for granted, to establish an analogy with another acceptable position, and to generalize. In spite of numerous attempts to define the process of reason, the most influential being Aristotle's syllogism and Descartes' and Leibniz's methods, there is no agreed process. The exercise of reason in one situation, conducting a scientific experiment for example, can bear little relation to the exercise of reason in another situation, such as negotiating a peace settlement.

According to commentators such as Lyotard and Habermas, reason provides a grand metanarrative in which we position ourselves and our beliefs.[106] It is a means of legitimating ourselves and our narratives, while delegitimating those outside our circle. Independently of how it is exercised, reason acts as a banner under which certain groups choose to stand, a means of structuring our view of society, or it serves as something to define ourselves against. Thanks to the Enlightenment legacy, narratives that invoke "reason" latch on to a powerful, highly privileged, and evocative term. Its use invokes a certain gravity of purpose. But it is possible to construct powerful narratives without appealing to reason and without diminishing the importance of the distinctions we wish to make. More radically, it is possible to construct narratives in which an appeal to reason plays no part and that invoke new metaphors, new definitions, and the rehabilitation of less prepossessing metaphors to construct important distinctions. For example, narratives can invoke concepts of agreement, without appeal to reason: as a confluence between modes of practice, as the fusion of horizons, as conformity to the norms of a community, as the recognition of difference, and as the construction of appropriate narratives. We can regard the narrative tactic of writers such as Foucault as a systematic attempt to set in motion, or continue, a series of discursive practices in which the preoccupation with reason is usurped by other concerns, such as power, technology, and the body. As I shall explore subsequently, the utopian aspects of digital narratives appear in a different light when we move beyond the Enlightenment preoccupation with reason versus unreason.

Counternarratives also exist that target the legacies from which digital narratives derive. The contradictions inherent in the nineteenth-century arts and crafts movement are well known. In spite of its appeal to the preindustrial, the arts and crafts movement was a bourgeois phenomenon, appropriated by the upper middle classes who had the time and the money to indulge a sophisticated romance with nonindustrialized products and life styles. It was of great regret to Morris that the arts and crafts movement catered to the upper middle classes, who appreciated the vernacular and the handmade more than the working classes who preferred not to surround themselves with rustic simplicity. Crafts are parasitic on industrial production, no less so than with computer "crafts." The "cottage industries" of software production depend on the stability of hardware, operating systems, and web browser technology. In fact, the software business seems to operate in webs of complex interdependencies, some of the linkages of which can be very fragile, as when a major operating system innovation compels small software producers to adapt their software to remain competitive.

Certain narratives pit themselves against the celebration of electronic communities. Critics such as Stallabrass identify the bourgeois dreams evident in cyberutopias and their narratives of universal, seamless networking (which he labels "New Age spiritualism and New Edge technophilia"), and how they are after all consistent with "modern corporate communications ideology."[107] Castells and others have demonstrated the implausibility of the egalitarian dream of the electronic cottage, and the linear extrapolations of changes that ignore the complex interactions between technologies and social and historical contexts.[108] In the twentieth century, the alliance between computer technology and the nonindustrial is an uncomfortable one, which suggests that the distinctions on which it is predicated are unsustainable. As various theorists of technology remind us, we are inescapably part of a technological world. As Heidegger indicates, technology points to a way of thinking rather than simply mass-produced products. Heidegger's narrative of the inevitability of our grounding in technology opens up certain modes of inquiry and constructs narratives in terms of praxis and "being in the world."

Certain narratives counter to digital utopianism present romanticism as inadequate for the task of accounting for the place of information technology in society. There is a general mistrust of both hyperbole about utopic IT futures and paternalistic warnings against IT dystopias. The romance with the machine has outlived its usefulness, and the post-Enlightenment reflections of contemporary thinking, including the philosophies of science and technology, are now permeating digital narratives in ways we will explore in chapters 6 and 7.

Certain counternarratives remind us of the political failings of both socialism and capitalism. Contemporary commentators such as Giddens regard the socialist versus capitalist or left versus right debate as unhelpful in accounting for the current state of political and social change. Truly countercultural movements are difficult to identify in late twentieth-century democracies, or rather the concept of "counterculture" is inadequate to describe what is happening in society and life on the Net. Giddens identifies what are commonly regarded as radical reform groups rather as "self-help" groups—articulate, reflexive groups of people who are bonded by common concerns, and who attempt to define the self in a way that has some significance in a nontraditional world.[109] Such groups include feminist and gay organizations, pro- and anti-abortion lobbies, support groups for ethnic minorities, and ecology groups. In this light, and in spite of the "soft socialist" rhetoric of certain IT commentators, the precommercial Internet phenomenon was not really an instrument of social revolution so much as a vehicle for highly articulate middle-class self-help groups. For Giddens, such groups function as alliances of individuals bent on the reconstruction of plausible narratives of the self in an increasingly unsettled late-modern world. What was formerly socialism is now caught in the ownership and trade of labor and capital. Self-help groups acquire capital and buy into the capitalist system to keep themselves viable. There is also profit to be made in paying heed to the opinions and influence of self-help groups, and much advertising is now directed at them and those under their influence. Internet communities are no different. Among other things, the World Wide Web is growing into a global interconnected supermarket of products and services. This is not to say that socialism has given way to rampant capitalism, rather that the worn-out socialist narratives are displaced by even more potent narratives that account for and reconstruct life in the digital age.

The notion of information as free property is countered by narratives of the inadequacies of network communications. Many people are guarded about what they put on the Net. Information is often out of date or it is broadcast after the author has maximized her return from it. New methods of commercial activity have also arisen on the network that the craft narrative is inadequate to elaborate. On the World Wide Web, partial information is often used as a hook to entice people to commit to a financial transaction later on. This is also a common practice with software development.[110] (Alternatively, the characteristics of IT crafts are disclosing what were there all along as properties of craft in the industrial age.) These new modes of commerce fit within a larger picture of the changing nature of corporations, which are wrestling with the tension between global and local operations. Furthermore, as is evident in the ambivalence of digital narratives, capitalism and revolutionary thinking are not so far apart. Capitalism readily embraces the concept of the revolutionary. The radical entrepreneur does not appear so different to the radical social reformer. The subtlety of such shifts is often lost in the soft socialist narratives of IT.

As a tool of digital utopia, systems theory provides various means of structuring and solving technical problems,[111] but it does not provide the universal theory that its early exponents hoped for. Its speculations are not produced directly by its methods, and systems theorists have to move outside systems theory narratives to develop their speculations and to construct "the big picture." "Imagination" is not a systems-theoretical term, though it is commonly used. Systems theorists commonly resort to romantic speculation when constructing the big picture, while appealing to the surrogate authority of systems theory to support their speculations.

Neo-medieval narratives of information technology's flirtation with the world of unreason, the mysterious, and the exotic other are situated in the Enlightenment tradition. Reason is never far from unreason, both of which are amplified in Enlightenment narratives.

Digital utopianism is a variant on the pervasive theme of unity and disintegration. The utopian quest is of a return to a state in which humanity and nature were united, when we were one with each other and with our technologies. The arts and crafts orientation of digital narratives reinforces this quest by inviting participation in the global village. The concept of genius has its origins in notions of access to the divine intellect beyond the material. The romance with systematization invokes concepts of a collective society of mind. Whereas rationalist and empiricist narratives trade in the computer as a tool of reason, romanticism also entertains the computer as a gateway into the world of unreason and the ineffable, a theme we will develop in the next chapter.

2

Cybernetic Rapture

The concept of cyberspace invokes exotic and Elysian narratives in which the computer age will eventually see a transformation to a new sensibility. The new age will be one in which the physical is transcended by information, providing opportunities for participation in a unity beyond the multiplicity and individuation of the material realm. According to Stallabrass's critical observation, the "enthusiasts of cyberspace let its propensity for dematerialization transport them into the realms of spiritual discourse."[1] I will show how cyberspace narratives continue the trajectory of the Neoplatonists, though it is a Neoplatonism variously transformed through idealism, the romantic tradition, and technorationalism.

The tendency toward such cyberspatial excess is evident in narratives that indicate how computers, networks, and mass communications subvert accepted notions of space and challenge the distinctions between mind and body, reality and unreality, organism and mechanism—that there are "simultaneous redefinitions of space, personal identity, and subjectivity that are emerging as the network grows."[2] The abandonment of traditional concepts of the body is an important component of this redefinition. For some narratives, with VR, "your conditioned notion of a unique and immutable body will give way to a far more liberated notion of 'body' as quite disposable and, generally, limiting."[3] Information technology seems to be affecting this subversion in tandem with other cultural and philosophical developments. According to Rushkoff, cyberspace, drugs, dance, spiritism, chaos theory, quantum physics, and pagan ritual conspire to ensure that there is a whole generation for whom "reality itself is up for grabs" and for whom reality "can be dreamt up."[4] Apparently the accepted reality is an arbitrary one, and to participate in cyberspace is to take part in "a movement that could be reshaping reality."[5]

Much seems to hinge on the issue of "reality" in technoromantic narratives. The term *virtual reality* already brings computers into collision with it. According to Heim: "With its virtual environments and simulated worlds, cyberspace is a metaphysical laboratory, a tool for examining our very sense of reality."[6] But to set out to examine "reality" is already to presuppose too much. Borrowing from Plato's legacy, phenomenology, and the terminology of Lacan,[7] I will examine cyberspace in relation to the concept of "the real," a concept that allows an inclusive, and even simpler, reflection on the nature of our being in the world and from which the concept of "reality" derives. As we shall see, the real implicates the concept of unity and the cybernaut's access to it.

The Real, the One, and the Many

The real can be pinned down to several tangible notions. Cyberspace narratives borrow substantially from fantasy literature, which similarly problematizes the real. In Carroll's *Through the Looking Glass,* Alice has an unlikely encounter with a Unicorn, in which the tables are turned. The Unicorn addresses Alice's companions as though Alice's existence were in question: " 'I always thought they were fabulous monsters!' said the Unicorn. 'Is it alive?' " Alice speaks: "Do you know, I always thought Unicorns were fabulous monsters, too! I never saw one before." With this confirmation that Alice is in fact alive, the Unicorn declares: "Well, now that we have seen each other, . . . if you'll believe in me, I'll believe in you. Is that a bargain?"[8] Alice is then asked to pass round the plum-cake. This simple parable reveals several important themes around which we construct the real. There is the concept of *proximity*. The two creatures have to encounter one another before their participation in the real can be confirmed, and even before it can become an issue. The real involves *sharing*. They strike a bargain about mutual belief in one another. The *body* is implicated, through the senses, quizzical looks, and the gastronomic communion around the plum-cake. There is also a *repetitive* element to the encounter. The Unicorn proudly repeats the Alice observation to the Lion, as if to confirm her existence. Finally, the whole encounter is infused with doubt, suspicion, and *ineffability*. People refuse to give straight answers to questions. There are things that seem to be inexpressible. Throughout the story, Alice is desperately trying to work out what on earth is going on.

Similar accounts around the themes of proximity, sharing, the body, repetition, and ineffability recur through narratives of the real. The expression "the real" captures the concept of what is within our reach, what is *proximal*, and what we can apprehend.[9] For the empiricist tradition, this is generally interpreted as what can be apprehended by the senses, but we do not need to presume the concept of the real as a source of data for the senses. The prosaic concepts of grabbing, holding, reaching will suffice for an understanding of the real. Apprehension of the real as proximal also implies a situation that is concrete and immediate. The real is what can be grasped here and now. The real also captures the concept of something *shared*. One of the criteria for deciding that something is real is that there is some agreement about it. So we participate in the real in a way that involves others, or community, or collectivity. We also participate in the real *bodily*. When we assert the reality of the world we frequently grasp the back of a chair or slap the top of a table, or to intimate that we are not dreaming we pinch ourselves on the arm. According to one commentator: "The world is real because when I bang my head against it, it hurts."[10] These are bodily actions, rather than verbal or contemplative ones. The notion of the real is intimately connected with embodiment. There is also an element of *repetition* in the appropriation of the real. We can go back to what was there yesterday and expect it to be there today. Experiments in a laboratory can be repeated to expose the reality of the same phenomenon, and to verify that something is real we expect to be able to experience it again. The vindication of a rule or formula is that it applies in a wide range of situations, it constitutes a regularity, its application can be repeated. Finally, the fact that we have to assert that certain things are real and others not, that there is debate at all about what constitutes reality, and that "virtual reality" has any currency as a concept—all indicate a pursuit, a longing, a preoccupation, even an angst, for the *ineffable*, the equivocal, the inexpressible, that is perhaps part of the human condition. In talking about the real as an issue, we are asking about the world and our place in it.[11] Cyberspace narratives commonly address the issue of the real in these terms, of proximity (being in data space, transcending the material), sharing (electronic communities, melding minds in data space, consensual hallucination), the body (wearing data suits, being out of body, electronic prosthesis), repetition (information and the primacy of pattern), and ineffability (the unattainable quest for digital utopia).

"Reality" is just one aspect of the pursuit of the real which, according to the analysis of critical theorists and of Heidegger, is an aspect of the real driven by the desire for control. We construct and define the world and our place in it in such a way that the whole ensemble is amenable to our manipulation, or at least in such a way as to sustain an illusion of manipulation. In other words, the way we manifest the pursuit of the real is as a technological pursuit, though it may not always have been, nor need it always be so, nor for everyone. To say that *we* manifest this technological imperative demonstrates the hold of the technological imperative even further. We could equally affirm that the technological imperative is disclosed to us—by whom or by what? Perhaps by technology itself, the real, or Being. Such is the power of the technological view of the real, that it may even conceal other aspects of the real. This is a line of argument presented within phenomenology and critical theory that I will develop in chapter 5, prior to considering its refinement and further development.

The concept of the real preserves these notions of the proximal, the shared, the embodied, repetition, and the ineffable. Following Derrida, we could say that these attributes of the real amount to the same thing. They are all concerned with presence, which is to say proximity, and each participates in various ways in a metaphysical discourse, the quest for ground and certainty. They also implicate unity and multiplicity. For certain technoromantic narratives, there is a striving to be closer to the unity (in information), casting aside the imperfection of multiplicity and individuation (materiality and embodiment).

Technoromanticism borrows from the powerful legacy of Plato's (427–347 B.C.) concept of the real, which first comes to light with his division of the world into (a) the realm of shadows, the sensible world (able to be apprehended through the senses), the material, which mortals inhabit, along with other particular things, against (b) the world beyond appearances. The latter is the realm of the ideas, universals and forms, the Intelligible world, the real, the unity. For Plato, ideas did not exist proximally to our everyday concerns, or even in minds, but in the supradivine realm that was "more real" than the material world. The realm of ideas is *the real,* which deals with perfection. In the sensible material realm we deal with imperfect circles, triangles, and specific instances of things that do not accurately fit the general description of their classes. These are just pale shadows of what exists in the ideal, which is to say real, realm of perfect circles, perfect triangles, and categories.[12] Plato's allegory of the cave describes a progression of states through which people come to participate in the real.[13] A person can be in the cave inspecting the imperfect shadows formed by the light of a flickering fire, or one can be outside facing the sun in all its blinding radiance, which is where the philosopher is situated.[14]

The authority of Plato's teaching on the residence of the real beyond the material is challenged in part by the influence of his student Aristotle's (384–322 B.C.) concrete, "empiricist" philosophy of matter and form,[15] with which it is often put in opposition. But Plato's concept (sometimes called "Platonic idealism") of the unity of the real beyond the multiplicity of the material has currency in popular culture today, including in cyberspace narratives, though it arrives here through various transformations.

The Other Real

Technoromanticism draws on this legacy of Platonic unity, particularly as developed by Plato's follower Plotinus, to be examined shortly. For Plato, the realm of the real is the unity, the world of Being opposed to the world of flux and becoming,[16] but there is another tradition in which it is the real that is the site of radical flux and indeterminacy. The real is the place of a paralogical tension between the one and the many.

Prior to Plato, the pre-Socratics had already developed narratives around the theme of unity and plurality. For Heraclitus, unity and plurality reside in each other: "Things taken together are whole and not whole, something which is being brought together and brought apart, which is in tune and out of tune; out of all things there comes a unity, and out of a unity all things."[17] That something should be the case and not be the case (whole and not whole) at the same time caused Aristotle some disquiet, and in *Metaphysics* he develops principles of logic, including the law of the excluded middle, to challenge the pre-Socratics.[18]

Thinking about the real as an ineffable, antagonistic, and paradoxical play between unity and multiplicity appears in many traditions, as outlined by Eliade in the case of rituals and legends. He notes that certain rites and beliefs have the aim of reminding humankind that "the ultimate reality, the sacred, the divine, defy all possibilities of rational comprehension."[19] They remind us that such realities "can only be grasped as a mystery or a paradox, that the divine conception cannot be conceived as a sum of qualities and virtues but as an absolute freedom, beyond Good and Evil."[20] Scholem offers a similar characterization of the writings of the Kabbalah, the medieval Jewish tradition that was sometimes at odds with philosophical teaching. The Kabbalah highlighted aspects of God that were "beyond rationality" and that become paradoxical the moment they are put into words.[21] The early German thinker Eckhart (1260–1329) adds weight to the agonistic relationship between unity and multiplicity in asserting that unity in multiplicity is never found anywhere but is understood: so "only in God are being and understanding identical."[22] The agonistic version of the unity myth appears in phenomena pertaining to the androgyne: ritual role reversals by men and women, and paying obeisance to beings who are both male and female. As outlined by Bakhtin and Eliade, the symbolic reversal of roles, the suspension of laws and customs under the pretext of carnival pranks, and orgiastic rituals seek "a reintegration of opposites, a regression to the primordial and homogeneous,"[23] but such activities are also "a symbolic restoration of 'Chaos,' of the undifferentiated unity that preceded the Creation."[24] Similarly, for the Pythagoreans, according to Aristotle, the number one is both even (which is limited) and odd (which is unlimited). So Unity consists of both the limited and the unlimited, the one and the many. For the Pythagoreans, the whole sensible universe is made up of numbers, so it is imbued with this conflict between unity and multiplicity.[25]

The real, as the site of play between unity and multiplicity, as a place of paradox, finds expression in Hegel's concept of the dialectic between being and nothing. It is a theme to which many recent thinkers have returned, from Nietzsche onward. It also finds expression in the tradition of verbal nonsense (*Through the Looking Glass*), to which surrealism was a party in the twentieth century. According to Esslin, verbal paradox is a metaphysical endeavor, "a striving to enlarge and to transcend the limits of the material universe and its logic."[26] Nonsense is not merely a foil to reason, but a further attempt to achieve unity with the universe by the destruction of language, which keeps us apart from the world by establishing the independence of objects. In exercising "the destruction of language—through nonsense,"[27] Lewis Carroll played with the arbitrary naming of things, a process which, according to Esslin, expresses "the mystical yearning for unity with the universe."[28] Esslin illustrates this conflict in his interpretation of Alice's journey through the woods "where things have no name." There she encounters a fawn, and Alice and the faun walk together through the woods until the normally timid animal realizes that it is a fawn, and that Alice is a human child, and bounds away.[29] For Esslin, the naming function of language therefore denies access to a harmonious unity, symbolized by the broken embrace of a human and a wild animal. Verbal nonsense restores this unity. As we shall see in chapter 6, the theme is also taken up in certain provocative narratives of the "cyborg," presenting the cyberspace phenomenon as dealing in the ineffable and contradictory.

But technoromanticism develops from a different trajectory. This is the monosemic legacy that equates the real, the Intelligible, the supradivine with unity, and the sensible, the material, and the earthbound with multiplicity. We are exhorted to escape multiplicity to be with the unity. This is the legacy of Neoplatonism, which finds recent expression in technoromantic speculations about digital ecstasis through "total immersion environments" and the collective rapture of humankind through computer networks.

The Ecstatically Real

Whereas technoromanticism speaks of the transcendental capabilities of information, for Plato and the Neoplatonics, access to the unity is by means of the soul. The soul is the immortal part of a living being. Plato describes the ascent and descent of the soul in the parable of the chariot (the soul), pulled by two winged horses (passions and sensual instincts), with reason as the charioteer. In its state of full freedom, the "soul has the care of all that is inanimate, and traverses the whole universe."[30] When the ensemble "is perfect and winged it journeys on high and controls the whole world."[31] But difficulty with the horses causes the chariot to fall, and the horses lose their wings: "one that has shed its wings sinks down until it can fasten on something solid, and settling there is takes to itself an earthly body."[32] The composite structure of soul and body is a living being.

Platonic ideas were taken up and developed in the teaching of the Neoplatonists, including Plotinus (204–270) and his followers. For Plotinus, the soul can gain access to the real but has to liberate itself from the world of matter through frequent *ecstasies*—that is, states of *being outside* (or *beside*) *oneself*. As explained by Marías, during ecstasy, "the soul frees itself from matter entirely and unites with the Deity, the One, to become the One itself."[33]

How does Plotinus regard the constituents of the real, in terms that we defined earlier, of proximity, sharing, embodiment, repetition, and the ineffable? According to Plotinus, participation in reality is by means of the soul, "which especially constitutes our being."[34] But what the soul participates in is the Intelligible realm or the "Over-World." The soul is more *proximal* than the body to reality: "Participation goes by nearness;

the soul nearer than the body, therefore closer akin, participates more fully and shows a god-like presence."[35] The language of proximity is translated into that of participating or basking in a divine emanation.

Participation in reality is a *shared* participation, but it is a sharing through mutual participation in the unity. According to Plotinus, we are "in sympathetic relation to each other, suffering at the sight of others' pain, melted from our separate moulds, prone to forming friendships; and this can be due only to some unity among us."[36] For Plotinus, it is a simple matter then to assert "the reduction of all souls to one."

Neoplatonism advocates a distancing from issues of the *body*. Rather than allow his portrait to be painted, Plotinus is reported as saying: "Is it enough to carry about this image in which nature has enclosed us? Do you really think I must also consent to leave, as a desirable spectacle to posterity, an image of the image?"[37] While he advocated looking after one's health, Plotinus maintained that people should allow their true character to show through, to wear away the tyranny of the body "by inattention to its claims."[38] The body is an impediment, and Plotinus provided instructions on how ultimately to enter into an ecstatic state of unbroken vision, where the self is "no longer vexed by any hindrance of the body."[39] The body is apparently the seat of mood and passion, which must be cast off, whereas the soul is disposed toward "the state of intellection and wisdom"[40] and can become "immune to passion." So the real of Neoplatonism denies embodiment.

The issue of *repetition* is manifested in the Platonic and Neoplatonic depiction of the intelligible realm in which there is no change or flux. Repeated inspection of the real brings us into contact with unity. Insofar as the sensible realm bears repeated inspection and reveals the same properties time and again, it participates in the real. Insofar as we identify the idea of a triangle while inspecting imperfect, sensible, material triangles of all shapes and sizes, we participate in the ideal triangle, which is to say the real triangle, the triangle resident in the Intelligible realm. The metaphor used by Plato and Plotinus to describe this phenomenon of participation in the real is of "the Idea impressed upon matter,"[41] which constitutes beauty. Repetition enters the story insofar as the things of the sensible realm are copies. They are clay ornaments from the same mold, imprints from a ring in a wax seal, which are all imperfect repetitions of each other derived from something more enduring.

The *ineffable* to which Plotinus appeals is not a yearning to grasp the sensible world as real but the yearning for unity, the universe "as one living organism."[42] The unity is above the material. The yearning is for retreat from earthly reality to a higher one, achieved through contemplation, ecstasis, or death: "The Soul of All abides in contemplation of the Highest, for ever striving towards the realm of the Intelligible and towards God."[43] Plotinus' philosophy is clearly not of engagement in the here and now. What is to be grasped is beyond what the body can accomplish. It is also beyond space and matter: "The escape is not a matter of place, but of acquiring virtue, of disengaging the self from the body; this is the escape from matter. The soul's 'separate place' is simply its not being in matter, not being united with it, not moulded in matter as in a matrix. This is the soul's apartness."[44]

Neoplatonism was in part a meeting of Greek and Christian thought. Its influence is extensive within the Judeo-Christian and Islamic worlds, through the Kaballists, Augustine (354–430), the scholars of the Middle Ages and the Renaissance.[45] Dante's (1265–1321) *Divine Comedy* tells of the ascent of souls through purgatory, souls who are amazed at how Dante's live body is able to cast shadows.[46] We have already reviewed the influence of the art of memory on concepts of the encyclopedia, which in turn seems to inform the current imperative to create vast interlinked multimedia collections of objects and artifacts. The sixteenth-century baroque memory theater, as explained by Yates, provided a spatial dramatization of the Neoplatonic world, with the tiers of the theater arranged to depict the progression of the soul to higher things, the object being "to store up eternally the eternal nature of all things which can be expressed in speech" by assigning them "to eternal places."[47] Milton's (1608–1674) puritan epic of the Fall in the Garden of Eden demonstrates various allegiances to Neoplatonic themes, including references to the "pendant world" hanging in a "golden chain,"[48] a reference to the hierarchical ordering of the universe in concentric spheres through which "body up to spirit work[s]."[49]

Ecstasis and the Enlightenment

Already we can detect resonances between Neoplatonism and technoromantic cyberspace narrative, the desire to transcend the constraints of the body and to enter another reality that is cyberspace.[50] But information technology narratives are also a party to modern or Cartesian thought that differs from Neoplatonism in several respects, and Neoplatonism comes to the cybernetic world filtered through Descartes' thinking. (Neoplatonism influenced Descartes through Augustine.[51])

Descartes raised the idealist prospect that we cannot know anything with certainty, that reality may be a deception—the nightmare prospect that everything is a dream, there is no reality, and an evil spirit rules the world. The only certain point is doubt and the entity undergoing the doubt, which is the ego, the self, the subject. For Descartes, the means to overcoming doubt, the means to improvement, is introspection through reason, and here Descartes and the moderns depart radically from Neoplatonism. According to Plotinus, reason is concerned with parts rather than unities and is thereby corporeal, of the body. For Plotinus, to grasp the intelligible, the real, one must employ means other than reason.

Descartes drew attention away from the soul and its desire to return to the unity from whence it came and on to our capacity for thought and reason. For Descartes, the goal of improvement is knowledge, which is the responsibility of the subject (the ego) who thinks. Contemplation does not take you out of the material world but draws attention to objects, things that are situated in geometrical space in the world.[52] On the other hand, for Platonic and Neoplatonic idealism, the real resided in the supradivine realm beyond human experience, in which one participated through study, meditative exercise, or the workings of symbol.[53]

So Descartes departs from Neoplatonism in his characterization of mental activity with consciousness[54] and his elevation of reason as the arbiter in all matters of judgment and perception. Cartesian idealism results in the subjectivization of knowledge;[55] the point of view is that of the first person singular: "I think therefore I am." By this reading, the Cartesian tradition is idealist but with the ego at the center.[56] Modernist narratives that implicate information technology inherit the legacy of this idealism, and for technoromanticism, rather than the soul winging its way back to its home in the unity, as we shall see, the stage is set for the autonomous, reasoning individual entering information space.

Returning to our theme of the constituents of the real, we find that for Descartes the real is what is *proximal* to the ego. Cartesian idealism/realism and its variants establish the concept of an independent reality, but the definition of reality presupposes an ego from which reality can be apart. The Cartesian, objective position is the first person singular position.

The real resists what is *shared*, as custom and tradition are sources of prejudice and error. The ego has a privileged relationship to the *body* but is independent of it. The mind is the prime means of appropriating the real. The body acts as a means of receiving sense data through which we can interpret the real, but through its frailty and susceptibility to persuasion it also acts as an impediment, and Descartes was at pains to divorce reason from the vagaries of carnal weakness and passions. *Repetition* as a constituent of the real is realized through method—repeatable procedure by which we establish truth, whereby anyone can reach the same conclusions by following the same method. The *ineffable* is Descartes' overcoming of doubt about the reality of his experience. Descartes did not settle the issue of reality but set up a puzzle, a quandary, and an anxiety. According to Bernstein, what Descartes raises, and what has persisted throughout the modern tradition, is anxiety about certainty—the "Cartesian anxiety"—apparent through various dualisms that focus around the issue of subjectivity.[57]

Neoplatonic and Cartesian idealism come together with most potency in the idealism of Berkeley, who appropriates the sensorium of the ego as the locus of the real, while at the same time denying the existence of matter, bringing us closer to the current cybernetic idealism. Berkeley was an empiricist. So along with Locke and Hume, as successor and antagonist to Descartes' rationalism and idealism, he asserted the primacy of sense experience and observation as the vehicles for knowing the world rather than introspection and pure reason. Berkeley was also informed directly by Neoplatonism, and one of his late works, *Siris*, includes a vision of the divine unity inspired by Plotinus.[58] This essay takes for granted the constraints imposed by the body: "We are embodied, that is, we are clogged by weight, and hindered by resistance."[59] Berkeley undertakes a spirited defense of the concept of the Unity developed by Plotinus and the ancients against charges of atheism, and he affirms the Neoplatonic progression by which one comes to appropriate the Unity: "Theology and philosophy gently unbind the ligaments that chain the soul down to the earth, and assist her flight towards the sovereign Good. There is an instinct or tendency of the mind upwards, which sheweth a natural endeavor to recover and raise ourselves from our present sensual and low condition, into a state of light, order, and purity."[60]

Berkeley is better known for his earlier work *The Principles of Human*

Knowledge, in which he explores the implications of the writings of his empiricist forebear, John Locke, and establishes that there are no material things (such as houses, mountains, and rivers) that exist independently of minds to think of them. Twentieth-century debate within analytic philosophy focuses on the logic of Berkeley's arguments as a test of empiricism rather than on his affirmation of a Platonic order. There is a lot at stake for the Cartesian and Enlightenment world views if the tenets of empiricism lead one to deny the primacy of the material world. Berkeley does not deny reality, or the reality of the world around us, but rather the reality of the *material*: "I do not argue against the existence of any one thing that we can apprehend either by sense or reflection. That the things I see with my eyes and touch with my hands do exist, really exist, I make not the least question. The only thing whose existence we deny is that which *Philosophers* call Matter or corporeal substance."[61] According to Berkeley, considering the logic of language, we can do without the notion of reality independent of our minds, which in turn solves certain philosophical problems, such as how "body can act upon spirit, or how it is possible it should imprint any idea in the mind."[62]

As everything is mediated by our senses, that is, decided by sense experience, we could never establish that a reality exists independently of us anyway: "If there were external bodies, it is impossible we should ever come to know it."[63] Locke had intimated that in our perception of the world there are first trees and then sense experiences of trees that "copy" or resemble the trees.[64] Our senses convey the qualities of "external objects" into our minds.[65] Contrary to this, Berkeley said there is just the sense experience of trees. If trees did exist apart from sense experiences, then we could not know about them; and since the concept of a tree as it is in itself independent of the experience of a tree would not make any difference to us, we might as well do without it. For Berkeley, all that exists are minds and the experiences of minds, and there are no independently existing physical objects to cause the experiences. We have an experience of a tree but then we do not need the concept of the tree as it exists unexperienced. The experience is enough; the second is excess: "It is evident the supposition of external bodies is not necessary for the producing [of] our ideas."[66]

How does Berkeley's philosophy participate in the real as we have defined it? For Berkeley, in *The Principles of Human Knowledge*, the *proximal* is what is received through the senses. The problematic is couched in terms of how it is that we each perceive the same thing, how is the reality a *shared* reality, and the empiricist answer is in terms of families of patterns. In his later work, Berkeley indicates his allegiance to Plotinus in asserting the higher reality of the world of the soul: "Plotinus acknowledgeth no place but soul or mind, expressly affirming that the soul is not in the world, but the world in the soul."[67] For Berkeley, the *body* serves as the vehicle for receiving sensations: "the things I see with my eyes and touch with my hands do exist,"[68] and the later Berkeley discloses a distancing from the body in keeping with Plotinus. *Repetition* is couched as an issue of continuity of experience and as the repetition of patterns. Berkeley's quest for the *ineffable* is to deny matter, though not reality, and to affirm what he regards as the commonsense view against the peculiarity of philosophers who have created various muddles by insisting on the independence of reality from the mind. Ultimately, reality resides with the Divine Mind with whom "things exist, during the intervals between the times of my perceiving them."[69] Later on, in *Siris*, we see the destination of Berkeley's idealism, as with Plotinus, as participation in a unity beyond the world of appearances and the body.

That the appeal to commonsense empirical evidence could result in the denial of matter and the valorization of the spirit realm has fascinated and horrified others of the empiricist tradition. For technoromanticism, Berkeleyan idealism provides further impetus to the quest for transcendence beyond the material. If we are always dealing with representations in the computer world, then we can dispense with and transcend the thing represented.

Romantic Ecstasy

Berkeley represents the idealism of the British empiricist tradition, which can be compared with the idealism of Kant's philosophy in Germany.[70] According to Kant, the way we understand the world is filtered by the way our minds are constructed. There is a real world, but it is unknowable to us other than through the peculiar construction of our minds. The real world is the "noumenal" world, which is unknowable (and ineffable). The world as experienced is the "phenomenal world."[71] According to Kant, how things are in themselves is outside the realm of thought.[72] The division between the unknowable and the knowable resonates with contemporary cyberspace narratives, not least for the prospect it opens up for multiple, subjectively constructed worlds. For Heim, even though Kant aimed for a "unity to regulate our construction of the world,"[73] he was the forerunner to a move that opened up the prospect of a multiplicity of worlds.[74]

Kant is generally regarded as the founder of German idealism, which established a particular problematic that was different to the English tradition of empiricism. German idealism continued with Fichte, Schlegel, Schelling, and Hegel,[75] and it was from German idealism that the quasi-philosophical movement known as romanticism emerged. As a reaction against the reductionism of rationalism and empiricism, romanticism later emerged as a major legacy of the Enlightenment.

As we have seen, the tenets of romanticism include the elevation of feeling, individuality, freedom, imagination, and the unity of human kind and nature. The nineteenth-century socialists were largely driven by the romantic impetus, which was also a reaction against industrialization. The proliferation of worlds other than the physical one, including utopias, fit the romantic imagination.

That contemporary romantic narratives should embrace computer technology and appeal to science is consistent with eighteenth- and nineteenth-century romanticism. Early romanticism elevated nature as a means to self-reflection. Nature is an integral whole with which humankind was once at one but is now separated. For romanticism, natural forces are analogues of life forces, nature is always in process rather than complete, and nature is personified as feminine.[76] For the romantics, science was the study of nature, and its methods were commended by some. For Shelley, the chemical laboratory symbolized joy, peace, and illumination.[77] Romanticism was uniformly against reductionism, individuation, and the presentation of science as independent of questions of value, though it did not necessarily reject scientific rigor. Romanticism was never undisciplined. According to Cunningham and Jardine, science also provided models and heroes and was itself informed by the romantic quest: "the self-image of the new 'men of science' was to be largely constituted by romantic themes—scientific discovery as the work of genius, the pursuit of knowledge as a disinterested and heroic quest, the scientist as actor in a dramatic history, the autonomy of a scientific elite."[78]

The romantics were also informed by Kant's and Burke's concepts of the sublime—awe and admiration at the various spectacles of nature that raise the soul above the vulgar and the commonplace, arousing emotions akin to fear rather than merely joy.[79] The sublime was manifested in the contemplation of raging cataracts, perilous views from mountaintops, the forces of nature, expanses of uninhabitable landscape, the infinity of space and time, but also breathtaking artificial structures and powerful machinery. According to Voller, the concept of the romantic sublime provided a substitute for Christian cosmology displaced by the growth of science. It was "an aesthetically grounded quest devoted to recovering intimations of the divine."[80] The romantic quest frequently discovered the sublime in the technological.

Romanticism also embraced Neoplatonism. Much of Plotinus resonates with romantic introspection—subjectivity as a means to truth. In pursuit of beauty, Plotinus said, "Withdraw into yourself and look."[81] As a sculptor cuts away bits of stone, we are to cut away all that is excessive, straighten the parts that are crooked: "labor to make all one glow of beauty and never cease chiseling your statue until there shall shine out on you from it the godlike splendor of virtue, until you shall see the perfect Goodness established in the stainless shrine."[82] Similarly, Fichte, in romantic vein, exhorted his students: "Heed only yourself: turn your gaze away from all around you, and inwards onto yourself; this is the first demand that philosophy makes of its apprentice. Nothing outside of you matters, but solely you yourself."[83] The Neoplatonic concept of the unity had a particular appeal to the romantic poets, who read Plotinus, and his influence is evident in Schiller, Wordsworth, Coleridge, Emerson, and Tennyson.[84] According to Schiller (1759–1805): "If every man loved all men, each would possess the All."[85] For Coleridge (1772–1834), the soul is to behold God: "Till by exclusive consciousness of God; All self-annihilation it shall make; God its identity; God all in all! We and our Father one!"[86] Turnbull identifies echoes of Plotinus in the essays of Emerson (1803–1882), who describes "that unity, that Over-Soul, within which every man's particular being is contained and made one with all other."[87] The romantics included the pre-Raphaelites, or symbolists, John Ruskin, William Morris, Oscar Wilde, Dante Gabriel Rossetti, and Arthur Symons, who eroticized various Neoplatonic themes, indicating a kind of union between the body and the soul: "I feel . . . that ineffable delight; When souls turn bodies, and unite; In the intolerable, the whole; Rapture of the embodied soul."[88] The pre-Raphaelite brotherhood, as some called themselves, was also fascinated with the occult, mesmerism, and spiritism, though Hönnighausen sees pre-Raphaelite mysticism to be driven less by a desire for union with a transcendent world than "a love of mysticism *per se*."[89] Such was the subjectivism of the romantic flirtation with Neoplatonism.[90]

The early socialists also invoked notions of unity and transcendence in the presentation of their utopias. According to Stedman Jones and Patterson, " 'socialism' began as an attempt to discover a successor, not to capitalism, but to the Christian Church."[91] Fourier (1772–1837) provides a highly elaborate theory of "passionate attraction" based on a "unified system of movement for the spiritual and the material world,"[92] so that "the attractions and properties of animals, vegetables and minerals might be coordinated on the same level as those of man and the stars."[93] The three aims of the soul are "Luxury of the five senses," "the progressive Series" and "Universal unity."[94] Various unity sects persist to this day, advancing little on the Neoplatonic theme. For example, Steiner's system of Theosophy culminates in "the seventh region, that of the actual soul life, [which] frees us from our last inclinations toward the sensory, physical world,"[95] where the soul is "absorbed into its own world, and the spirit, free of all restraints, wings its way upward into regions in which it lives only in its own element."[96] Steiner calls on Goethe and other romantic writers, and Theosophy informed modernism in art and architecture through expressionism. In contemporary cultural commentary, Jencks campaigns for an expressionist revival in architecture and art, appropriating quantum and chaos theories that "reveal the universe as a single, unfolding, creative event that is always reaching new levels of self-organization," the implications of which are "not only spiritual but architectural."[97] The stages of development of the butterfly are the new model for "cosmic evolution." In a similar vein, the sociologist Maffesoli asserts that "it is in rediscovering the virtues of Mother Nature that a feeling of wholeness is restored."[98] There is every indication that we are still in the romantic age.

As with the romantic appropriation of the medieval, we need not presume that soul, unity, and intellect mean the same to Enlightenment and romantic sensibility as they did in antiquity. Romanticism asserted the transcendence of the human spirit, the wholeness of Nature, and a better world beyond, against an emerging science of reduction in an increasingly industrial world. So romantic Neoplatonism was inevitably informed, or infected, by notions of the self as subject, the philosophy of individualism, genius, the bifurcation of feeling from reason, and the loss of symbol as a means of participating in the Intelligible.[99]

Disembodied Cyberspace

As we have seen, romanticism is the truly abiding aspect of the Enlightenment and emerges in popular culture, from Hollywood to the exponents of cyberspace. According to Cunningham and Jardine: "The political creeds which dominate our world, our views of human agency and morality, our arts and our conceptions of artistic creativity, all are rooted in the romantic movement."[100] Cyberspace narratives are informed by romantic (Gothic) fiction, as exemplified in Gibson's *Neuromancer*, in which cyberspace exhibits the trappings of the technological sublime. As such, according to Voller, the novel's characters "connect with an abstract geometry of data, a metaphor of the immaterial, risking life and mind in order to plumb the depths not of soul or psyche or mystery, but of the wealth and power immanent in data."[101]

Romanticism in Neoplatonic guise is a force in the late modern age. Writing in the early 1970s, Wallis, a scholar of Neoplatonism, draws attention to the "contemporary mystical revival" but regards it as an undisciplined Neoplatonism: "our present-day prophets, notably those of the psychedelic cults, too often ignore the necessity to the religious life of discrimination and self-discipline, and in this differ from the best mystics of all traditions, including the Orientals they profess to follow."[102] At around this time, McLuhan was proposing that a "current translation of our entire lives into the spiritual form of information" may "make of the entire globe, and of the human family, a single consciousness."[103]

Apart from their lack of discipline, cyberspace narratives conspire with notions of unity, ecstasis, and disembodiment to present a technological face to the Platonic universe. Their talk of lifting anchor from the constraints of a material world clearly resonates with the Neoplatonist tenets of distanciation and disembodiment. According to Benedikt, cyberspace fulfills the dream, "thousands of years old," of "transcending the physical world, fully alive, at will, to dwell in some Beyond—to be empowered or enlightened there, alone or with others, and to return."[104]

Cyberspace narratives also resonate with Berkeleyan idealism[105] the implication that we can inhabit an electronic world, and that the other world, where we sleep and eat, to all intents and purposes vanishes into unreality. The real is what we experience it to be, and if information technology constitutes part of that experience to give us new experiences, then reality is indeed altered. The real is what we think it to be, and through the mental prosthetics of the computer we can make it as we wish.[106] We can apparently violate space and time constraints in cyberspace by moving through objects, be in two places at once, handle numerical data as though it were physical, and merge with other minds. According to Sullivan-Trainor: "Today, network users can travel alone or with other 'minds' as they sit at their computers. Their minds enter cyberspace—a real place in which they can exchange ideas or emotions, but in which no physical presence or face-to-face contact is necessary."[107]

In his essay "The Erotic Ontology of Cyberspace," Heim directly addresses the meeting of the Platonic real with cyberspace.[108] He picks up Plato's theme of Eros (love or life instinct), which, when properly trained, provides the means of escaping from the contemplation of the beautiful in mere sensible things to contemplating the idea of beauty itself and the world of ideas (the real). Heim highlights the obvious reference in Gibson's *Neuromancer* of the "matrix" (the vast cyberspace network) as "mother," and how in both Platonism and cyberspace, "Eros inspires humans to outrun the drag of the 'meat'—the flesh—by attaching human attention to what formally attracts the mind. As Platonists and Gnostics down through the ages have insisted, Eros guides us to Logos [the real]."[109] According to Heim, ecstasis is at work: "Cyberspace is Platonism as a working product. The cybernaut seated before us, strapped into sensory-input devices, appears to be, and is indeed, lost to this world. Suspended in computer space, the cybernaut leaves the prison of the body and emerges in a world of digital sensation."[110] Heim indicates how cyberspace differs from the insensate world of Plato's ideas in that cyberspace is a fusion between the pure realm of cognition and the empirical data of sense experience, incorporating the "smallest details of here-and-now existence."[111] It is not necessary for us to decide here whether cyberspace exists or will ever exist as so described, or to decide on the implication that the concept of cyberspace is superior to Plato's concept of the real. The final transformation, or filter, through which this strand of idealism comes to us is clearly a technological one. The role of a technoromantic Neoplatonism in cyberspace narratives is exemplified in Heim's rubric:

"With an electronic infrastructure, the dream of perfect FORMS becomes the dream of inFORMation."[112]

In summary, how does this technoromantic, Neoplatonic idealism appropriate the real? For Plotinus, the *proximal* concerned a relationship between the soul and the real, which is beyond what the body encounters. The real is appropriated through ecstasis or through participation in symbol. For romanticism, notions of soul are transformed by concepts of the ego as self, subject, and individual. The techno-idealist adds computer technology to the relationship. What is real is what the ego is connected to on a computer: "real life is life on line."[113] The proximal is the electronic matrix, "the interfusion of nervous system and computer-matrix,"[114] appropriated by anyone anywhere by "jacking in."

For Plotinus, participation in the real is a *shared* participation, but it is a sharing through mutual participation in the unity. In Neoplatonic Kabbalistic doctrine, "God's essence is linked and connected with all worlds," and "all forms of existence are linked and connected with each other."[115] Romanticism embraces concepts of the unity of humankind, but there is always the threat of the collective, which represents mediocrity, and it is against this that the creative genius within has to assert itself. The starting point for participation is the ego and, at best, a collection of egos. In the electronic age, there is the expectation of ultimate participation in a united electronic consciousness in which we will all share. The translation of such a unity through the filter of the Enlightenment presents us with the prospect of a giant collective ego.

For Plotinus, the *body* is an impediment, and we are exhorted to enter into an ecstatic visionary state, where the soul is no longer hindered by the body. It is in the desire for ecstasis, and the mistrust of the body, that the confluence between Neoplatonism, romanticism, and the cyberspace theorists is strongest. Ironically, in cyberspace narratives the out-of-body experience is appropriated through technological means, which inevitably require connections to the body, and a complex material computer infrastructure, though here the phenomenon of connecting to a computer is thought to transcend the mere materiality of being a body connected to a machine. Vannevar Bush's article, "As We May Think," commonly taken as the inspiration for the development of hypertext and the World Wide Web, proposes that it may someday be possible to establish direct links from the record to the brain, bypassing the bodily senses: "With a couple of electrodes on the skull the encephalograph now produces pen-and-ink traces which bear some relation to the electrical phenomena going on in the brain itself. . . . who would now place bounds on where such a thing may lead?"[116]

The provision for ecstasis in the narratives of cyberspace is unbounded. Wiener, the founder of the science of cybernetics, was one of the first to seriously posit the prospect of reducing the body, as well as the mind, to information so that bodies might conceivably "travel by telegraph" as well as by train or airplane: "There is no fundamental absolute line between the types of transmission which we can use for sending a telegram from country to country and the types of transmission which at least are theoretically possible for a living organism such as a human being."[117] In an extensive summary of "cyberculture," Dery reports on the common theme of "discorporation" found in writing about cyberspace and provides numerous examples of the theme of bodily entrapment and eventual release in a new world through computers.[118] Apparently, we already have a prescience of this in that "growing numbers spend their days in 'static observation mode,' scrolling through screenfuls of data. Bit by digital bit, we are becoming alienated from our increasingly irrelevant bodies."[119] There is a "body loathing, a combination of mistrust and contempt for the cumbersome flesh that accounts for the drag coefficient in technological environments. . . . The software of our minds is maddeningly dependent on the hardware that houses it, our bodies."[120]

For Plotinus, the things of the sensible realm are copies or *repetitions* of each other derived from Intelligible things that are more enduring. Romanticism is a party to empiricist concepts of observation and repeatability, though it commonly reacts against notions of method, seeking rather a science that takes nature as a whole. For the techno-idealist, repetition as pattern is everything. Cyberspace appeals because it deals in patterns, the repeatable characteristics of everyday objects and contingencies, as exemplified in the writing of Moravec in the field of robotics. Moravec scorns what he regards as the "body-identity-position," which "assumes that a person is defined by the stuff of which a human body is made."[121] By way of contrast, he proposes the concept of "pattern-identity," which defines the "essence of a person" as "the *pattern* and *process* going on in my head and body, not the machinery supporting that process. If the process is preserved, I am preserved. The rest is mere jelly."[122] As every atom that makes up our body is likely to have been replaced through normal biological and chemical processes by the time we reach middle age, Moravec argues it is clearly only our pattern that stays with us. From this insight, Moravec posits a future in which we are able to transplant, copy, and merge the information of our bodies and minds in networked computers. Using the DNA code from the remains of those long dead, we will attempt to undertake "wholesale resurrection" through the use of "immense simulators."[123] (If we find these ideas uncomfortable, then apparently it is because we "are accustomed to looking at the world in a strictly linear, deterministic way," without taking heed of the "uncertain world described by quantum mechanics."[124]) We need not only put the code of our own bodies and minds into computers, but put that of all species, even those that are extinct. The outcome will be a "supercivilization, the synthesis of all solar-system life, constantly improving and extending itself, spreading outward from the sun, converting nonlife into mind."[125] If we happen to meet another expanding mind bubble, then we can negotiate a merger, requiring "only a translation scheme" between memory representations. According to Moravec, this process may already be occurring elsewhere in the universe and might eventually "convert the entire universe into an extended thinking entity, a prelude to even greater things."[126]

The *ineffable* to which Plotinus appeals is the yearning for unity, the universe as one living organism. For romanticism, the yearning is for unity with nature, the whole, freedom, and creativity. For the technoromantic, it is for such a degree of absorption into technology that not only the body but technology is transcended. The electronic matrix is something greater than the contingencies of individual components, their physicality and their failings. We become one with each other and with our machines. In a survey of the culture of "cyberpunk," Rushkoff indicates, with approval, the prevalence of belief in "the development of the datasphere as the hardwiring of a global brain," which is the final stage of the development of the earth as a living being, in keeping with the so-called Gaia hypothesis: "As computer programmers and psychedelic warriors together realize that 'all is one,' a common belief emerges that the evolution of humanity has been a willful progression toward the construction of the next dimensional home for consciousness."[127]

Such narratives are compelling in many ways but have their detractors and counternarratives. The theme of transcendent information technology generates narratives that position the materiality of information technology, that is, the spatial and temporal location of hardware and how we interact with it, at the margins. So there are the narratives of technical explanation that deal with certain aspects of the materiality of information technology—the engineering or ergonomics of information technology—and there are narratives of transcendent cyberspace. There is information technology as we encounter it in our day-to-day lives, and there is the transcendent computer that is not yet with us, the computer of the utopian visionary. In cyberromanticism, the configuration of cables, satellites, transmitters, receivers, computer processors, monitors, keyboards, and the body in front of the machine become subservient to the abstract system, the logic diagram, the utopian vision. The essential computer transcends the processor, the screen one sits in front of, and one's own body. Cyberspace narratives commonly speak of being immersed in a virtual world while making only passing reference to the materiality of the computer, the ergonomic environment, the clumsiness of the headset, or the inertia of the dataglove. The isolation of the information technology system from the material of which it is constituted does not comfortably integrate notions of technical breakdown, of ergonomics, or "human factors." Such narratives invoke what Grusin describes as the "trope of dematerialization,"[128] which marginalizes the world of practice—of material, human, and technological contexts. According to Markley, the rarefied idealism of such narratives leads us to ignore the "history of labor, of people building machines."[129] Cyberspace is "a fantasy based on the denial of ecology and labor."[130]

The usual antagonist of romanticism is empiricism, whose sober narratives of commonsense realism leave no space for the heady speculations of romanticism. The inadequacy of empiricism against the weight of technoromantic narrative is the subject of the next chapter.

Multiplicity

The Empiricist Tradition of Realism, and Its Critics

3

The Empiricist Legacy

Is empiricism the antidote to romanticism? If technoromantic narratives champion the issue of unity, then empiricist narratives speak ostensibly to the matter of categorization, breaking the whole into parts. If for technoromanticism the real is a transcendent unity, for empiricism the real is "reality," the material, natural world of particulars, subject to scientific scrutiny.[1]

We will examine empiricism and digital narrative through concepts of space. Concepts of space commonly appeal to notions of representation through spatial models, overcoming the resistance of space, the reduction of space to numerical systems, categories of real space versus space as experienced subjectively, and the suggestion of access to spaces in artificial worlds. Empiricism presents a range of spatial narratives: of space as represented, resisted, reduced, and divided. I argue that rather than countering romanticism, empiricism provides the conditions for technoromantic narratives to entertain the transcendent potential of computer space. Ultimately, empiricism's sober reflections on the ability of the computer to represent space takes us close to the romantic vision of cybernetic rapture, where, according to Csicsery-Ronay, "the computer represents the possibility of modeling everything that exists in the phenomenal world, of breaking down into information and then simulating perfectly in infinitely replicable form those processes that precybernetic humanity had held to be inklings of transcendence."[2]

The Virtually Real

Representation is a central theme in digital narratives. Architects, engineers, designers, planners, geographers, and the creators of computer games and simulators use computers to represent space, as in computer-aided design, virtual reality, and geographical information systems. Such systems employ databases in which numerical and other attribute data based on some coordinate system or other are stored, which can be manipulated according to rules of mathematics and geometry. Computer modeling is a party to changes in the nature of architectural, engineering, planning, and design practice, and it is becoming increasingly important in science, medicine, education, and training. Aspects of the entertainment industry appear to have been revolutionized through computer modeling, and the prospect of immersion in three-dimensional virtual worlds captures the romantic imagination. Computer simulations of three-dimensional objects of considerable complexity testify to the power of spatial representation by computer, particularly in computer visualizations, computer animations, and interactive, distributed, and immersive three dimensional environments.

Representing objects in space by computer implicates the concept of correspondence. That there is a relationship of correspondence between a sign and an object seems to provide a key notion within drawing practice, as highlighted in March and Steadman's key text on geometry in architecture: "In the case of a measured drawing from an existing building the draughtsman plots selected points of the real building, such as roof lines and the corners of openings, and 'maps' these onto his drawing. There is in this instance a one-to-one correspondence between the points in reality and their representation on the drawing, and vice versa."[3] In precision drafting, and in CAD models, the points and lines of the model in Cartesian space correspond to the corners and edges of the object in physical space. The correspondence can often be established by simple procedures and technologies such as the use of some measuring convention, surveying technique, or a camera, or even tracing outlines on a sheet of glass held between the observer and the object. The correspondence may also entail functions and algorithms of the kind outlined by March and Steadman for transforming and repeating objects defined in coordinate space. The obvious benefit of these mapping functions is that their effects are reproducible, so that at different times and in different contexts, one can achieve similar mappings following the same procedures and using the same technologies. The mapping between a representational model and the object can be achieved by several means. Objects can be mapped into a computer model manually (by "drawing" on the computer screen with a mouse, or by typing in coordinates) or by using scanners, photogrammetry, and digitizers. In turn, objects can be reproduced from computer models by the production of blueprints and instructions for builders, manufacturers, and fabricators, or by using CADCAM (computer-aided design and computer-aided manufacture) techniques, involving digitally controlled cutting machines or lasers for producing objects from the model that are identical to each other. We can also use algorithms to convert models to line drawings on a flat plane, or photorealistic visualizations. When we compare these computer-generated images with photographs of the objects, then we see that there is a further correspondence. The mathematical transformations used to generate the perspective view seem to correspond to the workings of light through the lens of a camera onto a photographic plate.[4]

The benefits of computer representations are well known and do not need to be catalogued here. But as anyone who has worked with documentation drawings (blueprints) or CAD knows, there are problems with the notion of representing objects in space in computer systems. Even though the principles of coordinate geometry and the workings of mathematical transformations are clear, there is no agreement on a universal graphical language for describing objects. There are as many ways of organizing and manipulating spatial information in a computer database as there are ways of describing objects in a manually produced line drawing, and not all are equally efficient or effective for all situations. There are limits to what can be simulated on a computer using a particular set of algorithms, and a computer system designed to handle one kind of spatial organization (a grid of columns, for example) may be unsuited to representing others (such as a mountain range). Furthermore, the numbers of points and edges in an object, and their locations in space, change according to how you choose to look at the object. To apply a geometrical schema is to engage in interpretation. Modeling a building, or any complex scene, requires one to ask what constitutes a corner and an edge in this situation, and, indeed, if it is the edges and corners that are of most interest? As is evident from experiments in "computer vision" and robotics, there are no algorithms for deciding and identifying which are the useful and important points and edges to be modeled in a particular object, such as a building, for all uses and for all contexts. The way you can represent an object is never exhausted.[5] These problems do not invalidate the importance of correspondence in representing objects in computer systems, but they indicate that correspondence is preceded by the more important notion of interpretation.

These problems and objections prompt some to say that what we are dealing with in a computer system, as in manual drawing or some other representational system, is a language. A language has to be learned, and, according to the structuralist view of language, elements in the language primarily bear a relationship to each other rather than to some thing referred to. The relationship between the signifier (the word or the entry in a 3-D database) and the signified (the scene being modeled) is arbitrary, according to Saussure, but decided by a language community. We will examine language and space more closely in chapter 4, and explore views of language counter to the structuralist one, but at this stage we can see that a structuralist linguistic view of computer representations affirms that there is no automatic mapping between a series of coordinates in a computer database and a "real world" situation. The idea of correspondence does not provide a good basis for a theory of language or for an understanding of computer modeling.

Digital narratives also often take the issue of *resistance* as a central tenet. If information technology concerns representing space and the objects in it, it is also concerned with violating the constraints normally associated with the world being represented. This ability of computer representation to supposedly overcome the constraints imposed by the physical world captures the romantic imagination. For Benedikt, "in patently unreal and artificial realities such as cyberspace, the principles of ordinary space and time, can, in principle(!), be violated with impunity,"[6] as is currently the case with films and cartoons. Apparently, there is no need to be constrained by physical laws in cyberspace. According to Mitchell: "Physical movement and phenomenal motion can now be disconnected; we teleporting cyborgs have found loopholes in Newton's laws."[7] Information technology provides a means of conquering the limitations of space and time by allowing us to be in contact with each other irrespective of spatial distance, thereby enhancing our freedom to do certain things. In this sense, space as experienced without the mediation of technology provides a constraint on our actions. Space is something to be resisted. The concept of cyberspace embraces the notion that we inhabit virtual spaces through communications media that transcend the constraints of physical space. Digital narratives commonly associate cyberspace with freedom. For example, Sullivan-Trainor states: "The broad availability of virtual reality systems across a network, accessible within the boundaries of our own homes and offices, will alter our perception of time and space and free us from the limitations of distance."[8] Cyberspace allows free movement through a vast, infinitely expanding virtual space. According to Rushkoff, cyberspace is "a boundless universe in which people can interact regardless of time and location."[9] Negroponte also predicts that soon we "will socialize in digital neighborhoods in which physical space will be irrelevant and time will play a different role."[10] Eventually, space will diminish to nothing at all, and the "digital planet will look and feel like the head of a pin."[11]

As well as providing resistance, to be overcome by information technology, space connotes freedom. To be free is to be given space, to be released from confinement. Cyberspace provides access to a larger, unconfined space, which is in turn analogous to enjoying freedom to roam the high seas and to discover new worlds. According to Heim: "The final point of a virtual world is to dissolve the constraints of the anchored world so we can lift anchor . . . so we can explore anchorages in ever new places."[12] Leary reminds us of the etymology of "cybernetics," which pertains to piloting rather than controlling. The cybernaut is continuing in the spirit of the great explorers and navigators in charting and finding their way about unknown seas.[13]

The computer as a means to freedom, either by overcoming the constraints of physical space, violating its laws, or presenting us with ever larger spaces in which to move, has its detractors and counter narratives. Anyone who has worked with three-dimensional modeling software knows that the creation of geometrical worlds is highly constrained. Computer networks promote new modes of communicative practice but do not reduce the world to the intimacy of a village. People still want to be together, in the same room, socially, and in work contexts. The World Wide Web presents new constraints (having to maintain up-to-date web pages for example), and browsing the net can involve sifting through masses of unusable data. If information technology overcomes space as resistance, or if information technology opens up new spaces, these effects are not experienced equally and for everyone. Benedikt implies that the violations of space accomplished through information technology are largely illusionary. He makes the distinction between actual violations of space and the representation of those violations, pointing out that "myth and fiction do not contain violations of ordinary spatiotemporal logic but *descriptions* of such violations."[14] The reduction of distance appears to be a practical—which is to say, contingent—matter. Our freedom is also at a cost, to others if not to ourselves.

Certain digital narratives also present the issue of *reduction* as a central theme. The view that information technology overcomes the resistance of space, or opens up to us ever new spaces, is sustained by the view that space can be reduced to its representation in a computer system. This is the view implied in narratives that look to mathematics and geometry, to the concepts of the point, line, and plane through which to construct an understanding of space. The point is without dimension but located in space on a coordinate system. A line is the shortest path between two points. A plane is bounded by lines, and volumes by planes. Other properties of objects in space, or different articulations of space, are either derived from these, are supplemental to them, or are imposed by our imaginations.[15] If we accept that the key to space lies in its mathematical description, then we may accept that the spatial experience can be captured in a virtual reality system, and such systems actually contain, reproduce, and re-present space. The distinction between physical space and computer-represented space thereby dissolves. If this is the case, then the realization of virtual reality systems that provide convincing spatial environments (of the kind that would persuade us we were on a holiday in the Bahamas, or even better) is a simple matter of detail. We need only provide more detailed descriptions of object worlds for virtual reality and better graphical routines for realism, faster processors, and more detail in the computer model to make it more realistic.[16]

Spatial reductionism also has its detractors and counternarratives. The success of computer simulations does not appear to depend on the replication of detail but on multiple representations, multiple views, heuristics, fudges, and other concessions to the context of use of the computer system. In the case of virtual reality systems, there appears to be no simple mapping from the world to the computer simulation via some simple description language and representation of the laws of physics. Arguments against such reductions are presented from phenomenology: that coordinate geometry is derived from spatial experience, not prior to it. Virtual reality provides another instance of what the phenomenological critic Dreyfus characterizes as being kept honest by programming computers. Spatial modeling and virtual reality, as with AI, "has called the Cartesian cognitivist's bluff."[17] If space was as Descartes suggested, then computers would have vindicated Cartesian reductionism long ago.

Certain digital narratives focus on the theme of the *division* of space, presenting the case that the problems of understanding computer modeling, cyberspace, and VR alluded to so far can be solved if we recognize that there are two kinds of space: space as it exists objectively in the world of matter, and space as we experience it—that is, practically or subjectively. The objective and the experiential do not always coincide. So there is "real life" (RL) space containing the objects we represent in computer systems and within which we have to eat, pay attention to health, keep up the mortgage payments. And there is space as a construct of the imagination, sustained through the technologies of three-dimensional modeling, the Internet, the World Wide Web, and interactive computer games. Computers give us access to new subjective spatial experiences. Space is thereby divided into the objective and the subjective, which is a distinction of some long standing.[18] Architects and geographers commonly distinguish between space and place. A space is reducible, can be described mathematically and on drawings such as plans and maps. On the other hand, place is memory qualified and imbued with value.[19] Some conceptions of place are subjectivist, invoking romantic concepts of sensitivity to the spirit of place. The information technology characterization of this dualism affirms that there is real space to which cyberspace is set in opposition.

There are narratives that run counter to narratives of space so divided, pointing to difficulties with the distinction between the actual and the experienced. Space presents to us in various practical contexts, including the laboratory, measurement, computer modeling, and cyberspace. The concept of objective against subjective space can be displaced by pragmatic concepts to be discussed in chapter 5. Furthermore, according to critical theorists, our insistence on an actual, objectively understandable, representable kind of space promotes certain political agendas. We will examine the concept of social space developed by Levebvre and others in chapter 5.

Then there are narratives of space *transcended* as outlined in the previous chapter. The full romantic transcendence into a cybernetic unity resides at the extreme end of the narrative spectrum of space represented, resisted, reduced, and divided, and builds on concepts of empirical reality.

The Empirically Real

Empiricism presents the real in terms of *reality*, a world differentiated into objects and categories. Digital narratives that trade in empirical reality develop around several themes: naive, representational, scientific, mathematical, and materialist forms of realism.[20]

I have indicated how the Neoplatonic pursuit of ecstasis informed empiricism, particularly through Berkeley, but the kind of empiricism that has been passed down to us casts the pursuit of the real (as proximal, bodily, shared, repeatable, and ineffable) in a different light. Empiricism presents the *proximal* as what is received through the senses. According to the empiricist, the real is defined as what is within our *bodily* grasp, through which the senses operate. Of course, our senses are not entirely reliable, and the reality and its appearance may be different. Locke distinguished between the primary or real qualities of objects, how they are in themselves and can be studied by science, and their secondary or apparent qualities, which is how they appear to the human senses.[21] A further way of casting the distinction is between physical and sensible properties of objects. For example, there is the real color of an object, as determined by scientific measurement, and there is the sensible color, as we perceive it.

According to empiricism, to the extent that we are able to recognize the same objects from our individual sense experiences, we *share* in the same reality. But it is the *repeatability* of sense experiences that provides the major foundation for understanding the world. Sense impressions are explicable in terms of patterns, which recur, even though the appearances of objects in our perceptual field change. The quintessence of repeatability is the scientific experiment. For a phenomenon to be real, it must be observed repeatedly or at least conjectured to be observable repeatedly, and the conditions for its appearance must be reproducible at different locations and times. Empiricism seeks to push back the bounds of the *ineffable*, that which is beyond explanation. It is interested in a state of complete knowledge, the scientific imperative being to know, which means to explain and to predict. Notwithstanding the Neoplatonic strand to empiricism, which resurfaces later with Comte and others, the empiricist tradition appropriates the real as an issue of *reality*, which is to couch our experience of the world in terms of an observing subject and an external world, and asks, "What is the nature of this external world, and how can we know it?"

So empiricist digital narratives, which trade in the use of the computer for modeling, representation, and simulation, distance themselves from technoromanticism. With computer systems, we can represent reality through data and process it to produce outputs, computer functions that make no explicit appeal to technoromantic excess. There are several variants of the empiricist narrative.

There is the theme of "common sense" or *naive empiricism*, which assumes that there is an object world of things, such as buildings, trees, and hammers, of the kind that we use and inhabit and that are able to be apprehended by the senses. Such a view is well articulated by the architecture critic Jencks: "Most scientists believe they are discovering something objectively real and that reality 'obeys' or is 'subject to' these laws. . . . I also believe this transcendent realm exists. Its importance is not only to set the universe in motion, and sustain it but, as far as we are concerned, to be a measure and standard for us—independent of us."[22] This is empirical realism. Thinkers from the analytic tradition of philosophy, such as G. E. Moore, have championed the empirical defense of realism, particularly against Berkeley's idealism, which Moore interpreted as positing that the world is "only spiritual." He attempted to refute Berkeley's proposition that we are only aware of what we can directly experience and to suggest that Berkeley's idealist position leads to skepticism about the physical world.[23] As an analytic philosopher, Russell also declared the independence of an empirical reality: "We may therefore admit—though with a slight doubt derived from dreams—that the external world does really exist, and is not wholly dependent for its existence upon our continuing to perceive it."[24] Popper also affirms realism as "essential to common sense," and adds: "Common sense, or enlightened common sense, distinguishes between appearance and reality."[25] Since appearances can be deceptive, the concept of reality is accepted as a reluctant necessity. Most empiricists agree that, whereas we can refute the particularities of Berkeley's arguments, we cannot positively establish the existence of the real world, either by incontrovertible observation or by logic. According to Popper, "Realism is neither demonstrable nor refutable . . . But it is arguable, and the weight of the arguments is overwhelmingly in its favor."[26] As a skeptic of modern empiricism Nagel also nonetheless concedes realism: "My position is that realism makes as much sense as many other unverifiable statements, even though all of them, and all thought, may present fundamental philosophical mysteries to which there is at present no solution."[27] For naive realism, the independent reality is what we can represent in computer systems. What is in the computer is only a representation. Putative encounters with cyberspace pertain to appearances and present no more of a challenge to reality than do the ephemeral appearances of dream objects.

Representational realism provides a further variant of the empiricist narrative, asserting that the world exists independently of us, and we can know it, apprehend it, and represent it to ourselves. According to Russell: "We may assume that there is a physical space in which physical objects have spatial relations corresponding to those which the corresponding sense-data have in our private spaces. It is this physical space which is dealt with in geometry and assumed in physics and astronomy."[28] As a critic of empiricism, Rorty characterize this kind of realism as the view of the mind as "the mirror of nature": "The picture which holds traditional philosophy captive is that of the mind as a great mirror, containing various representations—some accurate, some not."[29] According to representational realism, we have access to the real through our capacity for representation and introspection on that representation. The mind is like a powerful computer that contains representations of the world, including representations of space.[30] Representational realism brings the issue of correspondence to the fore.

How does representational realism interpret the real? The *proximal* is translated into a mapping between something designated as real and its representation. Proximity is established through words and signs that map representations to their objects. Different minds *share* the same reality by having the same representations of it. The real is an issue of *common* sense. The *body* is a conduit for sense impressions by which mental representations are formed, and presumably the organ of the brain is where the representation is stored. The world exhibits patterns that can be represented and *repeated* in the mind and also in computer systems. The achievement of a uniform, consistent picture of reality as it is in itself hints of the *ineffable*.

Narratives that implicate scientific or *metaphysical realism* (a further variant) assert that we can adopt a "God's-eye perspective" of whatever constitutes the real. We may not yet know what it is, and our understanding may at present be only partial, but there exists ultimately one correct description of reality (a further instance of the unity myth). In terms of our theme of proximity, scientific realism assumes that we can be outside the systems we are considering. Einstein supported such a view, claiming of physics that it "is an attempt conceptually to grasp reality as it is thought independently of its being observed."[31] For Einstein, in this sense "one speaks of 'physical reality.' "[32] Hawking also advocates one scientific world view: "I am hopeful that we will find a consistent model that describes everything in the universe. If we do that, it will be a real triumph for the human race."[33] Metaphysical realism seems to motivate the computerized encyclopedia project (CYC) of Lenat and Feigenbaum, in its goal to store what we take for granted about the world in a computer, to achieve automated commonsense reasoning from factual information, rule, and analogy.[34] As a critic, Feyerabend defines metaphysical realism as the view that "there are certain objects in the world and that some theories have managed to represent them correctly."[35] Johnson challenges metaphysical realism for depending on an impossible neutrality regarding observation and on outdated views that language works because of correspondences between words and things, which "in turn, requires an ahistorical, neutral rationality and a theory of meaning and reference in which words can map directly (unmediated by understanding) onto objective reality."[36]

Mathematical realism is a further variant on the theme of representational realism, developing the notion of space and objects converted (reduced) to mathematical description, as a means to establishing correspondence. Here the repeatability and predictability of operations with numbers and mathematical symbols determines the constitution of the real. For the Pythagoreans, numbers, figures, and geometry were the essence of things, and entities that exist are imitations of mathematical objects.[37] For Plato, the intelligible realm is also inhabited by number, and calculation is a means of training the mind to escape from the "prison house" of appearances. Those who are to take part in the highest functions of the state should study number until "by the aid of pure thought, they come to see the real nature of number."[38] The practice of calculation helps "in the conversion of the soul itself from the world of becoming to truth and reality."[39] Mathematical realism thrives in the current climate of scientific investigation, at least in physics, where the only "certainty" appears to be the behavior of numbers. According to Penrose, "some mathematical truths seem to have a stronger ('deeper,' 'more interesting,' 'more fruitful'?) Platonic reality than others. These would be the ones that would be more strongly . . . identified with the workings of physical reality. (The system of complex numbers . . . would be a case in point, these being the fundamental ingredients of quantum mechanics, the probability amplitudes.)"[40] According to Penrose, the mind mediates between the realm of mathematics and the physical world because of its privileged access to the real through the way it is structured: "With such an identification, it might be more comprehensible how 'minds' could seem to manifest some mysterious connection between the physical world and Plato's world of mathematics."[41]

According to mathematical realism, space conforms to the canons of mathematics and of logic. There are principles of coordinate geometry pertaining to translation, transformation, deformation, and combination, and there are the complex principles and theories of statics, dynamics, and the physics of particles, waves, and fields, concepts that seem to grant power to computation over space.

Materialism is a further variant on the theme of scientific realism. Materialism finds accord with the pre-Socratic atomists, who attempted to address the problem set by Parmenides about the existence of nonbeing. They settled this problem by positing that there is never a state in which there is *nothing*, but there are always entities—namely, tiny geometrical material particles that defy perception moving about at random under their own propulsion—which make up physical objects as well as supposedly empty space.[42] Plotinus found the atomists a ready target, particularly their proposition that the order of the universe is founded on random behavior:

> Certain philosophers (Leucippus, Democritus, Epicurus) postulate material principles such as atoms; from the movement, the collisions and combinations of these they derive the existence and mode of all particular phenomena, supposing our own impulses and states even to be determined by these principles. But it is an absurdity to hand over the universe to material entities and out of the disorderly swirl to call order, reason and the governing Soul into being. How can we imagine that the onslaught of an atom striking downwards or dashing in from any direction could force the Soul to reasonings, impulses or thoughts?[43]

Contemporary materialism defines itself within the context of a different problematic to the atomists. Materialism sets itself against what it calls Cartesian dualism, the implication from Descartes that there must be an independent, nonmaterial spirit that controls the body. For materialists, Descartes was responsible for promoting the myth that there is some ephemeral mind quality independent of the objective world. This is the myth of the "ghost in the machine" targeted by Ryle in *The Concept of Mind*.[44] As a resolute materialist, Dennett maintains that "there is only one sort of stuff, namely *matter*—the physical stuff of physics, chemistry and physiology—and the mind is somehow nothing but a physical phenomenon. In short, the mind is the brain."[45] According to materialism, science, as a unified and consistent discipline, will ultimately provide the means of explaining all human phenomena, such as consciousness: "We can (in principle!) account for every mental phenomenon using the same physical principles, laws, and raw materials that suffice to explain radioactivity, continental drift, photosynthesis, reproduction, nutrition and growth."[46]

For materialism, what is within our grasp (*proximal*) is matter, and we *share* this as we have the world in common. Our *bodies* include our minds, which are a kind of software for the circuitry of the brain as computer. It is even possible to conceive of thinking without a body. Dennett posits the conundrum of the brain in a vat, devoid of a body, but fed stimuli by scientists to enable it to imagine itself in a physical world—which he regards as "possible in principle" though technically impossible at the moment. For Dennett, this mind experiment is the starting point for exploring issues pertaining to human consciousness, which is a material brain phenomenon. For materialism, the *repeatable* patterns of electrical impulses provide a model for understanding consciousness. We are among the real things in the real world, entities produced through the repetitive operations of evolution, and products of our genes interacting with a material environment. For Dennett, the goal is the ultimate explanation of all things in materialist terms, including consciousness, the answer to which resides in information processing, and is illusive if not *ineffable*. But ultimately Dennett's materialism leads to a new metaphysics of natural selection. The current stage in evolution is the emergence of *memes*, which are self-replicating cultural units or ideas (like the idea of the wheel, wearing clothes, chess, or "Greensleeves") that thrive or expire depending on their place in the collective "memosphere" of human minds, but also of sayings, books, buildings, artifacts, and other "meme vehicles."[47] Empirical realism as materialism returns to a kind of ecstasis, the body transcended, and a transcendent collective meme pool.

Materialism presents as one of the most potent contemporary empiricist narratives, but it has its antagonists and counternarratives. The limits to materialism are not that it asserts that matter is all there is, but that it begins with a premise of reduction: that we can understand the world from understanding its parts, and the smallest parts at that. To deny materialism is not to embrace Cartesian dualism, the prospect of nonempirical ethereal entities that interface causally with the mind, nor to *deny* that matter is all there is. It is rather to reveal that such a postulate is just as limited and unproductive as saying that mind is all there is. Materialism engenders a particular discourse that leads again to the primacy of information, the repeating and replicating patterns of matter, mind, culture, and history. Materialism has difficulty with the proposition that the concepts of information and matter are cultural products as well.

The ways that space is handled in computer systems—as represented, resisted, reduced, divided, and transcended—participate in these empiricist formulations of realism to varying degrees. As we have seen, computer representations of objects, such as buildings, trees, and hammers, are commonly constructed on the assumption that there is an independent object world and a mapping, or correspondence, between an object and its description in a computer system. Some computer systems developers and theorists add the notion of the *model* as a further sophistication to mathematical realism. According to classical empirical model theory, constructing a computer representation involves some notion of the thing to be represented, a language for representing it, a modeling language, and functions for mapping the object language to the representation language.[48] By this formulation, it is formal languages that are being mapped onto each other through algorithmic processes, and the issue of relating a language to an external world is a matter of interpretation, which also involves the application of instrumentation and measuring technologies. Such an account opens the way for pragmatic understandings of computer representations to which we shall return in the next chapter.

Empirical realism is also evident in formulations about space as constraining. A Newtonian account of the physical world tells us that it requires energy to start an object in motion, energy is required to change direction from a constant trajectory, and movement through space, other than in a complete vacuum, encounters resistance that can only be overcome by expending more energy. Furthermore, we cannot perform action at a distance, so we need to be where we want to intervene, or we need technologies to extend our reach. Information technology seems to provide an essential mediating technology, and it overcomes the resistance of space through its ability to represent objects and intentions and to transmit these representations across distances.

What of space divided? According to empirical realism, there is the objective reality of the situation, the actual spatial location of people and objects. There are spatial and temporal principles to which people and objects conform, and the objective situation is that about which there can be uniform agreement. The physical rules of our existence in real space are generalizable and can be discussed rationally. On the other hand, empirical realism implies that there is also the situation as it appears, or is imagined, which may not be the same as the actual situation. Some empirical realists may therefore assert that cyberspace pertains to this realm where it appears that almost anything is possible. By this reading, the cyberspace phenomenon is perceptual, belonging to the varied and unreliable world of appearances. Spatial and temporal rules can be violated, and new rules invented and applied. The cyberspace experience is also sometimes idiosyncratic, peculiar to the person undergoing the experience. It seems to the person using computer communications that she is in a special space, cyberspace, and that she is spatially proximal to the people with whom she is communicating. From the point of view of the empirical realist, a further relationship exists between real space and cyberspace. The objectively real provides the metaphors through which certain information technology phenomena are understood. So we can describe the idiosyncratic, perceptual, and sometimes ephemeral world of computer mediation (cyberspace) in terms of location, movement, proximity, shape, extent, and other spatial metaphors. We use the language of real space to describe the perceptual world of the communications user. The position of scientific realism sets a sharp boundary between a physical reality and the experience of information technology and privileges the former.

But insofar as empirical realism points to information as the basis for all things, it also supports notions of the transcendence of space through information technology. As a technology for the reproduction, manipulation, and transmission of patterns, the computer provides a means of manipulating space and creating it anew, setting in place new spatial laws, extending the scope of space, and drawing us into it, or at least absorbing our essential, collective selves.

There are further narratives of the real that are heirs to the empiricist tradition, including *pragmatic realism*, which affirms the proximal, the shared, the embodied, repetition, and the ineffable, without appealing to the mind as the mirror of nature or dividing the universe into the real and the perceptual. Pragmatic realism does not require the objective existence of rules, depend on scientific procedure as the final arbiter of what is real, or rely on concepts of correspondences between knowledge and the world. It also leaves the matter of a divide between ourselves and an object world open and contingent. Moreover, it embraces the pursuit of the real in its discourse, acknowledging that it is just as interesting to examine how it is that we come to require the real as it is to discover what that reality is like. Johnson develops such a rehabilitated realism calling on the thinking of Rorty and the later Putnam: "Our realism consists in our sense that we are in touch with reality in our bodily actions in the world, and in our having an understanding of reality sufficient to allow us to function more or less successfully in that world."[49] The practical concern with the real involves our engagement as beings with bodies: "Our understanding is our way of being situated in our world, and it is our embodied understanding that manifests our realist commitments."[50] We will return to the elaboration of the pragmatic position and its relationship with phenomenology in chapter 5.

The Spatially Real

The issue of spatial representation in a computer system calls on empiricist narratives of space, about which more needs to be said. As we have already seen, Platonic, Neoplatonic, romantic, Berkeleyan, and technological idealism, along with the various forms of empirical realism, are a party to a range of binary distinctions. Whereas it is tempting to defend one side of a dichotomy, or refute the other, late modern discourse resists being drawn into these debates and taking sides. It prefers to assess the value and effect of the debates themselves and to assess the circumstances in which their narratives have currency. It prefers to move into new areas of debate, to generate new narratives that dispense with the old distinctions where they are "worn out," and to establish new distinctions. Besides, there appears to be little consistency within either the idealist or realist position.[51]

The distinctions between the objective and the subjective, rationalism and empiricism, and the absolute and the relative have been influential in understanding space, and they reveal much about the role of information technology. They also implicate concepts of the one and the many, unity and multiplicity. Moreover, they further indicate how the extremes of technoromanticism evident in cyberspace narratives are connected with the sober representational realism of computer modeling.

Objective Space

The notion that computer systems enable the representation of space assumes an objective spatial realm that can be represented in a computer, and from the concept of objectivity derives the possibility that there is both objective and subjective space.

Objective knowledge pertains to the world of things, independent of the thinker or observer and their state of mind, and subjective knowing pertains to the observer. Objectivity assumes we can move outside the phenomenon under examination, also characterized by Descartes as ridding ourselves of prejudice, a concept that is still current. According to Dennett, objectivity requires "a neutral way of describing the data—a way that does not prejudge the issue."[52] Objectivity participates in the real by positing the *proximity* of the ego, which stands aloof from the object world. We *share* this world insofar as we externalize it, throw our private observational insights to public scrutiny, while at the same time resisting the collective pressure of tradition and the effects of the *body*. The real is also disclosed through the operation of *repetition* as method. It should be possible for different people to perform the same observations or follow the same line of reasoning to arrive at similar conclusions, whether in the laboratory or through "thought experiments." According to Dennett, Descartes "clearly expected his readers to concur with each of his observations, by performing in their own minds the explorations he described, and getting the same results."[53]

On the other hand, subjective knowledge is the personal kind of knowledge, whether this is clouded by emotion, encumbered by the complaints of the body, or lucidly clear, impartial, and logical. A person may attach credence to the objective and the subjective in uneven measure, but rarely advocates one to the exclusion of the other. To this extent, the objective and the subjective are part of the same deal. According to the objective/subjective problematic, there is space as it really exists objectively, able to be measured, studied by scientists, modeled in a computer, and able to be reasoned about. Then there is space as subjectively experienced. Descartes defined space as measurable. But spatial experience is more than can be understood through a measuring system, and what was left out of Descartes' scheme of things constituted the subjective. Descartes' philosophy did not objectify space but effectively divided it into two, into the objective and the subjective.

Narratives of objectivity and subjectivity can also be retold in terms of the one and the many, unity and multiplicity. For objectivist narrative, the privileged position is that of the single overview, the correct view of the one reality, with subjectivity pertaining to many viewpoints, the vagaries of unscientific, idiosyncratic, individual points of view. There are many subjects, or individuals, in the world, and as many subjective understandings. On the other hand, for romantic narrative, which privileges the subjective, the one is presented as an issue of the unity, which is the unity of humankind with nature, beyond the individuation and classification considered necessary for objectivity. For romanticism, objectivity divides the world into many parts and is obsessed with individuation. The paramount example of individuation is the divide between objectivity and subjectivity itself. Empiricist narratives that attempt to reunite objectivity and subjectivity inevitably assume a romantic cast, as in Popper's "three worlds" proposition. According to Popper, objective knowledge pertains to the physical world, and subjective knowledge to the "world of our conscious experiences."[54] According to Popper, "Genuine or unadulterated or purely subjective knowledge simply does not exist."[55] Then Popper posits a third kind of knowledge that bridges these two worlds. This is "the *logical* contents of books, libraries, computer memories, and suchlike" upon which our subjective knowledge depends.[56] Nagel also addresses the problem of integrating the objective and the subjective in our conception of the real, asserting that reality "is not just objective reality."[57] For Nagel, realism has to accommodate oneself, one's point of view, the point of view of others, and "the objects of various types of judgment that seem to emanate from these perspectives."[58] Nagel also attempts a reconciliation, characterizing the objective position as one in which we step back from our initial view of some aspect of the world and "form a new conception which has that view and its relation to the world as its object."[59] In a series of moves reminiscent of the progression toward Neoplatonic ecstasis, Nagel proposes that there are degrees of objectivity, arranged in concentric spheres, "progressively revealed as we detach gradually from the contingencies of the self."[60] For Nagel, the object/subject distinction creates a problem, in that the two standpoints cannot be satisfactorily integrated. But for Nagel, the ultimate response is to hold to both sides of the opposition without favoring one element over the other: "Apart from the chance that this kind of tension will generate something new, it is best to be aware of the ways in which life and thought are split, if that is how things are."[61]

Certain digital narratives participate in this objective/subjective legacy.

To represent is to make explicit in an objective and universal way that can be recognized independently of the subject. In overcoming distance, we measure objective space, the time it takes to move through it by different modes of transportation, and compare these with the speed of electronic communication. To the extent that they assert or imply that space can be reduced to numerical coordinates, or that there is also a subjective dimension to the experience of space, virtual reality narratives affirm a Cartesian view. According to Benedikt, cyberspace is a particular realization of Popper's third, abstract kind of world that bridges the objective and the subjective: "It is the latest stage in the evolution of *World 3*, with the ballast of materiality cast away—cast away again, and perhaps finally."[62] Novak suggests that cyberspace is a meeting of the objective and the subjective, though it represents a triumph of the subjective. It is a "habitat for the imagination": "Cyberspace is the place where conscious dreaming meets subconscious dreaming, a landscape of rational magic, of mystical reason, the locus and triumph of poetry over poverty, of 'it-can-be-so' over 'it-should-be-so.' "[63] For Novak, with cyberspace: "Objective reality itself seems to be a construct of our mind, and thus becomes subjective."[64] Narratives that affirm that information technology is subverting orthodox conceptions of space, and even space itself, also participate in the Cartesian subject/object dualism. For subversion to be effective, a degree of credibility must already be attached to the orthodoxy, in this case objectivity. As we have seen, much supposed subversion of orthodoxy in information technology narratives is simply a reassertion of subjectivity.

Empirical Space

The distinction between rationalism and empiricism was developed by Descartes' empiricist critics. The Cartesian, rationalist, or "apriorist" position is that we can be sure of what we are able to derive through reason, understood as a process in logic, geometry, measurement and mathematics, independent even of experience. So Descartes' concept of space begins with space as extension, or measurement, in three dimensions along a coordinate system. In fact, according to Descartes, all properties of matter can be understood in terms of spatial and dimensional properties. The only essential property of matter is that it has extension: "The nature of matter, or of a body considered in general, does not consist in its being a thing that has hardness or weight, or color, or any other sensible property, but simply in its being a thing that has extension in length, breadth, and depth."[65] So a property such as hardness can be described in terms of resistance to movement through space. Furthermore, all properties such as hardness, color, heaviness, coldness, and heat can change. The one constant is that the object occupies space.

Descartes' ideas on the subservience of material properties to extension were contested by the empiricists, notably by Locke. To the question, where does the mind gain "all the materials of reason and knowledge?" Locke replied, "To this I answer, in one word, from *experience*."[66] Empiricism says that we do not derive an understanding of the world through some kind of innate reasoning ability but primarily through observation, or sense experience. In fact, we begin life with a "blank slate" onto which we imprint understandings developed through experience. So the properties of space are not derived primarily through intellectualizing about them, but by what our senses tell us: "We get the idea of space, both by sense and touch."[67] Locke then goes on to refute Descartes' insistence that objects and their extension are the same thing.

In the light of twentieth-century thinking, rationalism and empiricism are not very different. Rationalism seems to favor introspection, using the devices of logic. Empiricism prefers to see how things appear and to form conclusions from observations. Both appeal to reason as logic. Both advocate ridding the mind of prejudice in developing understandings of the world. Rationalism and empiricism are both objectivist. They seek to be clear about what is "out there" in the object world, and they both characterize and reject subjectivity as a prejudiced and inferior way of knowing. They are both driven by the imperative of ego-centered idealism.

Various movements of thought have taken their lead directly from empiricism, notably the pragmatism of John Dewey and the logical positivism of the Vienna Circle. According to Dewey, it was not until science became empirical, and therefore technological, that it overcame the rationalism of intellectual abstraction.[68] For Dewey, the strength of the empirical outlook of modern science is that it is essentially practical, grounded in instrumentation, technologies, and human practices. (We will return to pragmatism in the next chapter.) Logical positivism took a different tack than the pragmatists, affirming the centrality of empirically verifiable scientific propositions and their truth values.

The rationalist/empiricist debate implicates narratives of unity and plurality in various ways. Whereas rationalism seems to assert the primacy of one world view—the reduction of all spatial phenomena to extension, understood through reason—empiricism suggests a plurality of viewpoints dependent on the observational situation. As developed in logical positivism, empiricism seems not to presume a unity, the pursuit of which takes one into the realms of metaphysical speculation. But the French positivism of Saint-Simon (1760–1825) and Auguste Comte (1798–1857), on which logical positivism draws, is infused with the unity theme. Apart from its matter-of-fact dependence on the commonsense use of language, its embrace of science, and its rejection of metaphysics, Comte's positivism consists of several strands that reveal a form of idealism suggestive of Neoplatonism. For Comte, knowledge passes through three states, from a theological position to a metaphysical to a positive state.[69] (The idea that there are stages of knowing was built into many philosophical schemas, including the grand epistemological schemas of Kant and Hegel.) For Comte there is the theological phase, in which we animate the world by ascribing human characteristics to objects. Then there is the metaphysical phase, in which we abstract such beliefs into sophisticated systems of causation. The final phase is the positive one in which we rely on observation, independently of metaphysical schemas. The theological and metaphysical phases are inferior to the final phase, and they correspond to the primitive states in human societies, or the earlier stages in human development: "Now does not each one of us, when he looks at his own history, recall that he was successively a *theologian* in childhood, for his most important ideas, a *metaphysician* in his youth, and a *physicist* in his maturity?"[70] The positive or real state is the definitive state, where imagination is subordinated to observation.

The positivist creed advocated progress through science, the unity of the sciences, and it was against a dependence on concepts of causation, which it regarded as metaphysical. Comte is regarded as the founder of the science of society, sociology, and advocated that human phenomena, as with all phenomena, are subject to invariable natural laws. But in his later writings, Comte developed a religious aspect to positivism based on the putative principles of love, order, and progress.[71] There was no God in Comte's religion, rather the focus was on the Unity. For Comte, the ultimate Great Being is humanity itself: "one immense and eternal

Being."[72] Comte proposed the establishment of a church modeled on the Roman Catholic calendar and sacraments but with Humanity as the object of its worship.[73] Such ideas persist to the present among sociologists such as Maffesoli, who asserts, "When I am sitting in the café, eating a meal or addressing the other, I am really addressing the deity," which confirms "the link between the divine, the social whole and proximity."[74]

Positivism, without its quasi-religious aspect, was taken up by the group known as the Vienna Circle in the 1920s.[75] One of its leading lights was Carnap, who sought to bring "scientific strictness," understood through the "clarity" of mathematics and logic, to bear on philosophical issues.[76] According to logical positivism, the only meaningful propositions are those that can be empirically tested or for which we can imagine a means of testing. This is the verification principle, as summarized by Ayer: "The criterion which we use to test the genuineness of apparent statements of fact is the criterion of verifiability. We say that a sentence is factually significant to any given person, if, and only if, he knows how to verify the proposition that it purports to express—that is, if he knows what observations would lead him, under certain conditions, to accept the proposition as being true, or reject it as being false."[77] So deciding between realism and idealism, whether objects exist in reality or in the mind, or Berkeley's proposition about the nonexistence of matter, cannot be answered empirically. That is, we cannot imagine a way of testing the proposition such that we could arrive at a definitive judgment about its truth status. So the realist/idealist debate generates meaningless propositions, such as "there exists an independent reality," "our world has properties that we can never know about," "computers represent reality," or "the brain is a computer." The verification principle provides a means of arresting debate on a number of issues by declaring most utterances meaningless. According to Wittgenstein, in his positivist *Tractatus Logico Philosophicus*: "Most propositions and questions, that have been written about philosophical matters, are not false, but senseless. . . . And so it is not to be wondered at that the deepest problems are really no problems."[78] The closing words of the *Tractatus* attempt to obviate all speculation beyond what we can reasonably discuss, which is to say, to empirically test: "Whereof one cannot speak, thereof one must be silent."[79]

Positivism therefore addresses the issue of the real as the problem of realism versus idealism (and rationalism versus empiricism) by constructing its narratives through a process of exclusion in such a way that the issue of the real does not arise, at least to its own satisfaction. Though they are of interest as a means of exercising one's skills in propositional logic, the questions posed through the realist/idealist debate are metaphysical according to positivism, and therefore meaningless. According to Ayer: "We have seen that the dispute between idealists and realists becomes a metaphysical dispute when it is assumed that the question whether an object is real or ideal is an empirical question which cannot be settled by any possible observation,"[80] though the logic of the arguments put forward by Berkeley are of interest, as objects of analysis.

There are few serious positivist philosophers now, and Wittgenstein presents his refutation of his earlier views in *Philosophical Investigations*.[81] One of Popper's substantial contributions to modern thought was to refute positivism from the inside, and in its own terms, asserting that all empirical observations are in fact already theory laden and that we can never positively prove a scientific concept, only refute it, or refute its negation. So the positive assertion of matters of truth decided by impartial observation is not possible, only proof that something is not the case. Other critics of positivism such as Nagel, regard logical positivism as a form of idealism, in that it asserts that we cannot go beyond what we can think about,[82] thereby of its own admission limiting our view of the world to an ego-centered view. Critics of positivism have also pointed out the obvious circularity and contradiction in the positivist position. The positivist insistence on empirical evidence presents an unverifiable, metaphysical discourse (based on the verification principle) that one must conclude is meaningless by the positivists' own criterion.

But positivism lives on, and not only as a "straw man," in the narratives of its critics: romantics, neo-Marxists, liberals, phenomenologists, and poststructuralists. Logical positivism and the analytical school came together on the issues of language as primarily concerned with verification and truth statements, and this understanding of language persists, including in digital narratives.[83]

Positivism informed many of the people who are considered the founders of artificial intelligence, cognitive science, and systems theory. Allan Turing's test for intelligence (which he called "the imitation game") is an example of the verification principle at work.[84] That there is an empirical means of verifying how we can know that a machine is intelligent has helped preserve AI from the charge of metaphysics, at least until the arguments against the validity of the test by Searle and others are taken as persuasive.[85] Herbert Simon, a prominent figure in management science and artificial intelligence, was a student of Carnap and affirms the primacy of empirical testing as a means to vindicate the new "science of design."[86] Mitchell takes Carnap's semantics, and the use of logical rules for establishing general truth conditions to evaluate spatial geometries, and ultimately defines critical discourses in architecture in ways amenable to computation.[87] Positivism is also evident in the indifference of some AI and design theory research toward the "metaphysical question" of how AI models relate to human cognition—the correspondence or otherwise between calculative models and the way the mind actually works. As there is no means of testing the correspondence between models and some notion of reality, the question of the validity of the models is taken as meaningless. The characterization of language of Shannon and Weaver is also informed by positivism,[88] as is the characterization of language that looks to the primacy of "information content."

Propositional Space

According to the positivist model of language, computers provide special access to a means of forming descriptions of the world, and space, and manipulating them. Information technology revives and amplifies the positivist program. Information is what is there underlying the sentence, with the sentence as a carrier of information. In some respects this use of "information" accords with the etymology of the word to "inform," commonly taken as "to impart some essential characteristic to." The word is derived from the Latin "informare," which is "to give form to." To inform is to imprint a lump of wax with a form, in the sense in which Plato spoke of the forms of the Intelligible realm being imprinted on the material of the sensible realm. By this Platonic reading, one is able to inform by virtue of one's access to the Intellect, the realm of ideas. This forming is not superficial, as though giving a physical shape to something (a contemporary and somewhat contradictory meaning of "to form"), but it is to impart an essence. A thing's information content is its essence. To post-Enlightenment idealism, the intellect is no longer the intellect of the supradivine ideal realm, but the intellect of the individual ego. Information content is the thought, the idea, expressed as a sequence of signs, a sentence, by an intentional speaker.

Such narratives privilege the content of sentences over the word sequence. (The content pertains to the one, and the word sequence pertains to the many. There are many ways of expressing the same idea.) In interpreting a sentence, we have to get back to the idea beneath the surface phenomenon of the word sequence.[89]

According to the positivist view of language, the information content of a sentence can be extracted as a proposition. In other words, the proposition is the information content of the sentence.[90] Attempts have been made throughout history to codify a language of content. One such attempt is the language of the propositional calculus, or formal logic. It dates back at least to Aristotle.[91] Whereas sentences are very fickle and ambiguous, their information content, or propositional form, can be demonstrated to conform to regularities and can be analyzed as conforming to rules. In modern times, according to this bifurcated view of language, computers perform two major language functions. One is to store and convey sentences, to be interpreted by others, as in the use of databases of texts, word processing, electronic mail, and hypertext documents. The second function is to store, transmit, and also manipulate propositions. This latter function involves manipulating the relationships within and between propositions—exemplified in logic programming (with Prolog and LISP) and rule-based systems in artificial intelligence. Much of the romance with information technology generally pertains to this latter aspect of computers. Computers are thought to give us access to the content of sentences, of language, and thereby of thought.

According to this view, the idiosyncratic configuration of nucleotides on a DNA molecule is commonly understood as a code, which is in turn the information content of the DNA. As we have seen in relation to the techno-idealism of certain computer narratives, many claim that the DNA phenomenon is evidence of the ubiquity of information and think that its interpretation will give us access to the essence of life, and ultimately its transcendence. In the realm of social relations, other commentators such as Meyrowitz claim that the key to any human interaction lies with what is happening to information as it is passed from one person to another.[92] Not all of the information comes through sentences, but the many and subtle nuances passed through the various "channels" of the different senses.

The implication for Meyrowitz is that there is essentially no difference between being in a room with a person and being with them across a telephone or video connection. Normal conceptions of space are rendered irrelevant as far as our encounters with one another are concerned. It is the extent to which information can be passed from one person to another that is significant. In this sense, information transcends or conquers space. This view also supports the concept of space as a kind of constraint or encumbrance, which information, when allowed to flow freely, can overcome. There also follow the more extreme positions of Wiener of the prospect of the electronic "teleportation" of physical bodies and Moravec's expectation of the codification of all life past and present.

With the primacy of the proposition, space can also be described and defined informationally. Any location (point) in space can be defined in terms of three-dimensional coordinates, lines can be defined in terms of endpoints in space, and planes and volumes can be defined in terms of points and lines. There are also transformations that can be applied to represent the relationship between one configuration of elements in space and another. Other sophistications on the informational representation of space attempt to capture the natural or fractal properties of space in terms of recursive algorithms,[93] but such notions rely no less on reducing space to symbols, numbers, and relations.

As developed by positivism, empiricism leads to a concern with the primacy of language understood in terms of truth statements and their verification. What does propositional space add to empirical concepts of the real? *Proximity* is understood in terms of the relationship between the subject and the world appropriated through language, understood as dealing in truth statements, which is the only way that the world can be known. Language is a matter of communicating (*sharing*) thoughts and intentions, understood as information content, a series of propositions, between individuals. The *body* rarely features in positivist discourse other than as providing the organs of thought and communication. In common with all living matter, the body is after all the result of a code sequence. Matter is subservient, as is all of nature, including space, to the notion of information. For this positivist conception of the real, *repetition* is everything. What is repeatable about things in the world is their information content—that is, the sequences and arrangements of elements that can be reproduced. The quest is clarity and lucidity, sentences reduced to propositions, ready for verification, which denies the *ineffable*, though by one reading, Wittgenstein's "Whereof one cannot speak, thereof one must be silent" brings it to center stage.

Relative Space

To the modern mind, the greatest challenge to conventional concepts of space and concepts of the real comes from relativity and quantum theory. Modern physics seems to require modes of representation that in turn indicate extensions to the concept of space beyond what is understood through Cartesian schemas. In so doing, contemporary physics has become an emblem for a kind of quasi postmodernism, a romantic reaction against positivism, that licenses scientists to become autobiographers and poets in bringing together "material from a great number of different fields in the physical, biological, and behavioral sciences, and even in the arts and humanities."[94] The colonization of art, philosophy, and theology by physicists, mathematicians, and biologists seems to allow science to speak for all aspects of human endeavor, the irony being that it feels compelled to do so as the ability of its methods to provide unified accounts of nature reach their limit. How do these narratives bear on concepts of space and information technology?

To assume space as absolute is to take space as somehow given, objectively real, and a basis for understanding the object world. As defined by Descartes, objects have position and dimension, and they offer resistance to movement through space. The alternative is the relational view. For Leibniz, space is not itself an object but only emerges in a consideration of the relationship between things. Space is a derived phenomenon, dependent on the presence of physical bodies: "I don't say that matter and space are the same thing. I only say, there is no space, where there is no matter; and that space in itself is not an absolute reality."[95] The whole conception of space is relative to material objects. We only know of space because of objects and the relationships between them. Leibniz prepared the way for Einstein's mathematical formulation of space as a system of fields dependent on the relationships between matter, but also for thinking of space as derived from other considerations, as something other than a fundamental entity. So Leibniz's position on space demonstrates that what is normally taken as fundamental can be shown to depend on other things.[96] It was just that space was not one of them. Berkeley supported the Leibnizian view: "Concerning absolute space, that phantom of the mechanic and geometrical philosophers, it may suffice to observe that it is neither perceived by any sense, nor proved by any reason."[97] Kant took issue with Leibniz and sought to establish the status of space as absolute, invoking complex arguments about handedness. The only way that we can decide that an object is different from its mirror copy is that both can be compared within the frame of space as an absolute.[98] Much of Kant's philosophy of the real focuses on space and time.[99] For Kant, space and time are subjective spectacles through which we see the world. We cannot see the world as it is in itself, though space and time are empirically real, as real as are all objects revealed by our senses.[100] Through a complex line of argument, invoking intricate distinctions about judgment, Kant arrives at the paradoxical conclusion that space and time are found in our experience but are prior to perception.[101] In other words, our experience presupposes space: "The representation of space must be presupposed."[102]

The absolute/relative problematic is generally established by recourse to spatial metaphors. Descartes established the metaphor of the "Archimedian point," that single immovable point in space from which Archimedes thought that it might be possible to "draw the terrestrial globe out of its place and move it to another,"[103] and Leibniz established the "ground" as the datum on which things are fixed. The absolute is fixed to the ground, the relative is unstable, shifting, and indeterminate. These metaphors become even more pronounced when we move into the realm of measurement. There is absolute and relative space—space as decided by an absolute (or fixed) coordinate system, and space as defined by a relative (or variable) coordinate system. Newton assumed this distinction: "Absolute space, in its own nature, without relation to anything external, remains always similar and immovable. Relative space is some moveable dimension or measure of the absolute spaces; which our senses determine by its position to bodies; and which is commonly taken for immovable space; such as the dimension of a subterraneous, an aerial, or celestial space, determined by its position in respect of the earth."[104] According to Newton, the location of an object can be recorded in absolute terms on the global latitude and longitude coordinate system. It can also be recorded in relative terms using a local coordinate system, such as that found on a map. But the object could also be located in a coordinate system that is moving, as when it is located on a moving vehicle. The use of different reference grids is well known in cartography and in computer graphics and computer-aided design. In CAD, drawings and models are constructed using different coordinate systems, some of which may even be understood as moving relative to each other.

Such narratives also participate in the myth of the one and the many, unity and multiplicity. The absolute is a matter of the one, while the relative pertains to the many. Space as absolute asserts a unitary whole that is space, with everything fitting into it. Space as dependent on relationships between objects and matter implicates space as a multiplicity. The distinction is commonly characterized in terms of the homogeneity of space, the Cartesian view, and the heterogeneity of space, the Leibnizian view. (We will return to these themes in chapter 4.) Certain digital narratives also appropriate developments in scientific and mathematical conceptions of space to present the emergence of a new unity, an interconnectedness between nature, thought, and machines. But it does so by recourse to the notion of the relative rather than the absolute. Einstein is the authority on space and time in the popular imagination, and his theories speak of the eradication of the fixed point, even from that stronghold of certainty, early twentieth-century science.[105] Einstein's theories of relativity, which have been so important and successful in physics, have rendered "relativity" emblematic of progressive thinking.

The theories of relativity establish that space and time have different meanings depending on conditions, such as those present for a moving rocket relative to those on the earth. According to Whitehead, in the theory of relativity there is no unique present instant. Einstein conceived of the measurement of time as dependent on the motion of an object. In fact, space was to be replaced by motion as the basis for understanding the physical world. According to Einstein: "In the first place we entirely shun the vague word 'space,' of which, we must honestly acknowledge, we cannot form the slightest conception, and we replace it by 'motion relative to a practically rigid body of reference.' "[106] From this premise, the nature of space undergoes transformation.[107]

It is well known that Einstein's theories produce successful predictions in astronomy and atomic physics and are more accurate than Newton's theories of energy, motion, and gravity. Einstein's theories are mathematical abstractions of enormous potency. They unite concepts of space, time, energy, and gravitation.[108] The other great development in physics is quantum mechanics, or quantum field theory, which trades in notions of probability and indeterminacy, and with which Einstein's theories seem to be in conflict. Einstein made his objections to quantum mechanics well known, but the two systems conspire to present a particular view of the real, though the implications of relativity and quantum theory are not necessarily obvious, even to their originators.[109]

Quantum theory posed a serious threat to realism. According to Einstein, most of the theorists of quantum physics "do not see what sort of risky game they are playing with reality."[110] According to Fine, this risk put into jeopardy the traditional program of physics, which was to construct a model of an observer-independent reality.

In addition to its obvious technical and scientific utility, certain thinkers present quantum field theory as providing evidence for several important observations about the nature of the universe.[111] For Gregory, quantum theory abandons the rigid chain of classical determinism in favor of "a fundamentally probabilistic view of reality."[112] The various quantum equations do not describe the behavior of electrons but the probability of finding electrons at particular places.

According to Whitehead, Einstein's theories challenge scientific realism, "which presupposes a definite present instant at which all matter is simultaneously real."[113] The moment at which you can declare that two objects exist seems to depend on your own position and movement. According to Fine, quantum theory presents a schizophrenic attitude toward the issue of reality: "For the quantum theory of today is reputed to support realism with respect to a variety of entities, including molecules and atoms and electrons and photons, and a whole zoo of elementary particles, quarks maybe—and perhaps some even more unlikely things like magnetic monopoles. Yet that very same quantum theory is also reputed to be incompatible with realism."[114] Gregory indicates how relativity and quantum mechanics work against representational realism, making it difficult to maintain the convention of an absolute word-to-world fit: "Relativity, much to the discomfort of some people, shows that the truth of whose clock is running slower, mine or yours, depends on the frame of reference from which the statement is made."[115] According to Davies, the uncertainty of quantum mechnics is not due to the limits of our measuring tools, "but it is inherent in nature and not merely the result of technological limitations in measurement."[116] By this reading, insofar as we assume an independent reality called "nature," it is imbued at the microscale with indeterminacies, such that nothing can ever be said to exist with certainty. Heisenberg casts this problematic in terms of the involvement of the subject (the observer) in the experiment: "Now, this is a very strange result, since it seems to indicate that the observation plays a decisive role in the event and that the reality varies, depending upon whether we observe it or not."[117] Quantum theory indicates that the scientist is not set apart from a reality that is nature. Quantum physics provides evidence that the subject—that is, the scientist—enters into the observation to determine the existence of what is being observed, or more precisely the presence of the scientist is implicated in what is observed.

Relativity and quantum mechanics seem to indicate the empirical understanding of the real at its limits. The presumed distance between the observer and the observed seems not to hold at the finest grain of scientific observation. The observer and the observed are more proximal than assumed by empiricism. The repetition of patterns of behavior in what is observed and the predictive role of formulae assume the cast of probability. What is real is accounted for in terms of probability formulae. On the other hand, the more fine grained science's reflections become, the more idealist they appear. The explanations provided by the "new physics" seem as removed from the real, the real as proximal to our shared bodily engagement, as anything could be. Insofar as physics deals with the real, it seems to have constructed an ideal realm in which one can only participate by access to the most rarefied concepts, procedures, and formulae. According to Whitehead: "The new situation in the thought of today arises from the fact that scientific theory is outrunning common sense. . . . Heaven knows what seeming nonsense may not to-morrow be demonstrated truth."[118] According to Arendt, modern physics reveals the following: "The escape into the mind of man himself is closed if it turns out that the modern physical universe is not only beyond presentation, which is a matter of course under the assumption that nature and Being do not reveal themselves to the senses, but is inconceivable, unthinkable in terms of pure reasoning as well."[119]

The idea of information already infuses concepts of space through the empiricist tradition. Einstein's theories further implicate the notion of information transfer. The determination of measurements made within frames of reference that are moving relative to one another depends on observation, and comparisons of observations. The only way to compare what is happening to clocks on moving trains, or rockets, is to have someone observe them, and the fastest that these observations can be brought together is through electromagnetic radiation—light or a radio signal. For events to be observed as simultaneous, we have to observe them as such. Therefore, the fastest that we, or a recording machine, can receive an observation signal is the speed of light. This is how the speed of light enters into Einstein's equations. Two events may appear simultaneous from one frame of reference but not from another.[120] We can only compare rulers by transmitting signals.[121] By this reading, the primacy of information in the real resides less in replicable underlying patterns than in the artifice of experimental procedures that require the communication of signals.

Some commentators link quantum physics and computing through complexity theory, the exploration of the particular properties of numbers and algorithms by which patterns of order and chaos emerge and recede as if by complex laws, though there are none to be seen in the simple algorithms used. The most potent icons of complexity theory are the fractal diagrams of Mandelbrot,[122] showing infinitely recursive computer-generated filigree patterns.

Complexity theory throws a mathematical spin on the unity theme. There is the concept of "self similarity": that the whole and the parts are related through recursive patterns. So the patterning of the Mandelbrot set taken as a whole is evident in the details, though with variation: the strange bulbous shapes and filigrees of the whole are evident in the fine detail, analogous to how the small parts of a fern frond exhibit the general shape and pattern of the whole frond, albeit at a different scale. Complexity theory also intimates complexity in simplicity. The Mandelbrot algorithm is only a few lines of computer code, and yet is able to produce images with immense detail, limited only by the depth of recursion and hence computer processing time. Complexity theory also develops concepts of pattern in instability, the subjugation of chaos to the rule of order. For some parts of the Mandelbrot set, as a two-dimensional diagram, to move a little to the right or left presents little or no color variation. But around the edges of these stable regions, small movements can produce extreme variations in color, and within these regions some areas are more stable than others, and so on to infinite regress. According to complexity theory, certain systems can reach a state where very minor change in one variable (one small part of the system) can produce spectacular changes to the whole. So local, small-scale conditions can influence the movement of large pressure vortices in weather patterns, the so-called butterfly effect. Complexity theory develops the theme that systems can be unpredictable yet patterned. There are regularities in the Mandelbrot set, but there is no algorithm or rule by which you can predict the shape or position of the next filigree, other than using the program to generate it. There is a pattern there, though it is somewhat ineffable. Fractal diagrams seem to exhibit qualities of the romantic sublime. They reside at the precipice between ordered beauty and formless chaos, the unity in all things and their infinite individuation.

Penrose, Gell-Mann, and others link the fractal properties of number series to the concepts of order and disorder in nature, and also to quantum theory. According to Gell-Mann, by "reflecting on questions of simplicity and complexity, we perceive connections that help to link together all the phenomena of nature, from the simplest to the most complex."[123] As a convert to metaphysics, Jencks declares that quantum physics and chaos science contribute to the rediscovery of the "aesthetic and spiritual meanings of nature": "The more we discover via these new sciences, the more we find our connectedness to a creative and mysterious universe."[124] For Gleick, chaos theory is "turning back a trend in science toward reductionism, the analysis of systems in terms of their constituent parts: quarks, chromosomes, or neurones." For Glieck, chaos scientists "believe that they are looking for the whole."[125] This is the version of the unity myth presented in the narratives accompanying modern physics: the dialectic between simplicity and complexity. Fractal geometries could only be observed with the aid of computers, so the loop from information processing to the uncertainty of modern physics, and back again, is complete. The loop also brings us back to romanticism. In the narratives of cyberspace, relativity and quantum theory are presented as indicating the breakdown in conventional, Cartesian and Newtonian concepts of the real. They also bring us back to revived conceptions of unity and the interconnectedness of all things.

We have examined the ubiquity of idealist narratives and how they are infused with Neoplatonism, which they also transform. Both are caught in the same systems of thought. Contemporary empiricism ostensibly sets itself apart from such metaphysical extravagances. As modified through positivism, it claims a totalizing "commonsense" view of the world and is impatient toward the metaphysical, while forgetting its own metaphysical origins, including the religious motivations of its founders and Comte's thinly disguised Neoplatonism. Empiricism has fostered the dualism of the rational versus the irrational. The romantic Neoplatonism of Moravec (chapter 2) and the culture of cyberpunk is tolerated and institutionalized as an acceptable eccentricity, as an irrational diversion pertaining to popular culture. Yet technoromantic Neoplatonism is there, only thinly disguised, in popular science writing and elsewhere.

The claims that computers are altering our conception of space and of reality, and even altering reality itself, are sustained by the prosaic proposition that computers, drawings, and models are representations, understood as correspondences between codes, words, or images, and some reality beyond. If computers allow us to model, mimic, and represent reality, then they indeed allow us to alter perceptual fields, challenge and distort reality, and create alternative realities. If the world is essentially a matter of patterns, even infinitely recursive patterns of chaos and order, then these patterns can be placed into vast interconnected computer systems to create an electronically reconstituted unity. So rather than countering romanticism, empiricism provides the conditions for technoromantic narratives to promote the transcendent potential of computer space.

Space provides a good test site for examining the relationship between computers and the real. Computers promote further the privileged relationship between mathematics, geometry, and space, supporting the series of transformations through which space is represented, resisted, reduced, divided, and ultimately transcended.

How do we break the chain from representation to technoromanticism? For postempiricist and pragmatic narratives, the themes of idealism and empiricism are exhausted. There are other ways of understanding the real, notably through insights into the working of language. If we return briefly to physics, following the insights of the physicist Niels Bohr, it is possible to recast the problematic of the real presented by relativity and quantum mechanical theory in terms of language. Relativity and quantum physics equations can be seen as part of a pragmatic language game.[126] An acceptance of the ubiquity of language provides different insights into the real than those presented through idealism and empiricism.

4

The Symbolic Order

Rather than countering technoromanticism, empiricism seems to provide the conditions for it to flourish, through notions of representation and in the romantic reflections of popular science narrative. The chain from representation to technoromanticism, or empirical realism to techno-idealism, can be broken through pragmatic understandings of language. For empiricism and representational realism, the key component of language is the proposition, that succinct logical or mathematical sentence that captures precisely what is meant in an otherwise imperfect and ambiguous utterance. Though it has currency in computer narratives, this view of language has been mostly superseded now. Pragmatic concepts of language, particularly speech act theory, maintain that the concept of the proposition is severely limited. Even science does not operate through mathematical, logical, and linguistic propositions and their correspondence with the world. Structuralism and poststructuralism also argue against theories of correspondence, and see the construction of the real as a maneuver in language. Structuralism also elevates the role of myth, and, as we shall see subsequently, opens the discussion to potent variations on the unity theme and a critique of technoromanticism.

Pragmatics

Certain information technology narratives trade in the proposition as providing privileged access to the world, maintaining that it is possible to construct succinct, mathematical, and logical statements that can be stored in computer memory and manipulated, duplicated, and transmitted.[1] As discussed in chapter 3, propositions are the information content of natural language sentences.[2] Whether expressed formally or informally, propositions constitute the meaning of the sentence, and meaning is what is "intended" by the author, at least for positivism and its variants.

There are problems with the idea of the proposition as the information content of a sentence. It is possible to construct many different propositions for the same sentence, even in the highly formalized languages of logic. Whether one is using computers to manipulate propositions or sentences, in each event the entity dealt with is the sentence, in different forms. More strictly speaking, sentences are strings of symbols or tokens. It just happens that there are different kinds of sentences (or different contexts of use): natural language, and formal or propositional language sentences. Even though certain forms of symbol strings are able to be manipulated in various ways, as in the use of formal logic to prove theorems (and this can be accomplished under certain conditions in a computer), there is nothing in the string of symbols that carries meaning. Rather, we interpret such strings in certain contexts to mean something. The quest for the meaning or information content of a sentence as propositions leads us in an infinite regress, as every attempt to define the information content of a sentence leads to a new sentence, which in turn has information content, which is to say it can be translated further into other propositions. There is no end to the quest for the information content of a proposition. This indicates that to turn a sentence into a proposition is to translate one form of words into another, and the process is never exhausted, simply because we can never exhaust all the contexts in which the sentence, or the proposition, might be used. In spite of attempts to formalize the notion of context, it resists reduction to propositions. The propositional model of language also leads us to seek for origins and intentions behind sentences, and yet these origins and intentions prove forever elusive. To "state what you mean" is to embark on a process of interpretation and reinterpretation, a process that, in formal terms, is never exhausted.[3]

Narratives counter to the propositional view draw from the strand of language study known as "speech act theory," which emerged as a reaction within the analytical school against the primacy given to the proposition. Speech act theory also resonates with aspects of language theory in pragmatism, critical theory, and contemporary studies in hermeneutics.[4] As developed by Austin, speech act theory asserts that sentences do not so much report something, and thereby deliver propositions or information content, as make something the case by being uttered.[5] In order to cast doubt on the ubiquity and plausibility of the propositional view of language, Austin shows that there are sentences that do not seem to rely on conveying information at all. They do not describe or report anything. Neither are these sentences simply true or false. According to Austin, there are sentences for which "the uttering of the sentence is, or is a part of, the doing of an action, which again would not normally be described as, or 'just,' saying something."[6] In other words, there are sentences that are action sentences, or *performative* sentences. Uttering the sentence "makes it so." According to Austin, when a person says "I do" in the marriage ceremony, or "I bet you sixpence it will rain tomorrow,"[7] he or she is using a performative utterance and bringing some state of affairs into being. When the umpire says "Safe!" he creates a score. When the foreman of the jury says, "Guilty as charged," she creates a felon.[8] According to Austin, such statements are actually the norm. What we commonly regard as the traditional sentence, the sentence that has information content, is actually an abstraction or an ideal. It is derivative from the performative: "Stating, describing, etc., are just two names among a very great many others for illocutionary acts; they have no unique position."[9]

Speech act theory also opens language to the pragmatic concern with action in a context. Words and sentences mean what they mean by virtue of the context in which they are uttered, written, heard, or read. According to speech act theory, a context is a whole and cannot be reduced to parts, propositional, indexical, or otherwise. More precisely, the reduction of a context to propositions is itself a linguistic act and is carried out to some purpose within its own context. The notion of context also implies that a sentence is to be understood within a vast field, horizon, situation, or background, without which the sentence is not intelligible.

The idea of context also implicates notions of community. In an essay against the propositional or "conduit" metaphor of language, Reddy indicates how we often mistakenly attribute to libraries and databases a vast store of knowledge and information, but then asks what would that library be if there was no community able to read the texts or to interpret them?[10] Knowledge and information reside in situated communities of interpreters rather than in texts. Texts serve as a kind of "currency" in this interpretive community. In any communication, so much is unsaid or unwritten. Contrary to Dennett's and Dawkin's reductive concept of libraries and computers as "meme vehicles," what makes interpretation work is the irreducible, unwritten, even changing, context of norms, values, and practices in which those particular texts make sense.[11] In this light, the World Wide Web is best understood as a marketplace of words and images that fits into the emerging practices of the communities that use it, rather than as a repository of knowledge. In fact, the Web readily discloses how out-of-date, isolated, and unsupported texts act as an impediment, and cease to have currency, when there is no one interested in reading them.

When language is seen in this light, then notions of information are removed from center stage. The identification of information content as a series of propositions simply presents new sentences or utterance that fit within a particular context. The reduction of sentences to logical propositions[12] does not provide us with access to the content of a sentence, or of an argument; rather, insofar as logic has any relevance at all, the propositional form bears a relationship to natural language sentences established through a context of esoteric language practices—those of the logician. The logical manipulation of those sentences in turn occurs within the context of such practices. To reinforce the centrality of human practices, even in the use of logical propositions, it is worth noting the difficulty of adequately automating the processes of theorem proving as practiced by logicians. In the "knowledge base" of a hypothetical computerized expert system, there may be propositions pertaining to good design practice. The objective of the system would be to prove that a particular building design conforms to accepted practice. But it transpires that logical theorem proving is not an "automatic process" but relies on unwritten rules about ordering. When implemented in the logic programming language called "Prolog," certain logical problems occur that would cause no difficulty to a logician but that could in fact never reach a conclusion, particularly if the propositions are ordered in the "wrong" way. The automated proof process can get caught in infinite loops that would instinctively be avoided by the logician. So even abstract sentences, propositions, are to be understood within their own contexts—contexts that so far defy "automated interpretation."[13]

The propositional view of language deals in symbols and the relationships between symbols. Such symbols and relationships are repeatable patterns within various phenomena and signify their underlying order. On the other hand, according to pragmatic language theory, any depiction of a pattern is simply a tool for use in a particular context. There are certain configurations of symbols, and certain practice contexts in which such configurations of symbols are useful. According to the pragmatic view, the sequence of nucleotides in DNA (A, C, G, T, etc.) is a tool of science for prediction and is inseparable from the equipment, theories, laboratory, and other practices of the scientists who make use of the sequence in a particular context. In other words, the detection of a pattern is a matter of interpretation in a context. Even the similarity between sequences of words in different instances of sentences and propositions (as when we say that two sentences are exactly the same) is decided by interpretation. The practices by which we decide that two sentences are identical, even though they are on different pages and printed in different fonts, are fairly well established, and pertain to identifying words and word order. The means by which we decide that two sentences are similar, even though they have different grammatical structures, is also on the basis of certain practices—those of the grammarian. The pragmatic view is that pattern and order do not inhere within the phenomenon observed, as if there to be discovered, nor are they cast over the phenomenon by the observer. They emerge in practical contexts and are contingent on them.

That context is unable to be rendered propositionally or formulaically does not fit comfortably within narratives of language based around the proposition, which are intent on identifying underlying order to language, and operationalizing and controlling it, as in computer systems. Speech act theory is incommensurate with this endeavor and is perhaps of less interest within certain areas of cognitive science and artificial intelligence. Speech act theory focuses on language as a social phenomenon, and as such leaves the way open for a critique of the wider issues of language, and even of the study of language itself. From our point of view here, this pragmatic language narrative clearly deprivileges notions of information and the proposition and counters narratives in which information provides privileged access to the world through an ability to represent things operationally, consistently, and autonomously.

The Practically Real

Language features prominently in narratives of the real. Winch takes the pragmatic view of language, specifically that of the later Wittgenstein, to construct a pragmatic view of cognition and community. He provides the example of G. E. Moore's informal "proof of the external world": "He held up each of his hands in succession, saying, 'Here is one hand and here is another; therefore at least two external objects exist; therefore the external world exists.' "[14] But according to Winch, Moore was not so much enacting an experiment as "*reminding* his audience of the way in which the expression 'external object' is in fact used. And his reminder indicated that the issue in philosophy is not to prove or disprove the existence of a world of external objects but rather to *elucidate the concept* of externality."[15] Winch's view captures the tenor of pragmatism, that we *use* words. Words have currency and use value. They *make* things so by the utterance of them. To appeal to external objects (the real world and an independent reality) through statements in language *makes* those entities, not for all time in a metaphysical sense, but for a purpose and in a context of use. In other words, the real is contingent. This is not to separate language from the world, as if there is a real world imperfectly understood, and language that makes the world for us. The strong point of pragmatism is that language *is* our world, including the concept of reality itself.

This pragmatic narrative may appear familiar. It seems to accord with aspects of positivist views of language as expounded by the early Wittgenstein, but it goes beyond positivism. According to the later (postpositivist) Wittgenstein: "We cannot say . . . that the problems of philosophy arise out of language *rather than* out of the world, because in discussing language philosophically we are in fact discussing *what counts as belonging to the world*. Our idea of what belongs to the realm of reality is given for us in the language that we use."[16] The pragmatic position is that in our use of language we are more situated than we can ever imagine. Even the discussion of the real is a contextual language game.

How do these concepts of language impinge on the questions of the real posed through modern physics, which purports to have the last word on the constitution of the real? In his book *Inventing Reality: Physics as Language*, Gregory provides a detailed account of the development of modern physics from Newton to Einstein to quantum electrodynamics and quarks, as rendered formulaic by Davies, Hawking, Penrose, Bohm, Peat, Gell-Mann, and others,[17] but in a way that clearly brings out the dependence of science on language, indeed, its construction in language. He calls on speech act theory to do so, describing the history of science as a series of changes in the language of science. In this light: "The word *real* does not seem to be a descriptive term. It seems to be an honorific term that we bestow on our most cherished beliefs—our most treasured ways of speaking."[18]

Gregory explains how a way of talking about the world either works in some particular situation or it does not, and how nothing is added to our knowledge if we say the success or failure of a particular way of talking is due to a correspondence or lack of correspondence to the world.[19] An object such as a book is real because someone can ask me to bring them the book and the action can fulfill their expectation. Books are real "not because of some mystical connection between language and the world" but by expectations and fulfillments that are "made possible by our community of shared assumptions, conventions, and understandings—our shared language."[20] Words in language function as tools, and rather than looking for descriptions of what is "really there," "physics is about fashioning tools."[21] So objects of science such as electricity are no more independent entities than human traits such as anger, love, or respect. Electricity is a way of talking about how things behave.[22]

Entities in science such as space and time undergo revision as new languages emerge. According to Gregory, Einstein demonstrated the power of talking about space and time as though they were a unity and, in the process, showed that space and time are already human inventions, ways of talking about the world.[23] In the same way, particles and fields are different ways of talking, "not different subatomic tinker toys out of which the world is built."[24] This instrumental view of language accords with aspects of Dewey's pragmatism and with the writings of sociologists of science such as Kuhn and Latour.[25] It has also found support among some practicing scientists. According to Heisenberg, "what we observe is not nature in itself but nature exposed to our method of questioning,"[26] and Bohr noted that for observations and calculations about subatomic particles "no sharp distinction can be made between the behavior of the objects themselves and their interaction with the measuring instruments."[27] Bohr considered the implication and meaning of quantum mechanics, assuming physics as a way of talking about the world: "There is no quantum world. There is only an abstract quantum physical description. It is wrong to think that the task of physics is to find out how nature is. Physics concerns what we can say about nature."[28] According to Gregory, physicists are interested more in reliable predictions than in trying to understand how the world is, and the role of physical theories is to aid prediction. Quantum mechanics does not explain how the physical world works, but, through calculation, what will be the outcome of a particular experiment. So physical theories are ways of making calculations rather than building up mathematical pictures of the world.[29] The observation of subatomic particles by means of high-energy particle accelerators brings the role of instruments and experimentation into sharp relief. Subatomic particles are created through experimental conditions: "In the language of field theory they are particles that acquire enough energy to be promoted from virtual to real status. In this sense, physicists 'create' new particles rather than simply discover them."[30] Gregory sums up this pragmatic view of science: "Whether physicists talk in terms of classical field theory or quantum field theory depends on the problem they are wrestling with, much as the choice of a file or a saw depends on what a carpenter is trying to do."[31]

Against the charge that Bohr's language view is subjectivist, and that it promotes a view of the world as merely the subjective creation of individual observers, Gregory recognizes that there is nothing "subjective" about the methods of physics: "the flash on the screen that heralds the arrival of an electron is as objective as anything can be."[32] Gregory reflects on how the propensity we have to see the world in either objective or subjective terms shows how wedded we are to the idea that words picture an "objective" reality that is independent of the way we interact with it.[33] This "picture theory of language" is representational realism, a metaphor "we are so enamored of that being asked to give it up seems like being asked to give up reality itself."[34]

If science is subject to the workings of language in its conceptions of the real, then so is computer modeling. By this reading, geometrical computer models are not representations of reality any more than are sentences or formulas. The computer reveals through the methods of questioning it promotes (to paraphrase Heisenberg). Computer models are efficacious by virtue of what they enable you to do, the interpretive practices they support and disclose.

Pragmatic Space

If the propositional view of language provides insights into space by constructing its narratives around representation and correspondence, which implicate the computer, what does the pragmatic view offer? Space (and cyberspace) provides a good test case for pragmatism. Conceptions of space and time are mediated through technologies, practices, and communities. In science, this mediation is through measuring instruments, the conventions of the laboratory, the way that findings are reported, and the agreement of communities of scientists who uphold those practices. For the pragmatist, there is no *other* that is mediated. We can only know things scientifically as we measure them in practical and collective contexts. Einstein's theories of relativity show that the only fixed point in understanding the phenomena of space and time is the agreement about what constitutes the appropriate measuring device. Once we have settled on the kind of rigid measuring rule and the kind of clock, then there is no other space and time independent of what we measure. With this in mind, when making measurements under conditions of extreme variations in speed, mass, acceleration, or gravitational field, the results of the measurements do not accord with the usual geometrical (Euclidean) principles we apply, but with other principles.

The pragmatic conception of space also focuses on the issue of spatial differentiation. In its quest for order and pattern, the empiricist (and rationalist) view of space is a homogenizing one. The Cartesian view of space is that it is everywhere the same, defined through the uniformity of a number series along three axes of length, breadth, and width. Space is without qualities. The Einsteinian concept of space introduces an element of heterogeneity, though this is a heterogeneity analogous to granularity, variation of a single measurable parameter, like density, rather than of quality. Space is only "distorted" by the presence of matter. Subsequent to Einstein's theories, there is also the theory of "superstrings" that presents space as composed of intricately woven loops in multidimensional space—nine spatial dimensions and the dimension of time.[35] Insofar as these theories speak of space from the point of view of empirical realism, they present space as independent of practical or embodied concerns. Space is homogeneous to the extent that it is everywhere distorted by matter, here or at the farthest reaches of the universe. From the pragmatic point of view these properties of space are constructs revealed through particular practices and uses in language. Space discloses these heterogeneous properties in the context of practical concerns. Space discloses itself as curved by matter in certain contexts of measurement and calculation.

But from the point of view of empirical realism, science works within a realm of discourse that seeks to extend its reach beyond the laboratory, particle accelerators, and telescopes. Empiricist narratives have difficulty with the concept that space has certain properties at the point of collisions within a particle accelerator that do not exist outside it. The properties we apply to space apply here as everywhere. This is part of the language game of empirical realism, to abstract the particularities of an observation and say that they apply beyond the experimental situation and to affirm the homogeneity of the phenomena it observes, including space. The empirical view is of space as homogeneous, and the matters of quality, contingent judgments about use, or even practicality are of necessity marginalized in its considerations.

There are undoubted benefits in conceiving of space as homogeneous, and we are committed to certain technologized, instrumental practices in which this assumption works for us. Or at least our modern practices are so conditioned by the Cartesian, objectivist condition that it is difficult to see how it would mean anything for space to be otherwise. How can space be other than homogeneous? Henri Bergson (1859–1941) elucidated a pragmatic conception of space as heterogeneous, noting that it is a peculiarly human characteristic that we seem to be able to regard space as other than qualitatively differentiated. He notes that birds are often able to find their way across vast distances, presumably by sight, by smell, or by sensing magnetic currents, indicating that there must be an enormous range of spatial differentiation within the bird's experience. From this he deduces that space is not as homogeneous, nor as geometrical, for animals as it is for us. For animals, each direction appears to it "with its own shade, its peculiar quality."[36] According to Bergson we, as opposed to animals, have a capacity to defy all experience and regard space as homogeneous. He regards our ability to construct the conception of an empty homogeneous medium as far more extraordinary than the ability of animals to find their way in a nongeometrical world. It seems that we are able to react against "that heterogeneity which is the very ground of our experience."[37] We appear to have "a special faculty of perceiving or conceiving a space without quality."[38] Here Bergson is affirming that space is first and foremost differentiated. Everything in our practical experience tells us this is so, but we are able to break through this difference and see space in another way, as homogeneous.[39]

Prescientific or mythic concepts of space allowed for a range of spatial differentiations. As expounded in detail by Snodgrass and others, space discloses itself to the prescientific sensibility in terms of cardinality and in relation to celestial progressions and rhythms.[40] Life originates in the east with the rising sun, the west reveals darkness and death. Above are the heavens and beneath is the earth, both implicated in geometry. According to Platonic conceptions of geometry, shapes disclose space in different ways. The square and the cube resonate with the cardinality of the earth, and the circle and the sphere with the heavens. The Neoplatonic concepts of emanation from a heavenly source, the progression from Plato's cave to the source of illumination outside, and concentric spheres of awareness of the intelligible realm—all indicate the qualitative differentiation within, and the heterogeneity of, space and time. We participate in these differentiations through demarcations of space in the landscape, in architecture, and in ritual as well as through our participation in the cycles of the days and the seasons. These aspects of the mythic are metaphysical, claiming universality beyond the contingent, as does modern science, but from the pragmatic point of view we can see that they pertain to engagement between body, practice, and life. To build, make music, process, dance, and practice geometry, is to imitate the structures of the universe and embody a cosmology. According to Snodgrass, a built form "confines and determines the limits of the sacred, and therefore meaningful space from out of the unlimited extent of profane and non-significant space."[41] By its engagement with life practices, the mythic affirms space as significant and heterogeneous, which is to say qualitatively differentiated. Foucault makes similar assertions: "The space in which we live, which draws us out of ourselves, in which the erosion of our lives, our time and our history occurs, the space that claws and [gnaws] at us, is also, in itself, a heterogeneous space. In other words, we do not live in a kind of void, inside of which we could place individuals, and things. We do not live inside a void that could be colored with diverse shades of light, we live inside a set of relations that delineates sites which are irreducible to one another."[42] Space is lived space demarcated by sets of relations.

Following Lakoff and Johnson on the ubiquity of metaphor, it would seem that even our ability to turn space into a homogeneous medium pertains to our worldly engagement.[43] Our language use is imbued with bodily metaphors and image schemata that impinge even on our conception of the homogeneity of space. We see space in terms of coordinate axes through the bodily schemas of front, back, up, down, right, left, and space as pertaining to inside and outside, because we are so constituted in our engagement in the world. The notion of the scientific abstraction of space as a container, uniform, homogeneous, infinite in extent, and without qualities is an artifact of our bodily participation in a practical world of doing that includes to reach, grasp, ingest, stand up, move, and overcome obstacles. The pragmatic view suggests that if space is a party to this differentiated field of practical engagement at its most basic, then it cannot be grasped simply by coordinate geometry or by any other propositional system, in a computer. As with all tools, the value of the symbolic schemas we adopt will depend on the uses we wish to make of them. If science is subject to fields of practices and instrumentation, then cyberspace is equally prone to the contingencies of use. There is no system of "spatial representation" that we can put into a computer that is universal for all situations, that truly transcends the particular.

Pragmatic narratives of the real bring the *proximal* forward as an issue of engagement between language, tools, and practices. The *body* is also implicated, not just as a conduit for sense data, but in the metaphors through which we structure our understanding of the world. According to Johnson, the body is already in the mind. The identification of patterns (*repetition*) in our experience is a contingent matter that is explained neither as something that underlies the order of reality nor as projected onto the world, but as emerging contingently through our practices. The *ineffable* is this praxical field, which is not something we cannot talk about but is open to abundant interpretation. To identify the real as an issue of the proximal, the shared, the embodied, repetition and the ineffable, is to take a pragmatic stance toward the real. It is already to assume the real to be a matter of contingency. So too our emphasis on narrativity indicates a pragmatic orientation.

Pragmatism further diminishes the hold of empiricist concepts of reality that sustain technoromanticism. Once we see the real as contingent on practice (rather than the object of representation), then the problematic posed by virtual realities fades from view. VR does not "challenge the concept of reality," which is already subject to the praxical field. The most we can claim is that VR introduces new modes of practice and discloses aspects of our current practices.

The unity theme permeates the pragmatic space narrative as an issue of the indivisible whole that is the context, the network of equipment, against the multiplicity of interpretations and individuations that emerge in practical situations. Pragmatism hints at a dialectical tension between the whole and the parts. The homogeneity (unity) of space is in tension with heterogeneous, contextual differentiation (multiplicity), a project that is informed further by the tenets of structuralism and poststructuralism.

Structuralism

Empirical realism and propositional narratives of language also have to contend with structuralist language theory, which provides further implications for information technology. To the extent that narratives of cyberspace, VR, and computer modeling rely on notions of correspondence between symbols in a computer and the physical world, they are prone to structuralist critique. Structuralism constructs its narratives in ways that do not depend on the concept of correspondence, distancing itself from notions of reality.

Structuralism emerged as a reaction against nineteenth-century views of language, which were informed largely by empiricism and positivism and which held it as self-evident that language involved correspondence between word and thing, that there is an external reality of objects to which words make reference.[44] Nineteenth-century language theory was also concerned with classification, acknowledging the role of words as indicators of groups of things. It was the role of science, and linguistics as a science, to establish classification schemas, including grouping and classifying languages. Linguistics was also concerned with the lineage of language, how one language derives from another, in the same way that nineteenth-century botanists classified animal specimens and observed how one species appeared to evolve from another. Linguistics also sought after the "original language."

As a nineteenth-century linguist, Saussure (1857–1913) took these aspects of language for granted in his early teaching but later recognized a "crisis" in language studies, particularly over the elusive nature of the referent in language—the individual units to which words are supposed to refer.[45] It is not always possible to identify discrete, preexisting objects, particularly in the study of language itself. Linguistic categories, such as those used in grammar, appear as constructs, convenient fictions for some theory or other. But it was the limitations of history that most revealed the elusive nature of the object and therefore the limits of a correspondence view of language: "When a science offers no immediately recognizable concrete units, that means they are not essential. In history, for example, is the unit the individual, the epoch, or the nation?"[46] History still seems to function without deciding the matter. How important is correspondence then? According to Jameson, the crisis for history recognized by Saussure anticipated the crisis of realism in science that came with Einstein and then quantum theory. In the case of the conflict between the wave and particle theories of light, Jameson notes that "scientific investigation has reached the limits of perception; its objects are no longer things or organisms which are isolated by their own physical structures from each other, and which can be dissected and classified in various ways."[47] Aspects of Saussure's concern about the illusive nature of the referent resonate with those of pragmatism and speech act theory, but his linguistics took a different course.

Saussure's structuralism renounces the principle that words link to things, that language is a matter of names and naming, which according to Jameson is "the most archaic language theory of all."[48] According to Saussure, the linkage between word and thing is clearly arbitrary, though decided by the consensus of a language community. So the word *tree* bears no special relationship to that entity growing in the garden other than what convention allows, as is evident from the fact that there are different words available for the same thing in different languages. The relationship between signifier and signified is therefore arbitrary and the thing signified (the tree) is not so much an object as a concept. According to Saussure: "A linguistic sign is not a link between a thing and a name, but between a concept and a sound pattern."[49] Saussure's contribution to the issue of realism, or the real, is to provide a systematic way of studying language that does not require that language appeal to a reality beyond itself.

According to Jameson, this disdain for realism is still contested by what he calls the Anglo-American tradition, exemplified by Ogden and Richard's refutation of structuralism,[50] for whom, according to Jameson, "the most basic task of linguistic investigation consists in a one-to-one, sentence-by-sentence search for referents, and in the purification from language of non-referential terms and purely verbal constructs."[51] Ogden and Richards support the tenets of analytical philosophy (basic English, common language philosophy, and semantics as an organized discipline) against structuralism. They reconstruct the linguistic/semiological project in terms of correctness, causality, and adequacy. For this tradition, language is a barrier to accurately conveying intentions, and the problematic of linguistics includes developing means to overcome poor communication.

Structuralism does not deny reality, nor does it follow pragmatism in substituting the rhetoric of practice for that of realism, but it implies a resonance between language as a whole and the whole that is nature. According to Jameson, structuralism maintains that it is the whole language system that "lies parallel to reality itself"[52] rather than the individual word or sentence that represents or "reflects" the individual object or event in the real world. It is the entire system of signs "which is analogous to whatever organized structures exist in the world of reality."[53] Our understanding through language "proceeds from one whole, or Gestalt to the other, rather than on a one-to-one basis."[54] For Jameson and the Saussurean understanding of language, reality "is either a formless chaos of which one cannot even speak in the first place, or else it is already, in itself, a series of various interlocking systems—non-verbal as well as verbal—of signs."[55] Piaget (who has taken structuralism into the study of psychology) also maintains that in science the apparent harmony between mathematics and physical reality cannot simply be "written off" as "the correspondence of a language with the objects it designates."[56] He explains the phenomenon as a "harmony" between the system of "the human being as body and mind" and "the innumerable operators in nature—physical objects at their several levels."[57]

Saussure also departed from orthodox linguistics by proposing that the synchronic (time-independent, parallel) dimension to language is more revealing than its diachronic dimension. The diachronic, or historical, study of language focuses on the way languages change in time and how languages are derived from each other. To study the synchronicity of language is to look at the structures within any particular language, the relationships within the language, at a moment in time in its evolution, and compare it with the structures of other languages. There are structural similarities between languages, which seem to transcend the particularities of individual sound patterns and local grammatical differences. We learn more by examining a slice through language, or rather the multiplicity of the world's languages, at any moment in time than by looking at the derivations of languages.

As we focus on the synchronicity of language, it is the structures that become important rather than individual elements. According to Piaget, structuralism "adopts from the start a relational perspective, according to which it is neither the elements nor a whole that comes about in a manner one knows not how, but the relations among elements that count."[58] The idea of "the relationship" is very important in Saussure's linguistics. What are the constituents of these relationships? A linguistic sign or system of signs is made up of a signifier (word) and signified (concept). An instance of such a relationship makes up a linguistic sign, and the multiplicity of such relationships makes up a system of signs. According to Jameson, in abandoning an atomistic, empirical perception of isolated objects, Saussure had to posit another relationship of greater potency to account for the way language operates. He rejected the relationship between word and thing. He could also have considered the relationship between the part and the whole that became the focus of Gadamerian hermeneutics (interpretation theory), or he could have invoked the relationship of form against field, or figure against ground, of Gestalt psychology. The relationship that Saussure settled on was the simple relationship between sound patterns. What is the key relationship between the enormous collection of sound patterns that make up the lexicon of a language? Taken two at a time, sound patterns operate by virtue of *difference*. Saussure begins with the phoneme, the basic constituent of sound patterns, even more basic than the syllable. Language works because we are able to distinguish one phoneme from another: "The sound of a word is not in itself important, but the phonetic contrasts which allow us to distinguish that word from any other. That is what carries the meaning."[59] So the word *kin* is different from *tin* and many other words that are otherwise similar, because of the distinctiveness in this case of the initial phoneme.[60] What constitutes difference, or at least a meaningful difference, again depends on the conventions of the particular language community. Being able to speak and understand Mandarin relies on a certain set of phonemic differences that are scarcely recognized as differences by English speakers.[61] According to the idea of phonemic difference, only differences that have developed as important in the language are registered. Others might constitute peculiarities of accent or dialect, or go unnoticed.

Structuralism therefore trades in the primacy of binary opposition, appropriating the Hegelian concept of the indeterminate relationship between opposites (the dialectic).[62] According to Saussure: "*In the language itself, there are only differences*. Even more important than that is the fact that, although in general a difference presupposes positive terms between which the difference holds, in a language there are only differences, *and no positive terms*."[63] Utterances in language mean something by virtue of the totality of relationships with other signs around them—which speech act theory emphasizes as context.

Since terms do not relate positively to things and the terms themselves, the words, are arbitrary, one particular word would do as well as any other—if the conventions of the language community so adjust themselves. Jameson draws an analogy with money, where a unit of currency has the same function whether it consists of gold or silver coins, paper, or wooden tokens, "where the positive nature of the substance used is not as important as its function in the system."[64]

This aspect of structuralism reinforces the distinction between form and content, the empirical and propositional emphasis on pattern and information content, in a particular way. Words and their arrangement are the carriers of information, which depends on the relationships between tokens rather than on the choice of tokens themselves. So if we wish to make comparisons between computers and brains, as far as language is concerned, it does not matter that a computer deals in discrete electrical pulses and that the brain deals in electrochemical units. The replication of human neural processes by computer depends on relationships rather than on the choice of units or tokens within the system of currency. In each case, it is the structures, which is to say the patterns of relationships, that are important, and through which we may find concurrence. As we shall see subsequently, structuralism differs from empiricism in its focus on difference rather than on similarity as the basis for patterns of relationships. This opens structuralism to the subsequent radicalization of its theories.

The notion of language as a system of relationships, predicated on difference, suggested to Saussure that there are two aspects to language. There is the superficial or surface structure of language as it appears in any particular culture, the *parole* of language, language as it is spoken (or written) subject to local variation. But there is a structure underlying the different languages we encounter that is the *langue.*[65] Jameson regards this distinction as the key to the originality of Saussure's work. *Parole* is the active, *langue* is the passive dimension of speech. Issues of local accent, mispronunciation, and personal style are matters for the "science of *parole,*"[66] but Saussure is more concerned with the *langue*, that which exceeds the local, the structures beneath the peculiarities of specific uses of individual languages.[67]

One of the contributions of structuralism has been to open a space for considering language other than as spoken and written communication. Structuralism admits art, fashion, architecture, sport, and culture generally as forms of language.[68] Lévi-Strauss developed structuralism as a mode of research within anthropology to look at kinship patterns, taboos, culinary practices, ritual, marriage laws, and so on.[69] The approach requires that one look less at the origins and lineage of cultures and probe beneath the surface of cultural appearances. For structuralist anthropology, the equivalents of phonemic binary differences are phenomena such as clean/unclean, cooked/raw, male/female, in/out, young/old, heaven/earth, life/death, and so on. As with writing and speech there are cultural patterns that transcend the peculiarities of local conditions, the *parole*, and that constitute the *langue* that crosses local cultural boundaries. Lévi-Strauss provides the example of the kinship structure of uncle and nephew relationship in traditional premodern communities. Precisely who fills those roles is less important than the underlying structure, and the roles are filled by different people in different communities. The roles may be variously filled by father, priest, brother, and teacher.[70] The structures in myth provide a further illustration of the importance of langue. Members of a traditional village community may look at themselves in terms of their descent from animals, one group regarding themselves as descendants of bears, while members of another community see themselves as descended from wolves. According to the structuralist model, to dwell on the particularities of the choices of token, namely bears and wolves, does not get us very far. But to pursue the *langue* of this cultural system suggests a structure involving relationships between self versus others. Community A is as different from community B as are two different animal species. Bears, wolves, or eagles may equally suffice to establish this understanding of difference, which is caught up in the myriad ways community A deals with those outside the community. Lévi-Strauss calls this practice of taking whatever is available from the local conditions for the *parole* of myth "bricolage."

When anthropological study is opened up in this way, then we see that similar structural distinctions are made in modern societies, employing different tokens, with their own regional applicability, and in some cases equally dependent on distinctions between animal species, as in sporting allegiances.[71] From this identification of the *langue* of myth follows a theory of transformation. According to Lévi-Strauss, "mythic thought always progresses from the awareness of oppositions towards their resolution,"[72] and there are rules of transformation that bring about a shift from one variant of a myth to another.[73] To identify transformations of the unity story through Neoplatonism, romanticism, and technoromanticism (as in chapter 2) assents to the structuralist view of myth, though structuralism does not require that such transformations are time dependent.

Structuralism has constructed its narratives in such a way that the real, as pursued by empiricism as *reality*, is scarcely an issue. According to Barthes, language "is not expected to *represent* reality, but to signify it."[74] The code is everything. So computer databases do not *represent* spaces but *signify* them through systems of differences. The question of whether a computer image can correspond to reality therefore is not an issue for structuralism. Computer imagery participates in the play of signifiers, well explained by the workings of metaphor, to be discussed in chapter 5 (though there we take the insights of pragmatism above those of structuralism). We can see the implications of the structuralist position in the development of space as cultural artifact, and structuralism has provided an important contribution to understanding architecture, from traditional to modern, as imbued with codes. For example, in the case of a Romanesque church, one discerns a layout and symbol system pertaining to the cycles of the sun, the moon, and the zodiac. The statuary depicts the apostles, Christ the Chronocrator, and the Light. The organization of the building, including the disposition of entrances and windows, allows a certain orientation between liturgy and procession that is specific to a theology.[75] From a structuralist perspective, one could analyze these peculiarities of Romanesque churches in terms of an intricate matrix of oppositions: east/west, light/dark, past/future, time/space, life/death, heaven/earth, transience/eternity, and center/periphery. In so doing, we would find similar structural relationships in the layout of the Forbidden City of Beijing, a Hindu temple, or a Pawnee earth lodge.[76] If we look past the particularites of the tokens invoked, we see similar relationships of orientation, centrality, geometry, and references to the stellar and temporal. From a structuralist perspective, this common structure constitutes the *langue* of traditional or sacred architecture, of which the precise nature and name of the deities or notables involved, and the particular shape of the building elements and the ornamentation, pertain to the *parole*. Since there is no consideration of a mapping to reality, the question of the truth status of the objects depicted does not arise. The structuralist interpretation may investigate the transformations from one system to another that arise due to local conditions and history, and the way these relationships are implicated in myth, ritual, and life practices.[77]

In order to treat empirical notions of the real within a structuralist project, one would have to accord to empirical practices the status of a language, which is to treat the activities of science as cultural phenomena. So the language of physics as articulated by Gregory could be considered in structuralist terms. If science involves a series of transitions from one language system to another, or if there are rival language systems, then we would look for the oppositions within the various systems and, through them, discern various structures that transcend the vagaries of the particular theories under study. As yet, there appears to be little of this type of analysis under the rubric of structuralism. Structuralism has been developed as a science and looks to physics and mathematics as a model for its own methods. Structuralism does not presume to analyze science using structuralism's own derived methods. There is a field of study dedicated to understanding science from an anthropological point of view, as exemplified by Latour and Woolgar's book *Laboratory Life*.[78] They refer to laboratory practice as involving transformations of statements from one form to another, namely from highly qualified statements to statements of fact,[79] transformations being the operative force in structuralist concepts of mythmaking. But their study seems to owe more to critical theory and poststructuralism than to Saussure or Lévi-Strauss. In so far as Gregory's, and Kuhn's, account of science involves transformations from one system, schema, paradigm, or set of metaphors to another, it could be construed as structuralist—the implication being that science is less about progress, or discovering fundamentals, than simply about change, accommodating interconnected shifts in thinking and practices to take on board new contingencies.[80] A further account of the theories in science that could be construed as quasi-structuralist is to be found in the work of Lakoff and Johnson, though their work is ostensibly grounded in speech act theory, and they are critical of the structuralist approach to language of Chomsky, for example. From a structuralist point of view, Lakoff and Johnson have identified various oppositions that underlie thought, which they call "image schemata" or "gestalt structures." According to Johnson, he intends to demonstrate "that experiential gestalts have internal structure that connects up aspects of our experiences and leads to inferences in our conceptual system."[81] There is the in/out schema, the schema of force/counterforce, attraction/repulsion, center/periphery, and so on. He does not articulate all such schemas in oppositional terms, but it is a simple matter to make them into oppositions. Lakoff argues that such schemas lie behind all experience, including our understanding of mathematics. So mathematical concepts such as correspondence, order, equality, and operations can be understood in terms of such everyday experiential gestalts as linking, direction, balance, and agency.[82] One could extend this line of argument to consider certain scientific concepts of space, for example, as suggested earlier in this chapter. Mathematical concepts of space exhibit similar structures of left/right, direction, centrality, and so on.

Structuralism is important for our study of computer narratives because it opens the way for the consideration of difference and binary opposition, particularly as they permeate concepts of the real, notably the real and the symbolic, to be examined in chapter 7. Structuralist narrative suggests that to understand the relationship between computer models and the worlds they purport to represent, we would do well to consider the differences they establish before the correspondences. Insofar as computer descriptions make use of language, they work by way of difference, from the trivial precept that *X* and *Y* are useful variables in an algorithm by virtue of their difference, to the nature of variable types, which is as much about exclusion as about similarity. So in the case of interface design, the folder icon on the desktop is interesting and useful not first because it corresponds in appearance and function to a physical folder, but in how its behavior differs from that of other objects on the virtual desktop and in how it differs from a physical folder—it can be copied, the contents do not fall out, its contents can be displayed in different ways, you can have folders within folders, and so on. So too, VR worlds and immersion environments impress us with the differences they afford—apart from the obvious differences highlighted by the coarse granularity of images, inertia, and so on, we can walk through walls, change scale, and instantly transfer to a new location, opportunities that "reality" does not afford. Such systems of differences can be appropriated through the concept of metaphor, to which we shall return subsequently.

In summary, how does structuralism present the real? As developed by Saussure, Jameson, Piaget, and Lévi-Strauss, structuralism presents the real as a whole to which we relate through language as a whole system of signs. The *proximal* is the system of signs within which we are immersed. But there is a sense in which structuralism denies proximity as the basis of its structures. Empiricism presents meaning in terms of correspondence and gives priority to similarity, and networks of proximity between objects as a basis of classification, but for structuralism it is not similitude but difference that provides the basis of structure. The nonproximity of phonemes, objects, and concepts constitutes the relationships and therefore the structuring of the real.

On the aspect of the *shared* as a constituent of the real, for structuralism it is not so much sense impressions that are shared, but language, which works by agreements about meaning and difference, the conventions of the language community. The *body* emerges less as an instrument for the receipt of sense data than as an object implicated in a system of signs, as in Barthes' semiological description of wrestling, in which the "physique of the wrestler . . . constitutes a basic sign, which like a seed contains the whole fight."[83] Structuralism shares with empiricism its quest for patterns, the *repetitions* that lie beneath the surface phenomena, but patterns that are constituted on the nondeterminacy of difference are less instrumental and less amenable to predictive formulations than those of empiricism. Against this cost, structuralism offers the benefit that it is able to extend its theories into culture, language, and even science as culture, areas about which empiricism has less to say. The outcome is that structuralism has something to say on almost everything, as everything fits within some cultural context or other and can be discussed as a code system, from margarine to Einstein.[84] The discourse of the real is also subject to structuralist, which is to say cultural, analysis. The strategy I have adopted can be construed as structuralist—of identifying the proximal, the shared, the embodied, repetition and the ineffable as constituents of the real in different discourses of computation, realism, and idealism. So too, the transformation of the unity theme and Platonism through Neoplatonism, romanticism, empiricism, and the technological world view can be construed as an account of the transformation of a mythic structure.

Beyond Structuralism

Structuralism has its critics: narratives that take structuralism as their project. In spite of its rhetoric of difference, as with empiricism, structuralism implies a leveling of cultural phenomena and a neglect of differences that do not fit the theories. In identifying the *langue* of the customs, beliefs, life practices, and architectures of disparate communities, the differences that do not fit the deep structure may be glossed over. A structuralist account can engender a respect for difference or can trivialize difference by rendering everything the same, by virtue of an apparently similar structure of relationships. Furthermore, as suggested by Snodgrass, structuralism ignores the role of the symbol, at least as understood in the sense used by Plato, by which architecture and art are "formed, or in-formed, by imitation of and participation in archetypal paradigms."[85] According to Eliade, the examination of symbolic structures is a process of integration rather than reduction: "One compares and contrasts two expressions of a symbol not in order to reduce them to a single, pre-existent expression, but in order to discover the process by which a structure is capable of enriching its meanings."[86] As symbols address "the whole psychic life," we do not have to reflect on the symbols we use in order for them to work in us.[87]

In the case of architecture, structuralism has engendered a way of looking at buildings that implies that merit rests on how well a building can be read. To regard a building as text can present architecture as a heavily coded interreferential language game in which only the cognoscenti participate. Structuralism does not provide a satisfactory language for talking about practice, use, and function, let alone symbol and concepts of participation. This is a problem for structuralism generally, which seems to have limited its application to understanding computer systems. According to Jameson, structuralism has idealistic tendencies "which are already at work within the material itself, of encouraging the insulation of the superstructure from reality. . . . its concept of the sign forbids any research into the reality beyond it, at the same time that it keeps alive the notion of such a reality by considering the signified as a concept *of* something."[88] According to Jameson, structuralism "is essentially a replay of the Kantian dilemma of the unknowability of the thing-in-itself."[89]

In wrestling with the nature of cognition and human communities, Giddens further argues that structuralism has led to a rarefied play with notions of difference and an inability to engage the matter of context or human action. He thinks Wittgenstein and Winch provide a better account of the workings of language and society. At least logical positivism preserved an engagement with the world, or at least it preserved an engagement with agency that structuralism lost.[90] The postpositivism of Wittgenstein was able to translate this into a concern with *praxis*.

But structuralism has also institutionalized its own critique as poststructuralism,[91] which can be seen to radicalize structuralism's key insights, particularly through the work of Barthes, Foucault, and Derrida, among others. In *Of Grammatology*,[92] Derrida focuses on destabilizing structuralist concepts of language and by implication the empiricist, propositional, view that language can be understood in terms of sentences that have content, or in the language of structuralism that there is a *sign* (the sentence) and the thing to which it refers, the *referent* (the meaning or information content of the sentence). As Derrida argues at length, and as was well known to structuralism, it turns out that the referent constantly evades identification in any particular language situation. It is rather the case that signifiers appear to be referring constantly to other signifiers, and in turn these refer to other signifiers. So the word (signifier) *tree* may not simply refer to a single concept of tree but to my finger that is pointing to something, or the image of a tree, or an entry in a botanical classification, or a tree in a poem, or to solidity, life, or growth; and each of these signifiers refers to some other signifier. Insofar as meaning resides anywhere, it is in the trace left by the chain of signification.[93] The endless chain of signification is well illustrated in the case of computer imagery—in fact, in the whole milieu of the mass media. Images can be copied, reproduced, and recomposed in such a way that the original and what it referred to are never finally settled. If an image of a giant eye in a collage on someone's home page on the World Wide Web is a reference to the movie *Bladerunner*, from which it was "lifted," to what does the movie referent refer? The endless chain of signification features prominently in narratives of information technology, but Derrida's point is not that this is a new phenomenon of the digital age, but that language can only function, and always has, through the concept of trace. As Barthes says, "There always remains, around the final meaning, a halo of virtualities where other possible meanings are floating: the meaning can almost always be *interpreted*."[94] For Derrida (and Barthes), this endless referentiality is the norm, and any language situation in which we simply ascribe *this* signifier to *that* referent is a highly contextual and transient instance of ascription. For Derrida, the core of any phenomenon, such as notions of meaning in language, or that there are foundations to our knowledge, is in fact a nondeterminate phenomenon that is constantly in flux and play. It is not that there is no core or foundation to any phenomenon, but that the core or ground is unstable. It relies on what is supposed to be built on it. Alternatively, the foundation is always "foreign." It is something brought in from outside, by definition without justification, to found something new and different to what was there before. The concept of foundation seems to rely on instability to establish its status as a core.[95]

Poststructuralism therefore further challenges empirical narratives based on the proposition and information content as key features of language. Such notions are too metaphysical, grounded in ideas about correspondence between signifier and signified, denying the operations of difference and trace. But Derrida also challenges aspects of pragmatic concepts of language. In *Limited Inc.,* Derrida takes Austin's speech act theory to task.[96] While agreeing with much of what Austin says about the ubiquity of the performative utterance and the implications of that observation, he shows how Austin still depends on notions of the originator or author who brings intentions to the communicative act. Again, authorship is an elusive quest. Any text can be attributed to many participants. There are certain conventions about ascribing authorship to a text that are sometimes beyond dispute, but we cannot use the author function and some notion of intention as the basis for a theory of language without recognizing the nondeterminacy of the notion of intentionality.

There are many implications of Derrida's writing for information technology, though he does not discuss information technology explicitly at length. (Poster, Ulmer, and Luckhurst have attempted to tease out the implications of Derrida's writing on information technology,[97] as I have elsewhere.[98]) Some argue that Derrida's insistence on the disappearance of the referent and the constant interreference of signs is a property of electronic communications, and this contributes to the current age of uncertainty and disorientation. We can no longer be sure of an original source or original meaning as everything comes to us in digitally mediated form. But Derrida's view is radical precisely because he indicates that this is not a new phenomenon.[99] Rather, it is the case that language has always had this disorienting function. Derrida shows how many of the properties we commonly ascribe to writing have been present all along in speech "prior" to writing.

As a philosopher intent on destabilizing all metaphysical or foundational notions, Derrida provides no support for the view, which seems to sustain certain cyberspace narratives, that information or measurement allow us to represent, resist, reduce, divide, or transcend space. As I have discussed elsewhere, Derrida also trades substantially in spatial metaphors in his identification of the nature of metaphysics—that it relies on oppositions

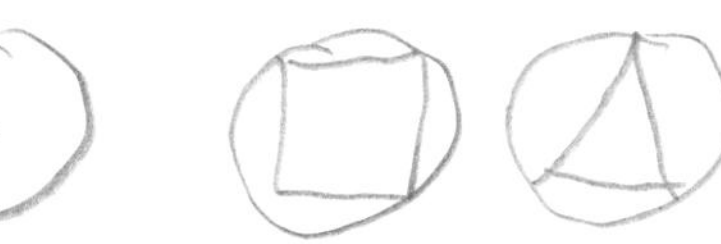

of presence and supplement or center and periphery.[100] The myth of unity opposed to multiplicity is a variant on the metaphysical schema of presence and supplement. The unity, the one, the ideal, the ground, the absolute, the homogeneous, speak of the privileging of presence. The many, the multiple, the individuated, and the heterogeneous speak of deprivileged supplement. The myth of unity and multiplicity is a metaphysical schema, favoring the former over the latter, though it is a schema that is variously radicalized in poststructuralist narratives. Poststructuralism extends the structuralist project, in that it identifies the quest for the real as subject to the important binary opposition of presence and supplement, which Derrida shows to underlie all discourses on the real, but that is itself the shakiest relationship of all.

A Critical Theory of Modern Myths

The tenets of structuralism have been adopted by critical theorists to open the discussion of language and the real to political interpretation and to construct critical digital narratives. Barthes' account of myth provides an example, though we must add that he is discussing *modern* myths, which are removed from what he regards as modes of authentic practice.

Myth is a feature of everyday life and language and is not restricted to grand narratives. According to Barthes, myth is simply a "type of speech,"[101] demonstrated in our dealings with ordinary things such as magazine covers, sports, and domestic appliances. Barthes provides many examples of contemporary myths, including the power of washing powders to liberate grime, chemical liquid cleaners to "kill dirt," the ornamental value of drinking wine, and reconstructing the adult world in miniature through children's toys. We may add the myths of information as a means to unity, progress through computers, emersion in cyberspace, digital utopias, and so on.

According to Barthes, not all instances of language or all cultural artifacts rely on myth. There are instances of language that pertain to the immediacy of use. So a comment about the weather from a farmer is a straightforward instance of the operation of the signifier/signified relationship pertaining to involvement in some task, whereas the same comment within other contexts is a mythic expression about one's mood, a foreboding, a general malaise, and so on. Myth is signification operating beyond the normal signifier/signified level of language. Barthes gives the example of a magazine cover showing a Pacific Islander in French military uniform saluting a flag. In signifier/signified terms, this picture may well indicate a person paying respect to flag and country, but the whole instance of the signifier and signified also carries a mythic signification, suggesting an empire in which all is well, where there is no discontent or discrimination.

Barthes only discusses *modern* myths in *Mythologies*, and myth bears a pejorative cast. Modern myths are not profound and deal in immediate impressions: "Myth hides nothing and flaunts nothing."[102] Myth, however, distorts, though it is "neither a lie nor a confession: it is an inflexion."[103] A myth is not improved or diminished by analysis nor by whether or not people see through it, but the peculiar feature of mythic language is that it appears as "naturalized language." In fact, according to Barthes, what allows people "to consume myth innocently" is that we do not recognize the myth as a semiological system. Myth conceals itself as indicating a causal process, and "the myth-consumer takes the signification for a system of facts."[104] So washing powders do drive out dirt, wine drinkers are more sophisticated than beer drinkers, children like to play at being adults, all is well with the empire, electronic communications are liberating, and computers are efficient. Myths are taken as what is real.

This brings the consideration of the real into a political framework. For Barthes, modern day myth is a middle-class, capitalist, which is to say bourgeois, phenomenon. One of the features of bourgeois life is that it is invisible to us as bourgeois: "The bourgeoisie is defined as *the social class which does not want to be named.*"[105] The link between myth and the bourgeois is that myth makes the contingent world appear natural and eternal: "Myth has the task of giving an historical intention a natural justification, and making contingency appear eternal,"[106] and this process is exactly that of bourgeois ideology.[107]

There are several features of the contemporary myth-laden bourgeois world according to Barthes. Society fabricates certain myths—"acknowledged evils," such as the avant-garde, eccentrics, street gangs, and scandals in politics and the church—to inoculate itself against the possibility of wholesale social transformation. The costs to bourgeois society of such tolerated evils balance the benefits of avoiding radical social transformation: "One inoculates the public with a contingent evil to prevent or cure an essential one."[108] In the computer world, we can see this in the identification of the issue of censorship on the Internet, the introduction of the "clipper chip," CIA conspiracies, computer hackers, cyberpunk, and software monopolists. It is also there in the fantasy and cyberspace literature, which are becoming increasingly difficult to distinguish from each other. According to Stallabrass: "The idea that one is participating in headlong, dangerous scientific progress imparts a frisson to the mundane use of today's technological equipment. Real apocalyptic tales are out there in profusion, but generally do not affect those who write these stories or read them."[109] A similar accusation can be leveled at cyberculture's propensity to elevate complexity: "Any attempt to grasp the actual complexities of the workings of global capital are abandoned to a literary spectacle of the sublime phenomena of complexity itself."[110] From Barthes' point of view, each of these myths is a diversion from structural bourgeois malaise. Baudrillard also identifies this diversionary function of myth in his account of artificial realities, as in the case of Disneyland theme parks. The apparent unreality of Disneyland is there to mask the actual "unreality" of Los Angeles. The same arguments would apply to computer simulations, games, and virtual reality. By virtue of their contrast with the rest of experience, they are a ruse to reinforce our belief in the reality of the rest of our experience, which is in fact prone equally, if not more so, to the workings of simulation, artificiality, deception, and domination.[111]

Modern myth also insures the neglect of history, presenting what society holds dear as eternal and independent of historical contingency. In the computer world, this is ignoring the historical contingency of the informational view, ignoring that the Enlightenment occurred in history and has a critique, not recognizing that there is a history to the use of technologies in teleological narratives, and "re-inventing" metaphysical schemas that point to the unity of all things through information. Myth also provides a means of dealing with objects that do not conform to the norms of bourgeois society. Myth allows instances of "otherness" to be treated either as the same or as "exotic" in some way. Hence, for empiricist narratives (and their IT variants), insofar as they pay it any heed, poststructuralist theory is subsumed as "subjectivity," Nietzsche is a "romantic," and the critics of artificial intelligence are simply warning us against excess. To the list of such institutionalized aliens as the computer "nerd," "cyborg junkie," and "techno-evangelist," we have the IT "critic," who does not fully understand but is admitted, entertained, and humored in the cause of open-mindedness.

According to Barthes, modern myth ultimately appeals to the tautology, such that when the resources of reason seem exhausted, then myth allows us to assert simply that things are as they are because that is how the world is, the most obvious examples being myths purveyed through proverbs and maxims: "knowledge is power," "think before you act," "science describes reality," and "the future is digital." According to Barthes, certain myths also work by reducing any quality to quantity. They "economize" intelligence, explaining reality "more cheaply."[112] This is common in empirical and computer narratives, where number, symbol, and formula provide the chief means of prediction. From Barthes' point of view, where these practical systems are extended to the realm of explanation, they then become myth.

From this angle the unity myth, and its variants, is a ruse. The prospect that we are entering into a new age of community—that IT is subverting accepted notions of reality, challenging outdated social and political schemas, and providing new opportunities for the liberation of our true selves—is a way of keeping people in their place, preparing us for new modes of commercial exploitation, and affirming the bourgeois social order. The prospect of cybernetic rapture presents the most pathetic deception of all. This impoverished manifestation of latter day Neoplatonism is devoid of theological and practical content. However premodern societies may have participated in the symbols of the one and the many, such symbols are now divorced from authentic life practices.

According to Barthes, in contrast to the workings of myth, there are ways of dealing with objects and language that are immediate, as in the use of mathematics in the laboratory and the academic paper, or when a farmer comments on the weather, a navigator describes a position, or an architect describes a door in a specification. Echoing speech act theory, Barthes says that such instances of language pertain to use. The other instances of language that are immediate belong to the oppressed. The oppressed person does not have access to mythic metalanguage: "Destitution is the very yardstick of his language."[113] As a bourgeois phenomenon, the duplicity that is myth "presupposes property, truths and forms to spare,"[114] but the oppressed have recourse to only one language, that of actions. Similarly, radical social reform involves the language of action and not myth: "Revolution announces itself openly as revolution and thereby abolishes myth."[115]

Barthes adopts the general neo-Marxist strategy which is to identify and expose the ways that discourses covertly reproduce their conservative agendas. As for Adorno, Horkheimer, Marcuse, and others of the Frankfurt School, the critical theory of Barthes trades in the language of social class, hegemony, domination, oppression, and ultimately revolution. Barthes adds modern-day mythologies, understood through the structuralist conjecture of signifier and signified, as a tool for the reproduction of capitalism and therefore a tool of oppression.

The culpability of modern myth finds support in Marcuse, who identifies the privilege granted to the *proposition* that pervades empiricism and its descendants as the purveyor of the reproduction of capitalism.[116] For empiricism, language is rendered inert and apolitical. The serious study of language is reduced to a consideration of the truth and falsity of propositions. What we actually do with language is presented as a concern isolated from the technological. Similarly, from Marcuse's viewpoint, scientific realism and positivism present the study and representation of things such as space without consideration of the social implications of those notions. Empiricism sees the computer as an apolitical tool to be used for good or ill. Social concerns are apparently beyond its technical agenda, but at the same time the propositional view conceals a political agenda. From Barthes' point of view, the proposition is the statement of fact that is the scarcely concealed myth. It is bourgeois myth masquerading as fact.

Marcuse characterizes the propositional view as "technological thinking," which is fraught with several difficulties. Technological thinking identifies, isolates, and then marginalizes the ethical. Propositions and the paraphernalia of logic (such as the syllogism) are put forward by individuals to settle a matter of argument and therefore to stifle debate. In actuality, the ethical is never "resolved" except by giving free rein to the indeterminacy of true dialogue. Technological thinking also trades in abstractions, indifference, and decontextualisation. So technological thinking specializes in alienation from the world. Technological thinking also trades in domination. According to Adorno and Horkheimer: "The general concept which discursive logic has developed has its foundation in the reality of domination."[117]

Structuralism embraces the indeterminacy of difference, but technological thinking makes no such gesture. It subjugates dialectical thinking—that is, the indeterminate thought that weaves its way through the various tensions inherent in any domain of argument and that permits the emergence of a new synthesis in the manner suggested by Hegel. In technological thinking, dialectical conflicts are tamed and rendered expendable and meaningless. Contradictions are thought to be the fault of incorrect thinking rather than harbingers of new thoughts. Concepts become instruments of prediction and control. According to Marcuse, technology itself (its objects and systems) embodies and reproduces this domination. It accomplishes this domination by virtue of its pervasive, totalizing presence.

Other more moderate critical theorists such as Habermas have extended the arguments of critical theory by focusing on language and how it can feature as an agent for reform. Echoing speech act theory, Habermas posits a "theory of communicative action," with communication intricately linked to practice: "Only the communicative model of action presupposes language as a medium of uncurtailed communication whereby speakers and hearers, out of the context of their preinterpreted lifeworld, refer simultaneously to things in the objective, social, and subjective worlds in order to negotiate common definitions of the situation."[118]

Critical theory therefore identifies abstract, propositional, and (we could add) informational concepts of communication, as well as myths, as perpetrators of domination. It shares with pragmatism a belief in the communal, situated, and contextual nature of language, the appropriation of which can lead to social reform. It appropriates structuralist concepts of language that provide a ready means of rendering issues of the real contingent, and it identifies myth as one of the ways that we construct the real.

Critical theory also provides direct insights into discourses on space. Insofar as concepts of space are informed by the language of the proposition, they are a party to the processes of domination, and according to Lefebvre we subdue space as a social phenomenon when we abstract space in mathematical and relational terms.[119] According to Lefebvre, the danger of abstract views of space is that we look at the *things* in space rather than space itself: "The dominant tendency fragments space and cuts it into pieces. It enumerates the things, the various objects that space contains." We fragment space into the specialties that deal with it, which obscures its important aspects. The major culprits are professional specializations that "divide space among them and act upon its truncated parts, setting up mental barriers and practico-social frontiers." For Lefebvre, "Architects are assigned architectural space as their (private) property, economists come into possession of economic space, geographers get their own place in the sun, and so on."[120]

Space is fragmented according to the division of labor, and the process reinforces the system of domination in which space appears as a passive receptacle that is independent of social relationships. According to Levebvre, instead of uncovering the social relationships that are latent in spaces and concentrating on the production of space and the social relationships within it, we treat space as an entity in its own right. We tend to "fetishize space in a way reminiscent of the old fetishism of commodities, where the trap lay in exchange, and the error was to consider 'things' in isolation, as 'things in themselves.' "[121] According to Lefebvre these "ideologies" of space have to be overcome so that we can reclaim space as the social realm. Derrida also echoes Lefebvre's concerns about social space. He supports the notion of space as social as opposed to space as inert and indifferent onto which we cast our social concerns—space as transcendent.[122] It is not that there is space, definable mathematically and absolutely, and that it is overlaid with social concerns. Rather, according to Derrida, these social concerns produce the "spatiality of space."[123]

As a theory of skepticism in the face of all notions of control, critical theory is rarely operational in the sense enjoyed by propositional narratives of space. Critical theory narratives do not provide operational guidance for the design of algorithms. Rather, critical theory assists in putting technological systems, and the theories we construct to support them, in their place—lest we are tempted to use their internal logic as a means of justifying and hence concealing conservative agendas under the disguise of impartiality. (We will return to the radical application of critical theory in Deleuze and Guattari's narrative of schizophrenia and information technology in chapter 7.)

Foucault develops narratives of technology and power that provide an indirect critique of the insights of structuralism, the critical theorists, and also bear on the issue of information. Some of the implications of Foucault's writing to information technology are presented by Poster.[124] According to Foucault, there is a human technology before there is a material technology. Society develops social means of reducing the wastage of violence and thereby embedding and promoting power. Foucault adopts the structuralist/poststructuralist understanding of the workings of the grand myth.[125] For Foucault, history operates as a series of transformations by which institutions arise to convert one manifestation of power to another. For example, the transformation from a feudal to a democratic system of power shows up as a transformation from a society in which torture is tolerated to one in which discipline is the norm. So the histories of prisons, schools, hospitals, and the military hold the key to an understanding of transformations of power. These social technologies in turn have their own material technologies of prison walls, textbooks, hospital beds, and rifles. Foucault's famous spatial example of how power is manifested in a material technology is that of Jeremy Bentham's proposed Panopticon, a prison building shaped in such a way that the guard can see every prisoner but the prisoners cannot see each other.[126] Interpreting Foucault, Poster sees modern centralized computer databases of the kind used by the Inland Revenue, police departments, and credit companies as a kind of "superpanopticon," in which surveillance by ordering, partitioning, and dividing (how the data is structured for any one individual) is complete.

But Foucault differs from the critical theorists in that he maintains we all participate in the technologies that surround us. Both guard and prisoner are the "victims," or subjects, of the panopticon. So in the case of computer databases, there is no one in control, and the people inspecting the databases are as much its subjects as are the rest of us. The computer is yet another instrument of institutionalized power. The computer is an outward manifestation, or a distillation of various transformations of power that have long been in train. There is a sense in which we see ourselves in the way the social technology has constructed the material technology of the computer. The Foucauldian view of technology removes some of the impetus of the critical theory argument. The means of dividing and ordering—be they myth, the proposition or the way in which we divide space—are not imposed by some intangible power elite nor are they some means of control against which we must revolt, but are part of a constellation of technologies by which we define ourselves and by which power has been transformed from one manifestation to another. Merely changing or rejecting the material technology will not necessarily bring about significant change.

But then for some commentators on technology, most notably Heidegger, to seek to implement change is just a further manifestation of technological thinking. The task is to let things be in their essence, which leads to a peculiar Heideggerian doctrine of letting entities, such as spaces, reveal themselves with their own characteristics and without attempting to impose on them universalizing categories. Attempts at interpreting Heidegger on this point have been developed at length by Borgman in the context of late-twentieth-century technologies,[127] and I have examined the relationship between Heidegger's thinking and that of the critical theorists elsewhere.[128]

The grip of empirical realism that sustains technoromanticism is released through pragmatic and structuralist narratives of language. The propositional narrative, the most primitive theory of all (according to Jameson), presents an extreme form of symbolization, the correspondence between word and thing. Pragmatic narratives of language concentrate on utterances in language as tools for action within contexts. Pragmatism contributes to understandings of science and displaces empirical and representational realism as the means of explaining how science operates. Structuralist narrative makes the break with reality, treating language as a self-referential system of signs. From the point of view of pragmatism, structuralism opts out of concern with agency and action. However, at the hands of critical theory, structuralism provides provocative narratives around themes of myth and political action.

This excursion into language exposes the deficiencies in narratives that trade in propositions and correspondence. In so doing, we have displaced the primacy of representation. Computers deal in symbol strings, propositions, formulae, algorithms, and sentences, but the view that their primary role is to represent is limited. From the pragmatic point of view, propositions are to be interpreted in contexts of practice. Their role as representing things, such as spaces, objects, and people in computer databases, is simply a shorthand for more complex processes by which, as toools and technologies, they fit into particular practices, pertaining to recordkeeping, computer modeling, visualization, simulation, and so on. There are clearly no all-purpose schemas of representation, which, if only we had the time and the computer memory, provide anything other than operations for particular contexts of use. In this light, the computer provides no access to a transcendent world by virtue of its privileged access to representation and the informational content of all things. Language does not work that way, and information content is a chimera.

Empirical realism has also been supplanted by pragmatism. Narratives that appeal to unified theories, asserting the preeminence of science and the primacy of fields of chaos and order, are metaphysical, if not ironical. Pragmatic views of space add further support to the contention that computers do not have privileged access to spatial description. Space is far too varied and contingent a matter to be represented, resisted, reduced, divided, or transcended in computers, as can be seen from constructions of space presented in critical theory and elsewhere. This is not to say that we cannot be immersed in computer environments, no less than we can be immersed in movies, reading a book, or contemplating an image. Such immersive experiences operate through the workings of interpretation and metaphor and through the provocation of difference, in romantic terms the workings of the imagination.

5

Pragmatics of Cyberspace

If empiricism begins with the primacy of the parts, phenomenology starts with "the whole." According to Heidegger, we look "through this whole to a single primordially unitary phenomenon."[1] Breaking a phenomenon into parts is a derivation to a rarefied context rather than a reduction to basic constituents. So to study space as a coordinate system, as in a three-dimensional computer model, is not to understand space at a basic level from which all other understandings can be derived but to move to a detached, rarefied frame that produces certain understandings in particular and limited contexts. According to Heidegger, to examine the unitary phenomenon is to engage in *ontological* investigation. To examine the things, the objects, as does science, is to conduct *ontic* investigation. In terms of our simple mythic schema, the ontological pertains to unity, the ontic to individuation and fragmentation, though, as we shall see, the ontological is the practical domain, with its vagaries and antagonisms.

Space and time provide good test cases for investigating the insights of phenomenology. Phenomenology allows us to build other constructions on the nature of space to those presented by empiricism and romanticism, and it provides accounts of the nature of mythic and utopian thinking. Phenomenology suggests that information technology is best understood through its involvement in various practices. The operations of computer systems that we interpret in spatial terms are contingent on practices, and narratives of digital utopias attempt to engage the pragmatics of anticipation, though at the level of the ontic rather than the ontological.

Pragmatics and Spatiality

Heidegger's thinking has touched most contemporary reflection on language and the real outside the empiricist tradition. It is worth studying Heidegger's phenomenology in some detail, as it represents a reorientation to the concept of the real (from empiricism), it is a position against which structuralists and critical theorists often parry, and it is gaining currency in technology studies and understandings of IT.

Heidegger's phenomenological treatment of space and the real follows several strategies.[2] In direct opposition to the Cartesian and empiricist systems, it does not presume the primordial and immutable divide between subject and object. Descartes assumed and reinforced the fixed nature of a self, a thinking thing (res cogitans), subject to doubt but sure of where that doubt resides. The things that are "out there" constitute the extended thing (res extensa), definable in spatial terms through measurement. The division of human experience, in terms of the subject and what is outside the subject, forms the basic premise of the various debates about realism versus idealism, objectivity versus subjectivity, the absolute versus the relative, and so on. The identification of subject and object as primary promotes contemplation as the principal means of understanding the real, which, as promoted by Plato and his successors, is a highly abstract activity. Contemplation suggests that the real pertains to the Intelligible rather than the embodied world of sensation and engagement. To exalt contemplation is to promote idealism. Insofar as the narratives of cyberspace and VR promote the concept of the cybernaut as subject presented with an array of three-dimensional objects in space for observation and manipulation, they invoke concepts of contemplation. The putative act of gazing in wonder at the spectacle of the data stream accords with the romantic's contemplation of the beautiful and the sublime.

According to the phenomenological position, rather than moments of contemplation, our most typical moments are those when we are engaged, absorbed in an undifferentiated world of involvement. We are busily going about our business of doing and making.[3] This is an experience with which everyone can identify: we are often most absorbed while working in the garden, driving, jogging, watching television, typing at a computer, or just engaged in the day-to-day routine of work. (In this respect, "immersion" is not a special feature of VR. A text editor, CAD system, computer game, or spreadsheet can be just as much an "immersive environment.") At such times, we are not contemplating who is this "I," what is "out there." Nor are we formulating plans and intentions, or maximizing the return to a "self" on effort expended. It is tempting to think of this engagement in purely empirical, psychological (ontic) terms, for which there is an objective world common to us all, and each of us becomes absorbed in parts of it to different degrees from time to time. According to this psychologistic view, we occasionally lapse into automatic mode and let our "unconscious" take over, or we succumb to habitual responses to situations without thinking of what we are doing. From the psychological point of view, we may also ask: what is this common thing that we are all absorbed in if it is not the real world? From the phenomenological position, to ask as much is to presume the subject/object divide sooner than we need to, and we have not achieved anything different from Descartes' formulation. In asking such a question, we are already presuming an isolated self. In order to obviate this problem, Heidegger adopts one of his many, very useful (and beguiling) linguistic strategies, and one that he adopts throughout his work, to replace notions of "self" with a term that can act as a placeholder for whatever the entity is that inquires after its own Being.[4] This entity is "Dasein," commonly translated as "existence," though preserved in the original German in translations of Heidegger's work to draw attention to his special meaning for it. Dasein is one of the concepts for which Heidegger claims *primordial* (ontological) status.

Cyberspace narratives of digital democracy trade in notions of community and solidarity. Dasein has several characteristics, which include the property of "being-with." There is a collectivity to Dasein that precedes the notion of many selves, society, or intersubjectivity. "Being-with" is another indication of the nature of human practice. We cannot act and do things in isolation. We always take over practices from within a community context. There is no such thing as doing and making in isolation. There is no such thing as a Dasein on its own. On this point the most ardent materialist would agree. Everything about human biology points to a shared being. We are so much creatures of our environment. Species develop as societies interacting within themselves, between each other, and within an environment. But according to phenomenology, we do not need to presume biology, behavior, evolution, or the existence of a natural environment before we can grasp Dasein as "being-with." We are already "with," and the construction that is biology (or even sociology) is a derived understanding to account for it in particular ways. Digital communities are not to be understood primarily as those formed from isolated selves communicating through networks, but there is already a solidarity, a being-with that is the human condition, into which we introduce various technologies, such as meeting rooms, transportation systems, telephones, and computer networks.

Cyberspace narratives promote concepts of digital worlds. Heidegger develops the key concept of "world." The other important property of Dasein is that it is in the world—it has the characteristic of "being-in-the-world." To be engaged unreflectively in doing and making is to be absorbed in the world. Here it is tempting to think of "world" in terms of something within which one is contained, a primarily spatial notion, and one that is again objectivist (and ontic). According to Heidegger, to think of world as environment, nature, or reality is to again jump ahead in the ontological hierarchy. In fact, it is the small word *in* that provides the key to a nonobjectivist understanding of being-in-the-world. The "in" of "being-in-the-world" is not a containment *in* but an *in* of involvement where there is no presumption of differentiation between entity and environment. Heidegger invokes various etymological arguments to demonstrate the primordiality of the noncontainment kind of *in*: We readily talk of being *in* love, *in* confusion, *in*volved, and so on. We are *in* the world in this kind of way before we are *contained* in an environment. If we have difficulty in describing being-in-the-world in ways that are contrary to objectivist notions, then this is because the philosophical tradition has all but covered over these primordial understandings. By this reading, we inhabit digital worlds, in an ontological sense, to the extent that we are engaged in practical activities, absorbed and involved with equipment, and not by virtue of assuming a viewpoint from some location in VR coordinate space.

It is also tempting to see the account of being-in-the-world described so far as involving a *fusion* between subject and object, as though the subject and object preexist but are fused when we become absorbed in a task. The phenomenological position is that there is *first of all* worldly engagement that we later articulate in various ways, including engagement in the distinction between subject and object.

What we have examined so far suggests something important about space. Already we can see that being-in-the-world carries different connotations to *being in space*, which invokes the *in* of containment. Furthermore, if there is a kind of *being-in* that is prior to a *contained in*, then space, or what we understand by space from the Cartesian and scientific view of space, is not an absolute. We do not begin with space as an objective entity and work out the place of humankind in it. We do not need to take the formulations of space of Einstein or quantum physics as a base from which to build an understanding of the world. Space is contingent and derived. It *builds on* primordial concepts of Dasein, world, being-with, and being-in. Another way of dealing with space is to reconsider its definition. In fact, this is what Heidegger does, stating that there is an entity more primordial than space and from which space is derived. This is the primordial (ontological) concept of *spatiality*. Spatiality does not presume the concept of a space that contains us but builds on concepts of engagement and orientation.

It is tempting to regard Heidegger's appeal to primordiality as a call for absolutism, but the phenomenological stance defuses notions of the absolute and the relative. To appeal to the primordial is not to appeal to an absolute, or if it is, it is to radically redefine the absolute. Heidegger makes liberal use of the term *essence* to describe the understandings he is propounding, but unlike the traditional use of "essence," Heidegger's essence of a phenomenon is not fixed but rather in a state of flux; it is indeterminate. So the way spatiality discloses itself is contingent on a practical situation. Spatiality is a pragmatic concept; to be spatial is to be engaged. Heidegger gives the example of leaving a room. Before I can move toward the door, I am already there in an ontological sense. I pervade the area of my concerns.[5] In the same way we would have to say we are already at the other end of the telephone line, we are with the participants of the on-line discussion group, we are already there at the receiving end of the video stream. It makes sense to speak of being there insofar as we are there in the sense of spatiality, though we see that there are many constructions on what it means to be in space in these cases: in terms of networks and connections, data channels, avatars floating in coordinate space in a virtual environment, and so on.

Heidegger's concept of spatiality highlights the importance of care: a further characteristic of Dasein is that we exhibit care. Care is not simply an individual disposition of philanthropy or a concern for people and things, but a disposition of Dasein toward the world as it presents itself to us at any moment. Within our experience of being in the world, there is a pragmatic understanding of proximity or closeness, and this closeness precedes any (ontic) notion of *measurable* distance. Distance is a function of our being concerned, or caring, about aspects of the world. So that which we care about the most at any particular moment is the closest to us. In our walking down the street, the pavement is as near as anything could be (measurably), yet it is remote compared with the nearness of an acquaintance one encounters several paces away. The commuter on her mobile phone is measurably close to her fellow travelers in the railway carriage but nearer ontologically to the person on the end of the phone. By this reading, computer networks are interventions into the fluid "networks of care" that already exist, though clearly each is implicated in the formation of the other. But even prior to the notion of caring for someone or something is the notion of care itself, as a disposition of Dasein.

So phenomenology builds an ontological picture of the primordiality of involvement. Things, objects, entities, are revealed to us through our engagement in the world at particular moments and in response to ruptures in the fabric of our involvement. Heidegger calls these ruptures "breakdowns," though he also says that the use of signification, or "pointing out," has the same character. To use language to signify is to withdraw from engagement and to focus on something. (In characteristic manner, Heidegger identifies that there is something before articulation in words, and this is the primordial phenomenon of "telling," which does not presume words.) When things are presented to us in this way, they appear to us immediately with certain characteristics. Examples of breakdown are where something presents itself as out of reach (the book on the shelf is too far away), the journey is too long (I am spending too much time in getting from home to the office), the paper is too large to fit in the envelope, and the computer file is in the wrong directory. In the moment of breakdown, things present themselves as too far, too long, too big, too heavy, too bright, and in the wrong place. In other words, we take things qualitatively and with value. Things are encountered as contextual, embodying our immediate concerns. We do not first of all perceive objects to which we apply qualities, or which we interpret and fit into a context, but we are already interpreting in the moment of perception.

Phenomenological study has focused attention on the issue of perception. When we hear a car engine, a bird call, a child crying, an ambient sound byte of running water in a computer game, that is what we hear—not first some noise, which we shape into something meaningful. The issue of encountering things as qualitative and value laden accords with aspects of Deweyan pragmatism. Dewey describes how the aesthetic always pervades our involvement. So the scientist inspecting a test tube, comparing samples, reading data in a table, or comparing computer plots is doing so under a horizon of concerns and values, many of which constitute an appreciation of beauty, order, simplicity, confusion, and so on. By this view, scientists in their laboratory routines are caught in the primacy of engagement, the consequences of breakdown, and the appropriation of things within contexts of value.

There are other ways of appropriating things in the world beyond how we encounter them during breakdown or as they are pointed out. These include the kind of dealings in which we attach measurements to things, list properties, identify rules for behaviors, and appeal to the dictates of logic. Such activities make sense in certain contexts because we are already able to appropriate things as having value. In fact, to measure is to move into a particular domain, a technological one, in which certain comparisons are made, uniformities constructed, and things become objects within certain practice contexts, particularly those of the scientist's laboratory. Heidegger does not develop the theme of laboratory practice as a party to everyday dealings in the world, as do those who study the ethnographic practices of laboratory life. Science is different and special for Heidegger, as it works with the Platonic and Cartesian myth of contemplation, for which everyday practical activity stops. Heidegger also gives credence to a further, and even more derivative, way of encountering things in the world, which is that of *pure* contemplation, where we strip objects down to bare sense data, and we encounter ourselves as self-sufficient subjects. The scientist stands detached before objects, and in a state of theoretical reflection. The cybernaut stands within the data stream of cyberspace. Here, the world consists of objects standing apart from an observing subject.

That there is something the world can be stripped down to appears as an admission that there is something essential to the world, able to be represented in computers, in an idealist and reductionist sense. To counteract this inference, it is worth considering that one of Heidegger's polemical tactics in *Being and Time* is to counter the Cartesian and Kantian notion that we begin to understand the world by withdrawing from it in a state of contemplation, that we begin with the concept of the pure subject standing off from an object world. He also counters the Cartesian method of reduction by which we build an understanding of the world from observations of the res extensa, building on coordinate geometry. One way to look at Heidegger's ontology is to say that insofar as we understand the world through method, the method proposed through the Cartesian tradition is the wrong way round. We should begin with the whole, and from then on our understandings are diminished, prone to the vagaries of particular contexts. Heidegger's arguments are in opposition to reductionism and materialism, and support the practical criticisms, as advanced by Gregory, of reductionism in science: "Once physicists pass beyond the simplest of systems, they find it incredibly difficult to make predictions by solving the mathematical equations describing the behavior of the world. Instead, they are forced to work with approximations and averages that sometimes allow them to predict quite well how things will turn out and sometimes lead to abysmal failures. Physics is the drive to predict the behavior of a world stripped of most of its complication."[6]

At face value, to assert that scientific understanding is a diminished rarefied understanding merely affirms the common romantic position, that science is appropriate within limited spheres of activity but is not the only view; that there is no one true description of ultimate reality. Science strips phenomena down to bare data. But Heidegger's vigorous theory of praxis absolves it from any such accusation, as do his profound reflections on the nature of technology. In his later writing, Heidegger sees the reductive aspect to science as symptomatic of the will to control, to reduce everything to number, rule, formula, a resource to be used.[7] The hydroelectric dam, the computer, and other devices of the modern world serve to reduce things to a common denominator, as "standing reserve," a potential to be measured, weighed, exploited, and used, thereby denying the individuality of things in their unique contexts.[8] He develops the provocative thesis that this will to control is the major way that Being discloses itself in the technological age, which began some time ago in early Greek thinking, and has been worked through in successive generations, notably with the thinking of Descartes and Leibniz, and more recently through rampant industrialization, consumerism, and mass communications. It is on the subject of technology that there is further accord with critical theorists, though they part company on the manner of dealing with it, and of course on the problem of Heidegger's personal political affiliations.

If we accept the phenomenological position, then the practical implications are that we cannot build any world understanding, or computer systems, on the premise of reduction. We cannot understand the way organisms work simply by looking at chemistry. Laying out the DNA code of an organism will not of itself tell us how the organism functions in its environment. We will not understand consciousness by examining neurons, or patterns of information. We will not understand how the human cognitive system functions and develops by observing and recording how human subjects behave in isolated conditions, no matter how repeatable the experimental results. We cannot build computer systems that are intelligent from sets of rules or neural matrices of weights and thresholds, whether in series or in parallel. Starting from coordinate geometry does not get us to what design is about. Patterning rules will not generate complex organisms, or life. Space as spatiality is not understood by appealing to coordinate geometry, Reimannian or Gaussian geometry, relativity theory, or quantum physics. With the reductions of science and mathematics, we are no wiser about how to get from London to Los Angeles, or where we are. Scientific reduction does not contribute to an understanding of the world in terms of our immediate primordial engagement, nor does it account for inconspicuous familiarity. This is not only because simple phenomena become enormously complex when we aggregate them, but because the reductive premise does not provide a truthful account of our being-in-the-world. It is not that the data, theories, and formulae are incorrect, but the premise is incorrect that we can build up to something like meaning, value, understanding, being-in-the-world, telling, temporality, and spatiality from components. So, according to this phenomenological narrative, it is appropriate to assert with materialism and scientific realism that science is getting down to fundamentals. It is beyond the realms of science to then assert that in doing so, it is able to construct the big picture.

Heidegger accepts that it is possible to decontextualize, to strip human experience of all properties of engagement, as in abstract contemplation or mathematical reduction. In analyzing Heidegger's approach to scientific realism, Dreyfus presents Heidegger as a plural realist who asserts that science is not the only way of constructing an understanding of the universe.[9] Science has access to a causal understanding of the universe, but this is not the only understanding. From the pragmatic point of view, we can refine this assertion: one of the ways science sets itself apart from other studies is in its engagement with a rhetoric of causality, its construction of causal narratives.

Heidegger's thinking also confronts the issue of the real and our preoccupation with it, by examining the quest for the real. This quest is one of the characteristics of Dasein. So the quest for proof that there is an external world independent of perceiving subjects is a problem that Dasein has set itself due to the way that Dasein is.

How does this pragmatism impinge on notions of the proposition and of information? As I have explored elsewhere, from the phenomenological point of view information can no longer be primary if we are to understand space through the concept of spatiality.[10] Space is to be understood by beginning with the phenomenon of *being-in-the-world.* According to the above arguments, care is prior to measurable spatial proximity, and we are with each other or with a thing insofar as we have concerns that require or involve the other or the thing. We are already proximal to what constitute our concerns. Electronic communications are among the other things that we are caught up in—the practical world of involvement, the world of inconspicuous equipment use.

This phenomenological view also places the structuralist project in a particular light. From Heidegger's position, the signifier/signified relationship is not primary but presents a derivative mode of understanding. Understanding begins with unreflective engagement. According to Heidegger, signs function as equipment. There are signs other than conventional language signs, such as simply pointing something out, and signifying occurs against the background of everyday involvement. If theories of signs are not fundamental, but preceded by the phenomenon of *praxis*, then is it appropriate to allow structuralism to speak of cultural phenomena beyond ordinary language, such as in Barthes' account of myth? Barthes suggests that there is an authentic, pragmatic use of language, by which the "mechanic, the engineer, even the user, '*speak*' the object"[11] without resorting to myth. But through its commitment to the code, structuralism appears mute on the subject of how authentic engagement operates as an issue of *praxis*. This is also detrimental to Barthes' argument, as structuralism does not admit *praxis* other than code into the realm of myth. Whether he means myth in a positive or negative sense, the workings of symbol as participation, forms of life, ritual engagement, and routine seem to be excluded from its workings. Myth presents only as a matter of reference, not of participation. Contrary to the distancing implied by structuralism, phenomenology commits itself to being-in-the-world as its point of departure.

How does the phenomenological position construct the real? The *proximal* is that which is primordial, immediate, pertaining to engagement in which there is no subject or object, and within this field of engagement we exhibit care toward certain things. According to Heidegger: "Reality is referred back to the phenomenon of care."[12] So, as concerned beings engaged in the world, the proximal is that about which we exhibit care, which is not to be confused with the psychological concepts of intention, will, wish, addiction, or urge: "Care cannot be derived from these, since they themselves are founded upon it."[13]

The *shared* embraces a precollective being-with. This is not the shared observation of many egos, as posited by empiricism. Neither is it the collective unity of Neoplatonism in which we participate through ecstasis. The world is already shared prior to such considerations. According to the phenomenological stance, our participation in the real is *embodied*. Heidegger's account of being-in-the-world is pervaded by references to the body, particularly the hand. He describes the way we are engaged in the world as that in which we encounter things as ready-to-hand. Contrary to this, the scientific, empirical, or ontic way of encountering things for study and contemplation is as "present-at-hand." But for Heidegger Dasein has a further characteristic that is prior to having a body. This is corporality, or embodiment.[14] There is a way of being-in-the-world that is oriented, a further aspect of care. One cannot be oriented equally to everything, that is, to all the equipment within one's sphere of concern or action all of the time. We face that about which we are concerned. Heidegger does not develop this theme at length but identifies corporality as a characteristic of Dasein that is prior to the notion of having a body. Again, this is not a reference to an idealistic, rarefied, otherworldly, disembodied kind of body but is one that is more situated and more engaged than the abstract body of empiricism, as a channel for sense data, or an object for dissection and study.[15]

For phenomenology, the concept of *repetition*[16] is preceded by practice, or *praxis*. Our being-in-the-world is pervaded by practical concerns. According to the empirical tradition, practice requires repetition, as in the practice of music, or in acquiring typing, drawing, or sporting skills. But this is a very limited view of practice that should be understood more primordially as a process of "taking over." We take over our various practices by virtue of being-in-the-world. We acquire the practices of using a computer system: the procedures for using a word processor, a CAD system, and communications systems as described in training manuals. But we also acquire myriad undocumented practices about file naming conventions, reading email at particular times of the day, organizing network aliases in directories, and methods of navigating through the web, and these interface with the other practices of the office: the use of Post-it® labels, filing paper copies of important email messages, using the fax machine, and so on. Furthermore, from the point of view of praxis, our determinations of what constitutes the real in any situation do not rely on repeated observations, but they rely on practices, which, if we are in science, include the practices of the laboratory, conducting experiments, calibrating, testing and using instruments, generating and interpreting computer output, documenting procedures, and writing up results. Heidegger summarizes this phenomenological position in *Being and Time* by referring to the ancient Greek word for "things" as *pragmata*, "that which one has to do with in one's concernful dealings (*praxis*)."[17] The primacy Heidegger accords to corporality and practice is echoed in Lakoff's pragmatic studies of language: "This 'being in touch with reality' is *all the realism we need*. Our realism consists in our sense that we are in touch with reality in our bodily actions in the world, and in our having an understanding of reality sufficient to allow us to function more or less successfully in that world."[18]

The phenomenological position embraces the *ineffable* in its narratives in terms of Being and anxiety.[19] Being comes before the real as a preoccupation of Dasein. Heidegger distinguishes between reality as pertaining to readiness-to-hand, that is, being-in-the-world, the ontological condition, and reality as concerned with the present-at-hand, the ontic or empirical way of looking at things. But Heidegger thinks that neither concept of the real does justice to the primacy of Being: "Reality has no priority," and "Reality is a kind of Being which cannot even characterize anything like the world of Dasein in a way which is ontologically appropriate."[20] Reality is a notion derived from the primal issue of Being: "Reality is ontologically grounded in the Being of Dasein."[21] So, for Heidegger, the issue of the real is displaced by the more important question of Being.

According to Heidegger, the ineffable aspect of Dasein's relationship to Being is anxiety, or angst (in the original German), sometimes translated as "uneasiness," and not to be confused with fear, which is not fundamental, being built on the concept of anxiety. Anxiety is a disposition of Dasein to reach out and find nothing, the "not-at-home," as Heidegger puts it: "That which anxiety is anxious about is Being-in-the-world itself."[22] By this reading, we organize our day-to-day practices to conceal this primordial anxiety, in other words, to render the world comfortable and homely. But chinks appear, as when technoromantic narratives transfer onto cyberspace various concepts of the sublime, the ineffable, and the transcendent. Such narratives attempt to mask our basic disposition toward anxiety by focusing on the objects of rapidly expanding computer networks and the infinity of virtual spaces. For Heidegger, such concerns are but a pale shadow of the primordial disposition toward anxiety, the seeds of which lie elsewhere.

Our formulation of the proximal, the shared, the embodied, repetition, and the ineffable as constituents of the real actually pertain to Dasein's preoccupation with Being. The proximal is addressed as being-in-the-world and readiness-to-hand. The shared is a matter of care and being-with. Embodiment is a matter of corporality. Repetition is an issue of praxis, and the ineffable is dealt with as anxiety.

Heidegger's phenomenology is a potent force in contemporary continental thought and has been highly influential to critical theorists, structuralists, and poststructuralists, including Foucault and Derrida, though it has left the world of the empiricists largely untouched.[23] Those who have been influenced by Heidegger are among his most enthusiastic critics. Heidegger's various dualisms, such as that between the ontological and the ontic, have been subjected to scrutiny by various writers. Adorno critically terms Heidegger's style as flaunting a "jargon of authenticity," that is itself rarefied and inflated.[24] Derrida takes Heidegger to task on many issues, including the issue of space. He maintains that Heidegger deprivileges space with his insistence on primordial notions of being-in-the-world that precede notions of spatial involvement. We will return to some of these criticisms and their implications for information technology in subsequent chapters.

Spatial Games

Computer games provide useful demonstrations of the grounding of Heidegger's concept of spatiality in praxis. Many games rely on the suggestion of movement through spaces, progression through labyrinths and corridors, perilous bridge crossings, climbing to places of prospect, seeking refuge, and so on. The spatial game of *Riven* (and its predecessor, *Myst*) is instructive in this regard. The game player is presented with an interlinked series of well-crafted, composed, and rendered perspective images. One "navigates" by clicking on different parts of the screen to move forward, look up, down, or turn right or left. Movement is by abrupt jumps from one position to the next. In spite of the primitive nature of this movement, to the extent that one "gets into" the game, involved with the activities of the game, one is caught up in spatiality. From a Heideggerian point of view, what makes the game compelling, and what makes the spatiality of the experience, is the activity the game engenders: solving puzzles, operating mechanical devices, opening and closing doors, pressing buttons, and boarding vehicles. Such games are also built around having one's progress impeded by obstacles, having to overcome resistance to progress, encountering breakdowns. Insofar as these breakdowns present themselves in terms of obstructions to movement, doors that will not open, draw bridges that don't budge, and incapacitated elevators, the game presents qualitatively in terms of space: inaccessible areas, dark corridors, vistas to distant islands, vertiginous prospects, lonely paths, and mysterious forests. As explorations of space, such games seem to rely on breakdown. One discovers that there are spaces to which one is denied access, that operating devices seems to both enable and restrict access to certain places, and that combinations of actions on devices in different locations have an impact on access. Insofar as the game presents issues of space, it depends on the extent to which the game participates in *spatiality*, which is the condition of pragmatic engagement that it engenders and through which breakdowns that provide opportunities for reflection and interpretation in terms of space (with qualities) can arise.

Until recently, such games relied on primitive graphics, or even just text, but game players would report on their sense of absorption in the game experience and talk about the experience in spatial terms. Though photorealistic rendering seems to contribute substantially to the experience, it seems to do so not simply by virtue of correspondences between the image presented to the visual field and real world objects, but rather by virtue of the plausibility, relevance, interest, and intrigue of the tasks set up by the game, tasks that may be enhanced in various ways by photorealistic rendering. The game sets up a field of practical activities that involves solving puzzles, that in turn involve actions that we interpret as moving around spaces, revisiting places, changing the point of view in looking at objects, trying to work out spatial connections, and thinking about geometry and topology. Of course, there are other things going on that pertain to praxis and that contribute to the game's success. The game is of a known genre. The human agency behind its creation and the folk lore surrounding its development are further dimensions of the praxical engagement. There is a communication, and a pact, between the game inventors and the player. The game cannot be so difficult as to be unsolvable (and it is easy to create a game that is too difficult given the combinatorics of switches and maze topologies), and not so simple that there is no challenge, the game is soon over, and the praxical engagement is terminated. It is rare to complete the game in one sitting, and few players would want to finish it in as short a time as possible. The engagement of the game is also characterized by the praxis of entering and leaving it, a certain nostalgia when away, an intimate knowledge, captivation, and affection for the game worlds. In sum, the game relies on praxis to function as a spatial game. When the game is complete, and all the puzzles are solved, it is possible to navigate through the various worlds and revisit places unimpeded, but by this time the game has lost its interest, and the high quality visual imagery is insufficient as an inducement to keep coming back. One could say that there is no more mystery once the puzzles are solved, but it is also accurate to say there is nothing left to do. Without the spatiality of praxis, its worlds cease to exist as spaces.

In the case of computer games, such as *Myst* and *Riven*, every move and every action has been anticipated by the game inventors and is exhaustible. It is worth comparing such games with other "immersive environments" such as C. Davies's *Osmose,* which is a virtual reality art installation that provides freedom of movement through a "virtual space," but a space in which there is nothing to do except contemplate, look, listen, and "float around." Based on the notion of relaxation through contemplation, for all its qualities as an artwork, it is difficult to see how this work presents as a spatial environment beyond its immediate fascination, because there is nothing to do. From a Heideggerian perspective, space builds on spatiality, which operates in a field of praxis, and if there is nothing to do that draws you in, then you become aware of other sources of breakdown extraneous to the focus of the system: the heavy headset, the low image resolution, the noises in the museum, the time constraint, the queue of people waiting to use the system, and so on.

The inhabited world outside the computer game experience has no such fixed constraints, though there are clearly modes of being in a place that facilitate different kinds of engagement. So the spatiality of living in a city is different to being a free-ranging tourist engaged in the praxis of way finding, taking photographs, covering key sites, and collecting travel anecdotes, which is different to seeing the city through the windows of a tour bus. In the same way that *Riven* ceases to engage us when the puzzles are solved and it ceases to be a space, but a series of perspective images on a computer screen, so for the tourist with nothing to do the city becomes a series of snapshots,[25] or the spatial world of the tourist contracts to domains for the micropractices of organizing luggage, attending to blisters, and looking for a clean lavatory. From a Heideggerian perspective, where there is nothing for us to do, then we construct our spaces as those in which there is a job for us with which we can be engaged.

If Heidegger's phenomenological narrative of spatiality and breakdown seems too complicated, we can simplify the formulation to the notion that the appreciation of space depends on practice. The lesson for the designers of computer-aided design and modeling systems, computer games, and VR environments (if they need any lessons) is that the issue of practice comes before the rarefied constructs of Cartesian geometry. The person at a computer engaged in the intricacies of modeling a building, animation sequence, or game world in three dimensions is involved in space to the extent that she is engaged in the practices of geometrical construction, applying texture maps, adjusting the lighting, working with the idiosyncrasies of the modeling system, constructing scenarios of use, and putting herself in the scene. The person inspecting three-dimensional graphical images sees them in spatial terms to the extent that she is engaged in interpreting them, looking for things, acquiring the language of the image, constructing narratives around the image, and placing herself in the picture, as can be said of the VR user spatially immersed in a model world.

There is still a great deal in Heidegger that can be applied to issues of information technology, and the application of his phenomenology is far from exhausted. Elsewhere, I have discussed at length how the ontological concepts of being-in-the-world, readiness-to-hand, care, being-with, corporality, and praxis disclose the cyberspace phenomenon.[26] In chapter 7, we will revisit Heidegger's concepts of the ontological and the ontic as issues pertaining to the real and the symbolic order, Lacan's insight on the unity theme. We will now recast the phenomenology of the real through the issue of metaphor to bring it even closer to our concern with IT.

Spatial Metaphors

We began chapter 3 by considering the issue of representation, suggesting that the primacy given to the correspondence view of language ultimately supports utopian narratives of cybernetic rapture and its variants (or at least the corespondence view is unable to halt technoromanticism's advances). If the correspondence narrative is supplanted by pragmatic, structuralist, and phenomenological narratives of language, then what is the relationship between space and codes entered into computer systems? Theories of metaphor provide insights beyond those of correspondence theory and also provide helpful ways of talking about the unity myth and the real.

Space seems to provide metaphors for the real, and space is implicated in metaphors that are ostensibly other than spatial. *Proximity* as an issue of the real pertains to distance in space. The Platonic concept of the Intelligible realm is of the real outside the cave in another space, or above the clouds, and there are concentric spaces through which we pass to get closer to the source of the Intelligible. Ecstasis is "standing outside," moving from the space of the body to a space outside it (in an "out-of-body experience"). From the empiricist point of view, space delineates what is other than ourselves. We are other than space and so the object world is the spatial world. We also stand apart from the world to gain an overview, the God's-eye perspective, as though located above the world that is laid out as a map. There is also the metaphor of the *self* contained within the object world. According to Kant's formulation, our experience of the real is mediated, or filtered, as if by spectacles. We cannot know the world as it is in itself, only as presented to us through our perceptual structures. The metaphors are of filter, impediment, structure, each of which involves an apparatus in space between us and the rest of the world.

Similar metaphors apply to conceptions of the *shared* aspect of the real. Individuals are distinguished from one another by their location in space, each experiences the world differently because of differences in spatial location—resulting in differences in perspective, and the real is what we recognize in common across those differences. The shared aspect to the real implies a network of entities connected by information flows in space and of people brought together across space.

Bodies are in space, with volume, orientation, and reach. Bodies move through space and conduct their activities there. According to Lakoff and

Johnson, space and the body are intimately connected through a series of metaphor structures pertaining to containment, movement, and force.[27] Even though their main concern is with how the body informs understandings of our perception of space, and other aspects of thought, we can see that the bodily notions of inside and outside, right and left, up and down, front and back implicate space.

Repetition (as a constituent of the real) is commonly regarded as an issue of time. If an experimenter follows the same method, the experiment conducted yesterday should yield a similar result today. But repetition also pertains to space. According to Bergson, we understand succession in time in terms of rhythm in space, and the most prominent metaphors of time are spatial. Repetition also pertains to the repeatability of observations in space. All conditions being equal, the experiment should yield the same results in this laboratory as in any other: the laws for the behavior of light apply on earth as on the moon. As we have seen, notions of the proposition, syntax, and patterns in language involve the production of duplicates that differ in their spatial location. Recurring patterns in DNA, and other patterns of relationships, are spatial patterns. Finally, the *ineffable* as a means of access to the real suggests there is something we strive toward, invoking metaphors of journey, negotiation, and movement, across space, as layers, concentric spheres, depth, elevation, or center.

But we also appear to understand space *through* various metaphors. There is the metaphor of space as resistance to movement, as though space is a viscous fluid. Then there is space as providing freedom, as though space is the open sea, or a huge vessel into which we can pour a small amount of liquid and watch it dissipate, or a conduit through which we pass information at lightning speed. Space also appears as ground, as an absolute, with other phenomena depending on its existence. According to the dictates of coordinate geometry, it is also a vast grid or mesh from which objects are suspended. When we move to modern physics, we see space as a taught fabric, distorted by the presence of matter, a transparent fluid of varying density, a finely meshed network, or an intricately granulated manifold. The master metaphor is number—space as a number series, equations, or variables in an equation. Of course, each of these metaphors in turn implicates space. How can we think of the open sea other than spatially, what is the ground other than something beneath us spatially, and what is mathematics without the metaphors of containment (set theory), balance (equality), and succession (series)?

The other great metaphor system for apprehending space is the language of the senses. It is often said that time pertains to the audile, and kinesthetic senses and space pertains to the visual.[28] According to various commentators, such as Dewey, McLuhan, and Ong,[29] the visual is the favored sense of the culture of print and of the Enlightenment. The visual sense is also linked to concepts of reason. To understand is to see clearly, to assume an overall perspective, that is, to see things laid out in space, even to lay things out spatially on paper as in lists, tables, and diagrams, following Ramism, and then the early Enlightenment encyclopedists.[30] We also apprehend space visually. The visual sense apprehends the world at a glance as objects laid out in various relationships to one another. Of course, the audile sense best captures space as an entity in itself, as ubiquitous and undifferentiated, with properties pertaining to ambiance, immersion, and homogeneity.[31] We are immersed in space as we can be immersed in sound. According to McLuhan, this immersion is a property of the electronic age. We are returning to the golden age of tribal life—unified, whole and engaged—in which the ear was the operative sense organ. Returning to our theme of unity, unity pertains to the aural sense and fragmentation to the visual.

Space also provides metaphors for metaphysics. As I have explored elsewhere, the whole metaphysical project as the quest for certainties and stability outlined and criticized by Derrida trades in the notion of center and periphery, presence as immediacy in space and time, and supplement as something extraneous, added on.[32] The metaphysical pervades intellectual discourses when we assert that the Intelligible is more real than the sensible, that the empirical presides over perception, that one sense is more basic than another, that speech is closer to thought than writing, and that spatiality precedes space. Insofar as the pursuit of the real is metaphysical, it sets up a series of binary oppositions: with certainty, the fixed point, the ground, and what is within reach and present on one side; and the marginal, excluded, appended, less privileged, excessive, and supplemental on the other. What is *in* is favored, what is *out* takes second place. Insofar as the discourse of metaphor trades in concepts of the literal and the metaphorical, it is also party to the same metaphysical distinctions. The discourse of metaphysics exposes an array of metaphors that implicate the spatial.

We seem to use metaphor in understanding space, and we use space as a metaphor for understanding other things. Heidegger's phenomenological account of being-in-the-world—the color and eccentricity of his vocabulary pertaining to space, body, the hand, and his later use of dwelling, earth, sky, gods, and mortals—seems to add further weight to the importance of metaphor. When Heidegger suggests that our being-in-the-world is to be understood as corporality, nearness, concern, care, and so on, he appears to be invoking spatial metaphors. Space provides a metaphor by which we understand our ontological predisposition to be near to one another and to things: concern *as* spatial proximity; orientation *as* facing something bodily.

Heidegger does not talk of metaphor in *Being and Time* but refers to metaphor disparagingly in his later work, where he makes it clear that his statements about dwelling and Being are not to be regarded metaphorically. But there is a thread to Heidegger's thinking that admits metaphor into the ontological framework. In *Being and Time*, Heidegger makes a distinction between "as" statements that are assertions, and the primordial hermeneutical "as." The "as" of an assertion corresponds to the common usage of metaphor, as in asserting "space as container," "computer as mind," and "body as machine." The hermeneutical "as" is that characteristic of Dasein that already understands circumspectively, that is, unreflectively. In the same way that Heidegger distinguishes between space and spatiality, time and temporality, the *in* of containment and the *in* of engagement, he distinguishes between the *as* of assertion and the primordial *as*. This is to make a distinction between metaphor and metaphoricity. (These are not Heidegger's terms.) Whereas metaphor is a feature of the assertion, a trope of language, with metaphoricity we hear as, see as, feel as, prior to any metaphorical assertion in language. We hear the sound *as* a piercing car horn, we see the picture *as* a beautiful portrait, we feel the tool *as* a hammer that is too heavy.

By this reckoning, it is inappropriate to say that Heidegger is merely explaining being-in-the-world through metaphors of space and the body. Neither is he making the weak point that being in space and having a body are characteristics of being-in-the-world. Rather, it is the case that spatiality and corporality are characteristics of being-in-the-world. Spatiality does not mean "being in space," and corporality does not mean "having a body." If there is metaphor in play, then it operates in the reverse direction. What we understand by space is derived from spatiality: space as spatiality. What we understand by the body is derived from embodiment: having a body as corporality. Space is derived from or dependent on spatiality, and the relationship is not simply metaphorical. But through "metaphoricity," the hermeneutical "as;" we are disposed toward interpreting the world in terms of space as we are first and foremost caught up in the issue of spatiality. Here issues of interpretation (hermeneutics), metaphoricity, and primordiality converge.

The discourse about metaphor since Heidegger is suggestive of this more embracing view of metaphor, as metaphoricity. The view of metaphor put forward by Ricoeur points to metaphor as a phenomenon of language that strikes at the heart of Being.[33] For Ricoeur, metaphor is not a superficial phenomenon, pertaining to the ornamentation of language, but deals with the interstice between Being and non-Being, the dialectical tension between the *is* and the *is not*. It is where imagination operates. In this light, to credit Heidegger with the presentation of a richly metaphorical view of being-in-the-world is not to diminish the claims of the phenomenological position.

Elsewhere, I have expanded on theories about the workings of metaphor and the controversies they entail.[34] In terms of structuralist and poststructuralist theory, to which Ricoeur, Derrida, and others accede, the power of metaphors is to disclose through difference—the *is* and the *is not*. Metaphors set up oppositions and tensions, and they seem to operate this way whether the metaphors are simple tropes, accounts of science, or reflections on Being. The *space as container* metaphor presents space as a vessel for containing objects. It also suggests that a container is *in* space (*container as space*), that space is what is outside the container. Considered as metaphor, the metaphor of space as container also prompts us to consider the prospect that space could be something other than a container for objects, a prospect that confounds most of us. To apply the concept of space to information and computing as in "information space" and "cyberspace," as metaphors, is to bring these tensions into play and work them through in different ways. Space as resistance suggests space as a conduit, and space as information entails information as space. Space as constraint also carries its opposite, the suggestion of space as openness, a superconductor, a vessel for emancipation. Following speech act theory, the way these metaphors and their entailments are worked out—which is to say their currency, their usefulness, and the way they are used—depends on their interpretive contexts, the fields of their application.

Gregory's explanation of moves within science as a language game can also be cast in terms of metaphor shifts. Through various shifts in theory, changes in instrumentation and laboratory practices, entities such as space appear variously as substance (the ether), field, continuum, and so on. According to Hesse, Rorty, and other philosophers of science, models are what drive science.[35] Models are metaphors in a particular guise, but unlike metaphor in prose or poetry, a scientific metaphor (a model) aims to settle its own ambiguities. Scientists think that a model, such as the wave or particle model of light, is satisfying insofar as it accounts for what is the case, rather than what is not, and a model is exhausted when the analogies within it no longer adequately account for the phenomenon under study. So new models emerge in response to disanalogies and failings of the old models. Of course, physics is now at a stage where there is a profusion of models. There is no unified, overarching scientific model to account for all aspects of space, time, and matter, despite the hope of Hawking and others. The Kuhnian account of change in science is of ruptured discourses, a trade in anomalies, where there is no unified picture.[36] The pursuit of the real in science seems to entail a metatension. Science is driven to settle the ambiguities of its metaphors so that they constitute statements about the real. The inevitable defeat of this quest, realized as physics touches the limits of empiricism, impels science toward a series of shifts and metaphor changes as rich as any encountered in literature. Hence the profusion of objects—quarks, antiquarks, superstrings, and leptons—and the growth in popular science writing, a resurgence of literary romanticism rivaling technofantasy and the cyberspace literature.

The construction that is metaphor supports pragmatic conceptions of space—the possibility that space is neither out there nor in here; neither an objective entity nor a subjective experience. Within the framework of metaphor, these constructions are contingent. Space can be different things in different contexts: in some contexts, it is nothing at all; in other contexts, there is no consideration of whether space is or is not. There are ways of dealing with the world in which space does not enter our considerations as an entity in its own right. The metaphorical view takes us back to words and their usage, which is not to trivialize words or our world, science, or the distinctions we make. It recognizes that we can never jump out of our word usage. We can only probe the edges of our usages for further disclosures, challenges, unsettlings. To claim the contingency of language, and thus of space, is not to reassert subjectivity or idealism, particularly as the pragmatic view of language shows subjectivity and idealism as also contingent. Neither is it the assertion of a new foundation, a new absolute called "contingency." Insofar as there is a metaphysics of contingency, the foundation is indeterminate and seen to be so.

The Contingency of Code

What are contingent, metaphorical narratives of space? There is a certain lexicon of terms we readily classify under the caption "space"—such as edge, contour, margin, center, periphery, form, pattern, point, line, and plane. Space is a convenient fiction, a rubric, for gathering together these disparate concepts, in a way that is similar to how we use "consciousness" to cover a range of phenomena—intention, knowing, autonomy, self, individual, and contemplation. In this light, we do not need to regard space as an independent, autonomous, "real" entity. Neither is it the case that space can be broken down into these elements, nor are they ways of describing space as something preexisting, an a priori.

It is instructive to return to the correspondence theory with which we began chapter 3, and the concepts of point, line, and plane that pervade conceptions of space in geometry, art, and computer modeling.[37] A point is a location with no dimension, a line is a path between two points, a plane is the path of a line, and a volume is the path of a plane. From these primitives and their interconnection, we build a picture of space as an infinity of points, a grid of lines, able to be partitioned and bounded by planes. There are also less formal spatial uses for these words. There is a point as a concentration, a mark, a center, an arrested movement, a location without dimension. There is line as edge, path, connection, barrier; and plane as surface, boundary, covering. These words are defined in terms of space; in turn, they define space, and the conception of space within which they sit is that of space as containment. In Heideggerian terms, they pertain to space (an ontic, or empirical concept) and not spatiality (which is ontological). But there are uses of the terms "point," "line," and "plane," other than those that refer to space. Following this theme, we could say that the geometrical point derives from the concept of pointing out as signifying, a corporeal notion like facing and being oriented. We speak of the point of a talk, and there is the fixed point on which others depend, which are not cases of a spatial metaphor applied to language but derivatives of the corporeal metaphoricity of being-in-the-world. "Line" relates to lineage as following on, descending from, succession, inheriting, and trajectory, which are references to temporality, that disposition of care toward the world that we commonly interpret empirically (ontically) as time.[38] The plane is that which is level and undifferentiated, the earth or foundation on which we build edifices from sets of differences. But the ground is the way the tradition, since Leibniz according to Heidegger, has concealed the primordiality of Being, that indeterminate ground that is also nonground. Point, line, and plane also pertain to embodiment more directly. A point is a central reference (it is where I am), a line is the boundary of my reach, and the plane is the ground on which I stand, extending to the horizon. Even in mathematical terms, where a line is the path of a point and a plane is the path of a line, there is an appeal to succession. A path is a transition through a succession of points in space. Ontologically, succession is related to temporality.

The technological imperative is to establish causal connections, and to introduce metaphors of control, that suggest other than difference and ambiguity. Seen technologically (ontically), point, line, and plane are causally connected. They are derived from one another through the concept of path: a line is the path of a point, a plane is the path of a line, and a volume is the path of a plane. Or they are connected through the concept of an axis system of one, two, and three dimensions, each dimension introducing a new object: zero dimensions, the point; one dimension, the line; two dimensions, the plane; and three dimensions, volume. From the pragmatic point of view, to manipulate space by appealing to such connections, as a formal exercise, is to participate in the technologies of drawing, geometry, and symbol manipulation, and their associated practices. The language that emerges, of space as a formal entity, is a convenient shorthand for a constellation of loosely connected and even disparate technologies and practices. The formal language of space also buys into a collection of highly privileged metaphors that promote and favor causation. From the point of view of the later Heidegger, and as taken up by some critical theorists, such privileging indicates a technological enframing that represents a distancing from the nature of things as they are in themselves. It is also inevitable.

However, it is possible to unsettle the presumed causal structures of space understood formally. To appeal to the path as the entity that connects point, line, and plane is to posit a further "fundamental" entity, that of the path, which cannot easily be explained without reference to points and lines, thereby producing circularity, or with reference to yet other entities such as succession, memory, or trace, each requiring further definition, and so on to infinite regress.[39] Furthermore, as a semiological structure, spatial formalism is built on notions of difference, in the same manner as any sign system, but also explicitly in the language of the empirical geometers. So the concept of a mapping, with which March and Steadman begin their exploration of geometry in architecture[40] assumes a difference between one system and another, as in the difference between the American and British ways of numbering floors in a building, the metric versus the empirical measuring system, the difference between a drawn plan and a building, or the difference between a plan and its mirror image. Mapping functions, procedures, and technologies presume difference, the identification and resolution of which is a contextual matter. As Saussure indicates, a language community works with the differences that constitute useful discriminations for it. That the configuration of lines on one part of a sheet of paper contains relationships that are different to those on another part of the paper, and that the difference warrants an explanation in terms of a mathematical mapping, is a matter of contingency. Such contingencies include the game of the geometer, that on this occasion variations in ink density and twists in the paper are unimportant differences, that differences pertaining to point, line, and plane are more interesting than color, weight, originality, effort, symbolic value, political condition, and social circumstance.[41] The consideration of difference opens the discussion to contingency and indeterminacy, which is not to diminish the value of formal geometrical studies, or their assumptions and simplifications, but to unsettle any claims to an internal consistency and determinacy independent of considerations of context, and hence any basis we may presume they provide for understanding space as appropriated pragmatically.

Spatial metaphors also extend to an understanding of information technology. The entities of network, node, link, coordinate, and access can be taken as spatial entities. We also understand information technology in terms that are prespatial, that is, pertaining to spatiality, in which network, node, link, coordinate, and access pertain to concepts that come before the notion of space. Such entities also assume the character of the technologies they are brought in to describe, pertaining to control and connection—the whole interconnected technological matrix.

We also understand space in information technology terms. Space is understood technologically through metaphors of control and manipulation. Space is one of the technological ways we encounter world, as is the distinction between the real and the ideal.

As we have seen, metaphors disclose aspects of the world through difference.[42] In the same way that buildings and artifacts placed into the world disclose that world, information technology discloses world. A sculpture placed in the urban fabric discloses the characters of the surrounding buildings, through contrast and difference. We often explain this in terms that presuppose something preexisting, but the capacity of things to disclose can be taken as primordial. This goes beyond the Leibnizian view that space is revealed by objects, that space exists by virtue of the objects in it. What does information technology disclose about space and the real? Information technology, and the spatialities it constructs—cyberspace and virtual reality—appear as the culmination of a primordial imperative. Perhaps the cultural imperative of cyberspace, with its ambiguous narratives of immersion, is technology's way of getting us back to an *in* that is not the *in* of containment, though it is a technological noncontainment.

There are several lessons from the phenomenology of space that aid in our understandings of IT and the real. Phenomenology provides further support in the case against the primacy of representation, and hence of the claims to transcendence of cyberspace. Phenomenology also replaces the language of correspondence with that of praxis, appropriated through the concept of metaphor, as a useful analytic tool. Phenomenology also takes us out of the self-referential language world of structuralism and reveals language as a practice, pertaining to contexts, communities, and the workings of interpretation. Finally, phenomenology sets up a particular problematic, that of the ontological versus the ontic. The ontological realm pertains to our being-in-the-world; the ontic pertains to the praxis of everyday systems, categories, and correspondences. This distinction proves useful in chapter 7 as an instantiation of the myth of unity and fragmentation.

The Mythic Function of Digital Utopias

Phenomenology also presents narratives around the unity theme through its reflections on time and interpretation. The digital utopia (chapter 1) is one manifestation of our embeddedness in time, or more accurately, our embeddedness in temporality, of which time is but an indication. The return to a golden age of unity involves a cyclical process of resolving the tension between the whole and the parts in a hermeneutical situation.

According to a positivist interpretation, utopias seem to fulfill a human propensity to predict, and thereby to plan, though prediction and planning are never dispassionate utilitarian exercises. Prediction is inextricably entangled with desire and dread, and the plan is often in mind with the prediction following. The obvious function of the utopia, like the Hollywood musical of the depression, is as an offer of escape from some inadequacy of the current times, such as the scarce or uneven distribution of wealth, overwork, monotony, exploitation, and fragmentation.[43] By way of contrast, the utopian world offers abundance, people are energetic, there is an intensity and excitement to life, people are honest and open, and there is a strong sense of community and social cohesion. There is also an escapist component to digital narratives. As with any hobby, surfing the net, communicating through on-line chat groups, and playing computerized role games and computer games can provide an escape from daily routine, the distress of unemployment, and unsatisfactory relationships. Discussion groups, MUDs, and role playing on the Internet,[44] as it is now and as it is projected into the future, seems to provide the ultimate fantasy world for escape. One can be an anonymous participant in all kinds of fantasies. The escapist aspects of IT are also translated into the hope of a better future.

But the metaphor of escape is inadequate as an account of what is going on in IT culture. Contemporary cultural commentators such as Žižek point to the obvious deconstructive redefinition of the escapist fantasy. We commonly regard a fantasy as a supplement, an inessential addition to reality, into which we can retreat, and from which we can counsel the lovestruck and the psychotic to extricate themselves. But on careful analysis, we find the fantasy bears all the hall-marks of reality. The fantasy is abiding, shared, and essential, and our reality shifts and changes to accommodate it. In chapter 7, we will consider the contribution to an understanding of IT utopias of psychoanalytic theory as radicalized by Lacan, Žižek, and others, for whom the phenomenon of desire plays a major part.

Utopianism also exemplifies certain kinds of mythic thinking,[45] particularly that which exhibits the pattern of paradise, fall, and salvation—the state of unity, individuation, and the return to unity. According to Plattel, the mythic pattern parallels aspects of Hegel's dialectic of thesis, antithesis, and synthesis. So for McLuhan, there was tribal society in which people were immediately engaged with the world, and in which there was priority to the aural sense. The fall came with manuscript and print culture with its emphasis on the visual sense, which spawned notions of detachment and objectivity. The final stage, the synthesis, is to a retribalization and unity brought about by electronic technology. The paradise is tribalism transformed. For Morris, there is the paradise of a craft-oriented medieval society, the fall of industrialization, and eventually salvation through revolution. In the progression, the new world is always more than simply a return to the way things were. The fall is a kind of excursion by which something is brought back to transform the original situation.

The same pattern recurs among writers who are not overtly utopian. For Dewey, paradise was tribal society in which there was an immediacy of doing and thinking. Greek thinking separated doing and thinking, rendering aspects of life abstract and unconnected. The Copernican revolution reintroduced the primacy of practice and an engagement with equipment and instrumentation back into life and produced modern empirical science. Bakhtin's characterization of the world of the Roman Saturnalia and medieval carnival invokes a golden age united in base and earthy laughter, lost to the modern age. The implication is that we could recapture this unity, if only we could learn to laugh. For Heidegger, the old order was the pre-Socratic world of unity in thought, the fall came with later Greek thinking in which the technology of making dominated all thought. Being meant "being made," and society is now dominated by instrumentalism. "Salvation" comes through a kind of contemplation of things that lets them be in their essence, including an understanding of how being is implicated in technology. The mythic pattern is ubiquitous. Even positivist world views modeled on progress commonly hark back to a state of being in the past, reflection on the current climate of discontent, and projection of a hope. Descartes's rationalism presents a return to a state of pure thought unencumbered by cultural and authoritarian prejudices. Logical positivists remind us of the best of Greek thinking, particularly Aristotle's exclusion of the middle ground in reason, or the more recent reflections of the empiricists. They point to the current demise of thinking, which is imbued with misguided metaphysical notions that have deflected philosophy from its true analytical path, and project a world that corrects the reasoning errors of the present and the past—a scientific world. According to A. J. Ayer: "There is no field of experience which cannot, in principle, be brought under some form of scientific law, and no type of speculative knowledge about the world which it is, in principle, beyond the power of science to give."[46] There is the hope of mythic salvation through science in this speculation.

This mythic interpretation of utopianism shows that teleological narratives are, after all, the continuation of a tradition that is prior to the Enlightenment rhetoric of detachment and objectivity. The metaphor of myth also opens digital narratives of progress to a consideration of the nature of thought itself, as elaborated through the philosophical tradition of Hegel and appropriated by twentieth-century pragmatism, structuralism, and phenomenology. In Hegelian terms, the mythic pattern is that of a dialectic, an indeterminate process. The concepts of the past glory, the present distress, and the future salvation are informed by one another. In the case of Morris's romanticism, hope for the future is cast in terms of a medieval past, the conception of the medieval is informed by dissatisfaction with the present and hopes for the future. The present dissatisfaction is understood by comparison with what once was and with what could be. The past is constructed and interpreted to reflect current preoccupations and future hopes. The hope for the future lies in some kind of integration. The parts are brought together again, implying that there was once something that was whole and has since become disintegrated. The simplest metaphor is that of progress. There is a forward progression to something better. But looking forward is also looking back. The past provides something against which to assess how much better the future will be, but the better future is a return to a more complete and integrated primal whole.[47]

Digital narratives with their variants reflect the same pattern. Inheriting the preoccupations of the Enlightenment "myth," commentators point to the reconstruction of a communal past, the reestablishment of Enlightenment hopes of a free, active, and informed civil society. In keeping with the preoccupations of systems theory, some commentators look to a reconstruction of an ordered world, brought about by the application of scientific methods. The mythic pattern is also woven into the methods of systems theory. The undesirable present is the current state in which there is some kind of problem, the future projection becomes a goal, the past is an unprejudiced, objective state to which we have to return before we can set out to accomplish the goal.

The mythic pattern outlined here is not static but is part of an ongoing cycle of repeated excursion and return.[48] The past, present, and future are continually transformed. Plattel recasts the romantic conception of the utopia in terms of the horizon. The Nietzschean concept of the horizon is also taken up by hermeneutical scholars. A horizon is not fixed but shifts, depending on your point of view. The mythic pattern also follows the structure of the hermeneutical circle proposed by Dilthy, Heidegger, and Gadamer. In any interpretive situation, the experience of the past is brought to bear on a current situation to project into the future.

The utopia, including the IT utopia, fits within this cyclical pattern of a play between past, present, and future. However, the pattern is not necessarily time dependent. Utopias can be other than future worlds. The most radical insights into the phenomenon of utopian thinking come from phenomenology that seeks to extricate the hermeneutical process from the presuppositions of conventional understandings of time. According to phenomenology, it is not necessary to foreclose on a temporal interpretation of the utopian phenomenon in order to understand it. Heidegger offers further provocation on the nature of time and hence of digital narratives.

The Pragmatics of Anticipation

Digital narratives, particularly utopian and progress narratives, are clearly caught up in the notion of time. We often think of time as the last fortress of certainty and the absolute: time exists and advances inexorably; events follow one another in succession; everything is subject to the authority of time; history is the interpretation of time that is past; predictions and utopias are about times that are to come.[49] Some grant that we can interpret time in many different ways, but the phenomenon of time itself is an absolute. History may be a construction, there is no returning to what actually happened in the past, and memory plays tricks on us, but our inability to capture precisely a past moment does not dissuade us from affirming our existence in a present with a past behind us and a future ahead.

If time is the last stronghold of certainty, then it is a very elusive one. It turns out that time is extremely difficult to grasp. As Bergson points out, metaphors through which we describe time are commonly spatial.[50] The notion of time behind us and ahead of us borrows from the concepts of spatial orientation: we orient ourselves spatially with a front and back. To describe time in succession, we commonly appeal to repetitive patterns, which we draw out spatially in charts and tables, time lines, circles, or with reference to clock faces. We position our hands in various ways in space to describe our experience of time, and we use spatial terms such as "succession," "sequence," "one after the other," "in front of," and "behind." In referring to time in the abstract, we commonly appeal also to the behavior of sound, such as the ticking of a clock, or a heartbeat, or progressions within music, and there is a history of associating space to the sense of sight and time to the sense of hearing. One of Bergson's major contributions to an understanding of time is to render the concept of time elusive, if not indeterminate. Derrida provides a similar service in unsettling concepts of time. He identifies the privileged concept of the present, by which we commonly define the past and the future. The past is a present that is no longer, and the future is a present that is still to come. However, this phenomenon of the present eludes capture and representation, as the present is that infinitesimally small interstice between the past and the future.[51] To Derrida, this is the unsettling nature of all "absolutes." On analysis, they turn out to be dependent on the very structures that are built on them. "Absolutes" are floating, indeterminate entities. In every case, at the core of our supposed certainties, are indeterminacy and uncertainty.

Bergson also introduced two concepts of time—time as it is experienced and time as measured. Time as experienced is heterogeneous. One moment is different from the next, the rate at which time passes depends on circumstances, we experience time cyclically, and so on. Measured time is the homogeneous time of clocks and other machines, the time of experimentation and science—the regularly pulsing rhythm to which we relate our measurements.

Heidegger radicalizes Bergson's reflections on time further by distinguishing between *temporality* and time. According to Heidegger, our engagement with temporality is part of a basic human condition that precedes our identification of concepts of time. Heidegger does not assert that there is real, objective time, to be opposed to time as we experience it, but that there is a basic phenomenon (which he calls "temporality"), more basic even than time, and which we occasionally give expression to in terms of sequence, past and future, cycles, time, and so on. A good way to grasp Heidegger's concept of the temporal is to consider the character of a *situation*. According to Heidegger, we are always situated. Situations reveal themselves most tellingly (in their essence) when we are fully engaged in them, unreflectively. As we have seen, as a philosopher of the pragmatic, Heidegger presents the essential or typical situation as one of action not of contemplation.

What is the character of a situation? According to Heidegger, within any situation there is that which is given, what we take for granted, our predisposition. We don't need to articulate them, but there are things that we take as given in any situation, what we would normally regard as our experience prior to the current situation. There is also a comportment or orientation within the task in hand.[52] We have already discussed this as an issue of care, which precedes the notion of spatial proximity. We care about what is going on in a situation, which comes through in our actions rather than our reflections. The artist wants to see the drawing completed, as evident in the care with which she applies the paint, the motorist cares about the journey, the traffic, and safety. The computer programmer endeavors to write a program that will run bug-free. In walking to work, I care about the landscape I pass through, evident in the route I take and what I stop to look at on the journey. I anticipate the destination, through my care to make a living or even my care toward what my work entails. The computer game devotee cares about discovering the code to unlock the domes (*Riven*). All of this points to an attitude of care and concern in what we are doing, though this is not a reflective or contemplative state. We are simply involved in doing.

Of course, there are distractions from unreflective doing. There may be a breakdown, as when the artist has to stand back and contemplate an unresolved placement of color; a pedestrian unexpectedly steps onto the road; there is something strange and new in the landscape; the secret door in the computer game will not open. In the event of distraction from our concerned dealing with the world, we articulate our desires and expectations. At this moment, care is translated into anticipations that are unmet, desires that are unrealized. Such moments reveal the "not yet." The "not yet" takes many forms,[53] a sense of unease, incompleteness, anxiety, dread, and it is commonly realized as a concern with what we call "time" and "future." Something that is not and that should be resides in what we call the future.

It is difficult to discuss Heidegger's concepts of temporality without getting caught up in psychological and subjectivist concepts, and without appealing to the everyday concept of time. Language is so caught in the ordinary concepts of subjectivity and of space and time. However, Heidegger's position is well summarized in a passage in *Being and Time* that rewards close scrutiny. In describing temporality, Heidegger states: "In our terminological use of this expression, we must hold ourselves aloof from all those significations of 'future,' 'past,' and 'Present' which thrust themselves upon us from the ordinary conception of time. This holds also for conceptions of a 'time' which is 'subjective' or 'Objective,' 'immanent' or 'transcendent.' Inasmuch as Dasein understands itself in a way which, proximally and for the most part, is inauthentic, we may suppose that 'time' as ordinarily understood does indeed represent a genuine phenomenon, but one which is derivative."[54] The concept of temporality does not foreclose whether we are speaking of a personal, idiosyncratic, or subjective experience, or on whether we understand time objectively. Temporality comes before psychological interpretations of time, or time as measured.

Following Heidegger's argument, the concept of future is one fabrication we create to house the "not yet." What constitutes the future depends on the situation. As we have already seen, there are many futures, utopias, futures that are reconstructions of the past, and residences of the unrealized promise of information technology. The past is a construction to account for a range of phenomena, including remembering, assessing what is given in a situation, the prejudicial horizon from which we make judgments, and the background of experiences. One of Heidegger's terms for this aspect of temporality is "primordiality." As we have seen, phenomenology gains substantial mileage from distinctions based on the notion of primordiality. For example, the concept of the *thing* is more primordial than the concept of the *object*, unreflective involvement is more primordial than detached theoretical analysis, and temporality is more primordial than time. Heidegger does not mean that one concept came before the other historically or chronologically, though we may be able to put that construction on it. There is a way of preceding something that is more basic than a chronological succession.[55] Dreyfus explains Heidegger's use of "primordial" as indicating our most direct or revealing encounters with things. One interpretation is also more primordial than another if it is more complete, detailed, and unified.[56] So Heidegger's concept of temporality accords more directly with the way we encounter situations at a practical day-to-day level than the usual concept of time affords.

If Heidegger's concept of temporality focuses on actions, and is pragmatic, then we are at liberty to ask what practical difference it makes how we regard temporality or time. Is it the case that Heidegger's pragmatic view of temporality is just another way of looking at situations and at time? According to the phenomenological method of inquiry, we cannot appeal to any empirical criteria of truthfulness in deciding between different concepts of time. However, it does make a difference in the case of IT's utopian and teleological narratives.

Heidegger's concept of temporality brings certain features of digital narratives to center stage. It is worth elaborating on these to ground further discussion. If utopian and progress narratives are tired and unrevealing, then what do pragmatic and phenomenological narratives offer? It is not that something startling and new necessarily emerges but that ideas otherwise marginalized by the hyperbole of narratives of progress are allowed their place in the sun.

First, narratives of progress tell us a great deal about the situations generating them. Science fiction has to resonate with the current situation to appeal, perhaps by exploring the implications of extending current practices.[57] Utopias, including cyberpunk fiction and its ironic narratives of disjunction and fragmentation, excite interest in the here and now. Utopic and dystopic constructions are part of the discursive practice that surrounds a technology; that provides impetus for its development; that allows the developers, marketers, and consumers of computer systems a sense that they are involved in something important, and critics of IT justification for their misgivings.

Second, predictions and utopias are expressions of hope, excitement, misgiving, or despair. To use the language of logical positivism, they are value laden. According to Heidegger's phenomenology, our comportment toward a situation is one of care, and distractions from our unreflective involvement present themselves qualitatively. So predictions embody qualitative assessments of a situation. To assert that soon we will have automatic computer agents trawling computer networks for information suitable to our needs is not a dispassionate statement of a future state of affairs, but a commentary on the current situation. As a qualitative conjecture, it appears as an expression of hope or regret. It operates as a performative utterance: the conjecture will make it so, or perhaps it will ensure that it will not be so.

Third, different concepts of time apply in different situations. There is no construction of what time is other than in a context of concern. Time as the regularly pulsing rhythm of a clock is time as it has become in the context of particular measurements, the compilation and use of timetables and calendars, and calculation in scientific experiments. So too, the march of time and the concept of progress in digital narratives abstract the phenomenon of technological change to notions of increased speed or decreasing size linked to some moral imperative. The concept of progress is part of the Enlightenment legacy, and progress itself is contingent. We can always ask, progress for whom? progress at what cost? The concept of progress is too contingent to serve as an overarching explanation of the nature of IT change.

Fourth, different views of time are not equally applicable to all situations and carry more or less authority depending on the situation. In the case of technical accounts, prediction is more authoritative than recounting dreams, telling fantasies, or presenting scenarios. Certain narratives carry weight through the authority of the people making the predictions, providing the assurance that someone is in control of technology.

Fifth, we cannot help but exhibit care toward the world. Notwithstanding the imputation of control, progress narratives are demonstrations of care. Heidegger regarded such technologizing as misguided and inauthentic, but we can look for what is authentic in aspects of digital narratives. The quest for a more equitable world through IT, the veneration of networked communications, and the quest for labor-saving computer systems are affirmations of concern about the world, to be in touch with one another, and to have more leisure to spend time together. Digital narratives reveal these concerns and open them to scrutiny in particular ways.

Sixth, one characteristic of IT is rapid change. We find ourselves in a multiplicity of situations that are widely different. The trauma caused by technological change need be understood not only in terms of continua of faster, smaller, better, worse, less personal, and more connected, but also in terms of discontinuities within our experience and our ways of coping. The misgivings, where they occur, about current technologies are often that we have not established modes of practice in which these new technologies are integrated. During some failure within our habitual patterns of day-to-day work, objects become conspicuous from the general field of equipmental engagement. In Heideggerian terms, a breakdown occurs in our habitual and concernful dealings in the world. What emerges in the breakdown are technological objects: computers, modems, networks, the Internet, and cyberspace. Given our propensity for care and control, we construct futures in which these objects are reintegrated into the web of our involvement in the world.

Seventh, new, as yet unrealized technological situations are best understood in terms of human engagement, practice, and interpretation. We find ourselves in a complex web of technologies and practices. Predictions and utopias are exercises of the imagination: complex responses to diverse situations. In other words, utopias are *interpretations*, taking account of cultural, social, and technological factors, and competitive practices between manufacturers, the division of the production process, scale economies of production, agreement about standards, the extension of network infrastructures, and developments in microchip design. As situational and contextual, IT predictions come under the purview of hermeneutics, and any attempt to map a prediction against the moment is an interpretation.

Enlightenment and romantic narratives speak to a different situation than our own, with a particular set of conditions, variously interpreted, constructed, and deconstructed. So far in this book we have seen how every attempt is being made to consign the Enlightenment to the realm of the other, to render it strange and unfamiliar as we grapple with new social and technological conditions.

For phenomenological narratives, space and time are ontic phenomena, pertaining to reduction. Spatiality and temporality are ontological, pertaining to Being. Phenomenological investigation seeks an understanding of the whole and from there proceeds to understand the contingencies this brings to light. The process of interpretation involves this play between the whole and the parts, whether at the level of the grand narratives of IT visionaries or the micropractices of reading an email message. Phenomenology also provides potent arguments against the representational paradigm of empiricism and thereby further weakens the chain supporting romantic digital transcendence. The phenomenological position also opens up other narratives of IT. Emphasis shifts from the mysterious, enchanted realm of computer-mediated spaces, new realities, and idealized digital futures to use, social practices, and equipmental contexts. We will see how these themes are transformed in the light of the investigations of surrealism and its progeny.

Ineffability

How Contemporary Narratives of Fractured Identities Challenge Technoromanticism

6

Oedipus in Cyberspace

In this chapter, we examine how the theme of personal identity, as developed in the context of computer systems, further advances our understanding of narratives of unity under the trajectory of technoromanticism. In the process, we examine how narratives of rupture, paradox, and nondeterminacy further loosen the hold of technoromanticism.

Certain digital narratives suggest that networked computing hinders or enhances our involvement in the public realm, democracy, freedom, and other articles of the Enlightenment, romanticism, and modernism. The threat or promise of losing or enhancing our identity is articulated as one in which "we singularly lose our orientation, we enter a desperately unpredictable world out of our control, in which our wills are subsumed by the system."[1] But according to Pearce, the future also holds "potential to free us from hierarchical structures, allows for individual expression, and enables the ultimate definition of our individual and collective humanity."[2] Stefik indicates the power of IT to transform identity by renaming the information superhighway the "I-way," which entails "a search for ourselves and the future we choose to inhabit."[3] According to Franck, networked communities, games, and virtual worlds privilege the private world of the imagination, offering release from our mundane existence, presenting an "endless array of possibilities to be imagined and created."[4]

Other narratives identify the dislocation engendered by information technology, claiming that IT allows a multiplication, fracturing, and dislocation of selves, that is at the vanguard of a new postmodern sensibility. In computer-mediated worlds as Turkle finds them, "the self is multiplied, fluid, and constituted in interaction with machine connections."[5] In electronic role games, she meets characters who put her in "a new relationship" with her own identity.[6] For Poster as well, the new media "cultivate new configurations of individuality."[7] For such commentators, this understanding is at odds with the modernist one, which apparently misses the cultural significance of IT. Whereas modern society fostered "an individual who is rational, autonomous, centered and stable," according to Poster, "a postmodern society is emerging which nurtures forms of identity different from, or even opposite to those of modernity,"[8] and electronic communications technologies may be enhancing "these postmodern possibilities."[9]

The rationalistic, the romantic, and the postmodern are often entangled or confused in digital narratives of identity. Turkle has no problem conflating the aims of artificial intelligence, romanticism, and postmodernism:

"A classical modernist vision of computer intelligence has made room for a romantic postmodern one."[10] For Turkle, the computer systems of today "often embrace a postmodern aesthetic of complexity and decentering,"[11] and computing is a great boon to postmodern theory whose abstract ideas it has rendered concrete: "Computers embody postmodern theory and bring it down to earth."[12] Turkle's technoromanticism embraces a transformed artificial intelligence, which now assumes a personable, egalitarian countenance: "Mainstream computer researchers no longer aspire to program intelligence into computers but expect intelligence to emerge from the interactions of small subprograms."[13] The new, emergent artificial intelligence represents a "turn to 'softer' epistemologies that emphasize contextual methodologies."[14]

In 1970, Turkle published on the influence of the psychoanalytic theories of Freud and Lacan in French philosophy. She subsequently turned her attention to computing and developed the provocative theme that what we say about the computer sometimes tells us more about ourselves than it does about the computer, in the manner of the Rorschach ink blot test. Her more recent book, *Life on the Screen*, attempts to develop the psychoanalytic theme in the light of personal computing, networking, communications, theories of complexity, artificial intelligence (AI), and artificial life (AL). According to Turkle, as Freud and Darwin employed dreams and beasts as the test objects for their modernist theories, the computer is the new focus,[15] providing objects that "exist in the information and connections of the Internet and the World Wide Web, and in the windows, icons, and layers of personal computing."[16] For Turkle, the new AI, including the theories of connectionism and neural networks, invokes a language of links and associations that "evokes the radically decentered theories of Lacan."[17] The computer and psychoanalysis come together in that the Internet experience helps us to develop models of "psychological well-being" and indicate the extent to which "our reality is constructed." Such models "admit multiplicity and flexibility," acknowledging "the constructed nature of reality, self, and others."[18]

Other narratives cultivate a more radical position than Turkle on the theme of IT and identity, intimating the primacy of rift and division. According to Haraway, a new subjective entity has emerged in the electronic age, namely the "cyborg." The cyborg is apparently the product of three rifts. The distinction between human and animal was breached with

Darwinian evolution and its variants, exemplified by recent concerns for animal liberation, and the ecology movement. The second breach is of the boundary between organism and machine, where advanced prosthetics and computers "have made thoroughly ambiguous the difference between natural and artificial, mind and body, self developing and externally designed, and many other distinctions that used to apply to organisms and machines."[19] In fact, for Haraway machines often appear to be more alive than we are: "Our machines are disturbingly lively, and we ourselves frighteningly inert."[20] The third rift is of the boundary between the physical and nonphysical. Our machines now are computer programs and intangible data systems "made of sunshine; they are all light and clean because they are nothing but signals, electromagnetic waves, a section of a spectrum, and these machines are eminently portable, mobile."[21] But whereas information and the cyborg entities constituted by it are fluid, people are not, "being both material and opaque."[22] In contrast to people, cyborgs are "ether" and "quintessence."[23]

"Cyborg" is a term derived from "cybernetic" and was invented and reinvented both by NASA scientists[24] and by science fiction writers to describe a hybrid human-machine. Now there are many people dependent on synthetic body parts, prosthetics,[25] life support systems, and mechanical aids. Haraway speculates that perhaps "paraplegics and other severely handicapped people can (and sometimes do) have the most intense experiences of complex hybridization with other communication devices."[26] But Haraway declares that there is a sense in which we are now all cyborgs: "We are all chimeras, theorized and fabricated hybrids of machine and organism; in short, we are cyborgs. The cyborg is our ontology; it gives us our politics. The cyborg is a condensed image of both imagination and material reality, the two joined centers structuring any possibility of historical transformation."[27] Haraway uses the term "cyborg" as an ironic provocation to explore contemporary social conditions and, in particular, the place of women. Cyborgs are "the illegitimate offspring of militarism and patriarchal capitalism, not to mention state socialism," but she adds: "Illegitimate offspring are often exceedingly unfaithful to their origins. Their fathers, after all, are inessential."[28] Haraway approves of the cyborg and the challenge she/it poses to the male order and rationalism.

Turkle, Haraway, and other theorists of cyberspace appeal to the psychoanalytic theories of Freud and Lacan in their explorations of identity. I will show what psychoanalytic theory reveals about digital narratives. I see Haraway's use of the cyborg as the introduction of an *object* into a discourse in a way designed to provoke. Such narratives participate, to varying degrees, in the legacy of surrealism, which developed particular concepts of the object and appropriated aspects of psychoanalysis. Surrealism brings the IT theme of the real versus the imaginary into sharp relief, and leads to a renewed investigation of the myth of unity and fragmentation as the myth of Oedipus.

Surrealism and the Computer

Surrealism was a movement in art and literature that began in the 1920s and officially ended with the death of its founder, André Breton, in the 1960s.[29] Surrealism sought modes of expression that claimed to call empirical reality into question.[30] The tenets of surrealism resonate with those who expound the IT world as vertiginous, full of strange juxtapositions, complexities, layerings of meaning, and contradiction. Surrealism drew on several legacies, including the culture of the absurd. As the Enlightenment achieved full flower, the absurd found potent expression with thinkers whose business was otherwise order and rule—mathematicians such as Lewis Carroll and the naturalist Edward Lear who provided posthumous inspiration for the surrealists. But notions of the absurd can be traced back to antiquity. In chapter 1, we examined Bakhtin's concepts of the premodern carnival as involving the suspension of conventional morality and the reversal of hierarchies, and Eliade's study of paradox in unity myths. Esslin presents similar accounts in his study of the absurd in theater identifying the largely unrecorded mime plays of Greek and Roman antiquity, which feature the clown as the *moros* or *stupidus*, whose "absurd behavior arises from his inability to understand the simplest logical relations."[31] In such plays, the unities of time and place were violated, and the workings of dreams, hallucinations, and paradox prevailed. There was a mingling of realism with highly fantastic and magical elements. Such elements recurred in medieval mystery plays (devils, personified vices, the court jester) and are integral to Shakespeare's themes (from the fantasy of *A Midsummer Night's Dream* to the exalted madness of Ophelia in *Macbeth*), the Italian Renaissance theater (*commedia dell'arte*), the puppet theater, vaudeville, silent movies, the Marx brothers, W. C. Fields, Jacques Tati, and eventually the avant-garde theater movement of the 1950s, the theater of the absurd

(Samuel Beckett, Albert Camus, Harold Pinter, etc.). Surrealism drew from this tradition of "verbal nonsense."[32]

Lyotard links surrealism to the romantic concept of the sublime,[33] as articulated by Kant[34] and Burke.[35] The sublime is an aesthetic category in tension with beauty. It implicates the sense of terror from a safe distance, as when standing on a precipice, sheltering from a violent thunderstorm, and beholding a raging cataract. The romantic artists sought to evoke the sublime in nature: raging rivers, spectacular mountain scenery, the struggle for survival within the animal kingdom, the lion attacking the horse (more recently replaced by vertiginous immersive environments, fractal landscapes, and violent computer games). But for Lyotard, the essence of Burke's concept of the sublime resides in the distortion of taste, imperfection, the shock effect, privation, and terror: "privation of light, terror of darkness; privation of others, terror of solitude; privation of language, terror of silence; privation of objects, terror of emptiness; privation of life, terror of death. What is terrifying is that the *It happens that* does not happen, that it stops happening."[36] For Lyotard, the aesthetics of the sublime "outlined a world of possibilities for artistic experiments in which the avant-gardes would later trace out their paths."[37]

Surrealism was a movement of the avant-garde in art and literature, but it clearly influenced popular forms of culture. Dalí was the great popularizer of surrealism and lived near Hollywood in the 1940s. He is known to have collaborated with Walt Disney,[38] and Alfred Hitchcock was clearly influenced by him.[39] Surrealism, variously transformed through avant-garde theater, the tradition of the absurd, and the comic tradition, informed the bizarre theater of groups such as the Cambridge Footlights in the 1970s, spawning influential television entertainments such as *Monty Python's Flying Circus*. Surrealism also informed the psychedelic cults of the 1970s, and links between the Beetles of the 1970s and the avant garde are not hard to find,[40] and in film, music videos, styles of advertising, and new age and techno-music that trade in appropriation and "mixes" of sound patterns.[41] Surrealism also informs contemporary literature, including science fiction. Gibson identifies the mutant breed of science fiction "that mixes surrealism and pop culture imagery with esoteric historical and scientific information."[42]

The tenets of surrealism seem to pervade aspects of digital narrative, particularly in the way it presents the mass media, computing, and the medium of the Internet and the World Wide Web. Digital narratives inherit the legacy of surrealism in popular culture but also amplify and add to it. Commentators such as Boyer[43] have already established the connection between information technology and surrealism, particularly in relation to the new formations of the city. I will introduce the tenets of surrealism through their manifestation in cyberspace narrative. These tenets promote an antirationalist aesthetic, nonsense as opposed to reason, and a surreality that surpasses the real.

Certain digital narratives suggest that when immersed in the web, we are barraged with data from various sources juxtaposed in apparently random ways and out of context. The Web is formed much as the surrealists considered montage or collage—as the juxtaposition of objects out of their normal contexts: the torso of a woman, a tuba and a chair as clouds floating above the ocean in a painting by René Magritte, or a gramophone with legs protruding from the horn in Domínguez's *Never*. Max Ernst wrote that when the "ready-made reality" of an umbrella is placed with that of a sewing machine on an operating table, the occasion provides the possibility for "a new absolute that is true and poetic: the umbrella and the sewing machine will make love."[44] Coupling disparate realities in this way "render conditions favorable" for new transmutations and the expression of pure acts, such as love. In a similar vein, a commentary on cyberspace and architecture by Novak highlights "morphing," which he links to alchemy, as the operative mechanism in the computer age.[45] The combining of elements as in alchemy, particularly the transformative, alchemic function of words, is a strong surrealist theme, and Breton drew parallels between "the Surrealist efforts and those of the alchemists."[46]

Surrealism shares with structuralism the Hegelian appropriation of dialectic and difference, realized through the juxtaposition of image against image. In his *Manifestoes of Surrealism*, Breton records: "The image is a pure creation of the mind. It cannot be born from a comparison but from a juxtaposition of two more or less distant realities. The more the relationship between the two juxtaposed realities is distant and true, the stronger the image will be—the greater its emotional power and poetic reality. . . . The emotion thus provoked is poetically pure, because it is born free from all imitation, all evocation, all comparison."[47] Surrealists eschewed the literal expression of comparison (the use of "like," "such as," "just as"),[48] preferring the suggestive power of grammatical usage or "grammatical indeterminacy."[49] Chénieux-Gendron provides several examples of such evocative juxtapositions from surrealist poetry: "The hand holds the night by a thread," "The hourglass of a falling gown," "The ruby of champagne." Collage, as explored by Max Ernst, functions in much the same way, involving "irrational" juxtapositions of ready-made elements. Such theories resonate with the tenets of structuralism, and of metaphor as a creator of meaning, which "is not limited to revealing preexisting forms still unknown to us."[50] The IT world trades in the conjunction of ready-made terms no less obscure than in surrealist poetry (web, cyborg, I-way). Gibson refers to collage in the writing of *Neuromancer*: "There are so many cultures and subcultures today that if you're willing to listen, you can pick up different phrases, inflections, and metaphors everywhere."[51] This mode of writing in turn provides a metaphor for cyberspace. The Web presents us with elements, images, sounds, and a variety of media in ambiguous juxtaposition.

González also links the creation of the cyborg with surrealist montage, which allows "representations of body parts, objects and spaces to be rearranged and to function as fantastic environments or corporal mutations,"[52] reminiscent of the mechanical heads and bodies portrayed by the surrealists Hanna Höch, Raoul Hausmann, and Eduardo Paolozzi. Surrealist works were also "multimedia" works, and surrealism was concerned with art in all its manifestations: poetry, painting, sculpture, photography, and film. Surrealist exhibitions would include all of these, combined and interacting in diverse ways.

To search the web is to explore a vast "city" within which one stumbles across strange objects and encounters surprise—the net surfer is a flaneur.[53] The surrealists were excited by this aspect of cities such as Paris, as evident in Walter Benjamin's *Arcades Project*[54] and André Breton's novel *Nadja*, where Breton reports that he often went to the Saint-Ouen fleamarket "searching for objects that can be found nowhere else: old fashioned, broken, useless, almost incomprehensible, even perverse" objects.[55] The concept of the "object," found, transformed, out of place, and a provocation, features prominently in surrealist art. To valorize the notion of the decontextualized (or recontextualized) object may seem to contradict the subjective focus of surrealism, but the surrealists were interested in instantiations and not classifications, objects placed in the "wrong" categories, and the real subsumed within the surreal rather than the real against the ideal.[56]

Alexandrian provides a "classification" of the objects of surrealist art, indicating the strangeness of surrealist taxonomy: the found object, the natural object, the interpreted found object, the interpreted natural object, the readymade, the assemblage, the incorporated object, the phantom object, the dream object, the box, the optical machine, the poem-object, the mobile and mute object, the symbolically functioning object, the objectively offered object, and the being-object. Examples include old-fashioned manufactured objects, "whose practical function is not evident and about whose origins nothing is known"[57] (a found object such as an old mangle or a fencing mask), a bed of stones (a natural object), a collection of small crystal cubes, balls or feathers in a striking arrangement (a box), a device for presenting optical illusions (an optical machine), and the bizarre ceremonial costumes of Dalí, dolls and statues of human forms as symbols (being-objects).

Histories of computing are populated with "objects" (the abacus, Babbage's Difference and Analytical Engines, the Jacquard loom, the Turing Machine) as if in compensation for the apparent "objectlessness" of cyberspace that attempts to transcend the constraints of hardware. The computer world is also populated by synthetic objects: trash cans, folders, virtual pocket calculators, and magnifying glasses. Three dimensional object models and object-oriented programming amplify the object fixation of computing as a surrealist enterprise (as well as a reconstruction of the baroque irrational cabinet[58]). The computer game *Myst* ("the surrealist adventure that will become your world") is populated by objects in unfamiliar contexts some of which apparently have no function in the game plan—telescopes, crystals, boxes of found stones, wondrous mechanical devices—and the makers of the sequel, *Riven*, published new objects on the World Wide Web as they were being rendered prior to the game's release. The games also present familiar objects such as switches, taps, clocks, pipes, pilot lamps, bolts, clamps, and light fittings as strange by rendering them with exaggerated photorealism, larger than life, and in settings where we are unaccustomed to seeing them: caves, mines, temples, and submarines. In *Neuromancer*, Gibson installs Duchamp's *Large Glass* in a space station, because he liked the piece and "wanted to get it into the book somehow."[59] Haraway's invocation of the "cyborg object" is typically surrealist, with the object or device transposed to the genre of cultural commentary.[60] The cyborg is presented neither as a class of objects, an instance, nor as a metaphor. It is presented, described, and analyzed as object, a being-object, analogous to one of Dalí's costumes. According to Haraway: "The cyborg point of view is literal, material, and technical; it is built, located, and specific—like all meaning-making apparatuses."[61] As we shall see subsequently, the use of the "object" is common among philosophers and commentators informed by surrealism.

In accord with the surrealists' interest in the world of imagination and dream, computers (the Web, computer games, and electronic role playing) seem to provide ample opportunity to celebrate the marvelous, dreams, fantasy, and the labyrinthine. Gibson describes his fictional cyberspace as a "consensual hallucination,"[62] where, as reviewed by McCaffery, "data dance with human consciousness, where human memory is literalized and mechanized, where multinational information systems mutate and breed into startling new structures whose beauty and complexity are unimaginable, mystical, and above all *nonhuman*."[63] For Csicsery-Ronay, in cyberspace "hallucinations and reality collapse into each other"[64]; for Franck, virtual reality presents "a fluidity and speed of movement that are more akin to dreams than waking life."[65] Stefik's book *Internet Dreams* trades on the mystery and gravity of dream theory, saying that when we invent systems following the metaphors of "the keeper of wisdom, the communicator, the trader, and the adventurer . . . we are traveling along paths of development worn smooth by ancient and modern dreamers."[66]

Surrealism saw reality as multifaceted and found inspiration in the fantastic sixteenth-century art of Hieronymous Bosch, primitive art, and art produced by mystics and the psychologically disturbed. The surrealists thought that the latter provided greater access to the workings of the unconscious mind. They saw correspondences between the production of art and the dream state and valorized the imagination, fairy tales, and the state of childhood. Surrealist art was inspired by the symbolism of dreams, the latent content of which could be revealed through psychoanalysis. But according to Alexandrian, in all this, surrealism is not fantasy. More accurately, it seeks "a superior reality, in which all the contradictions which afflict humanity are resolved as in a dream."[67] Breton believed in the future resolution of the two contradictory states of dream and reality "into a kind of absolute reality, a *surreality*,"[68] further understood as the unity of the unconscious and the conscious.[69] Breton amplifies the dream quality of surrealist art. In his *Manifestoes*, and quoting Baudelaire, he claims that surrealism "acts on the mind very much as drugs do. . . . It is true of Surrealist images, as it is of opium images, that man does not evoke them; rather they 'come to him spontaneously, despotically.' "[70] Dalí was known to enter deliberately induced psychotic hallucinatory states to further promote the dream experience.[71]

Aspects of IT culture, cyberpunk, and the Web seem to trade in the importance of unconscious processes, the carnivalesque, and "unreason." At the very least, the Web treats information and imagery in a fluid and nondeterministic way. Breton wished to create objects of the kind one might encounter in a dream, that are "indefensible from the standpoint of their utility." Such objects might "throw greater discredit on 'reasonable' beings and objects."[72] The surrealists were interested in the emergence of creativity from random behavior, as exemplified in early experiments in automatism (automatic writing): the generation of random streams of words to gain access to the unconscious.[73] The surrealists were also interested in machines, "cleverly constructed machines that would have no use,"[74] as a means of generating chance occurrences: "Absurd, highly perfected automata, which would do nothing the way anyone else does."[75] Some of the interest in the generative capabilities of computers could be construed as surrealist in intent, such as the three-dimensional "organisms" of Latimer or, in the world of architecture, the generative experiments of Novak[76] and Frazer.[77]

In the computer world, ray tracing and radiosity software create synthetic photorealistic imagery, image manipulation software is used for presenting and experimenting with the photographic image, and virtual reality purports to present synthetic environments.[78] Computer imagery presents tensions between the real and the imaginary in ways that would have delighted (or horrified) Dalí. The cyberpunk literature suggests that the computer world challenges various empiricist "conventions," including assumptions about reality. Virtual reality has caused us to "reassess reality," according to Frazer, and "the transcendence of physicality in the virtual world allows us to extend our mode of operation in the physical world"[79]—claims echoing those of surrealist art. According to Alexandrian: "Writing, painting and sculpture became aspects of one single activity—that of calling existence into question."[80] For Orenstein, the theater of surrealists such as Antonin Artaud sought to " 'alchemicaly' transform the spectator so that he [or she] may awaken to a new vision of reality and experience the marvelous as only the surrealists have defined it."[81]

The Web is currently populated with home pages in which individuals assert themselves as dreamers, free spirits, and fountainheads of original ideas that can at last be published to the world. Surrealism is also grounded in such romanticism with its valorization of individuality, selfhood, and subjectivity. Surrealist writing also trades in the first person singular and celebrated its continuation of the romantic tradition with Breton's catalogue of their romantic forbears (Swift, Hugo, Poe, etc.).[82] Surrealism attempted to transcend the strictly logical, emphasizing the role of genius, the subject, the power of the imagination, and, echoing Rousseau, the celebration of childhood as providing authentic, "risk-free possession" of the self.[83]

We have already considered the links between romanticism and early socialism, the continuation of this legacy in the computer world, and its claims to radicality. As certain commentators see the promotion and defense of computer networking as a form of political action, so for surrealism. Surrealism broke away from, and outlived the anti-art movement known as Dada, which sought simply to shock and scandalize. According to Alexandrian, surrealism declared the rights of fantasy against "a world racked by war, with boring dogmas, with conventional sentiments, with pedantry, and the art which did nothing but reflect this limited universe."[84] Surrealism was an inflammatory movement with a radical concept of freedom, according to Sarup, aiming for "the liberation, in art and in life, of the resources of the unconscious mind."[85] Breton defended the practical, revolutionary role of surrealism against charges of social irrelevance. He saw no limit to the surrealist's application of Hegel's dialectical principle, "to the exercise of a thought finally made tractable to negation, and to the negation of negation."[86]

But the radical ambitions of surrealism were in conflict with its popularization. In arguing against the use of surrealist technique as a set of conventions, Breton highlighted the danger of emphasizing the "picturesque quality" of the strange and antagonistic juxtapositions of surrealist art, "rather than usefully revealing their interplay."[87] Collage and automatic writing were to be means to an end, and the artist and writer should make the effort to observe what was "going on inside themselves"[88] rather than mimic a style. Breton saw the work of surrealism in the realm of art as a "preparation" for developments elsewhere. Surrealism provided an opening for "what had been thought to give way at last to the *thinkable*."[89] But as soon as the "thinkable" is commonly thought, the time has come to move on. Ultimately, surrealism could not withstand the effects of its popularity. Furthermore, according to Lyotard, the art of the avant-garde was so tainted by the capitalist economy that it became impotent as a radical movement. It even relinquished its prime aesthetic category, that of the sublime, for "sublimity is no longer in art, but in speculation on art."[90]

Surrealism and the Real

How does surrealism address the issue of the real? The surreal embraces, subsumes, and exceeds the real. The surreal is what is within the (*proximal*) grasp of subjective experience, is resident in the gap between truth and illusion, and is accessed primarily through the unconscious. According to Breton, "the world of dream and the real world are only one."[91] The real world dips into the surreal, "the current of the given,"[92] to reveal differences in intensity between truth and illusion, which are of course inscribed in one another, "our mental equilibrium seeming obviously to hang upon this very precise distinction."[93] If cyberspace narrative wishes to add weight to its claims for the challenge, ubiquity, and transforming power of cyberspace, then it could do worse than identify it as a manifestation of surreality—a technosurreality—that subsumes our everyday concepts of the real. Through technologies that allow the everyday to dip into the realm of cybernetic "consensual hallucination," we encounter the clash between truth and illusion and confront the delicate balance between sanity and madness, ideas which I think capture the tenor, if not the substance, of Heim's otherwise enigmatic essay *The Erotic Ontology of Cyberspace*[94] and Haraway's invocation of the transcendent (but instantiated) cyborg identity. In any event, what the surreal presents as within our grasp exceeds the object world of empiricism, and even the humble *object* is recast as a provocation to our sense of truth and illusion.

The surrealist's concern with subjectivity, the unconscious, and the world of dreams ostensibly disqualifies the surreal from concern with the public realm, claims to empirical validation, and the rhetoric of communal praxis (the *shared*). But in spite of this subjectivist focus, Breton was concerned with coaxing writers from their pedestals and looked for "a democratization of the artistic function."[95] Automatic writing was one means of invoking shared participation in the surreal. According to Chénieux-Gendron, the seances conducted by surrealists during which automatic writing was practiced presented a conflict for Breton, whose thinking alternated between the "affirmation that each person's individual heritage is brought to light through these techniques (whether one is talented or not) and the assertion that the intellectual material obtained during these seances is communal to a number of people."[96] Surrealism participated in the Enlightenment problematic of the residency of creativity in the private, the communal, or the public sphere and clearly favored the former. Participation could not go too far, and surrealism required exclusivity to maintain its impact. According to Breton, this exclusivity was its lifeblood and maximized its potential to provoke: "The approval of the public is to be avoided like the plague. It is absolutely essential to keep the public from *entering* if one wishes to avoid confusion. I must add that the public must be kept panting in expectation at the gate by a system of challenges and provocations."[97] Similar dilemmas are presented by ubiquitous networked communications for the narratives of cyberpunk and other movements, which seem to trade in the clandestine, and for whom the Internet was only attractive as long as it was accessible to the few. In any event, advocates of both the surreal and the "consensual hallucination" of cyberspace do not admit to requiring the validation of shared experience to appropriate their objects.

The *body* assumes varying roles in the appropriation of the surreal. As an object, the body is placed in unusual situations and juxtapositions. For example, in one of his paintings, Dalí arranged the elements of a room to make up the features of Mae West's face. There is Vitor Brauner's strange use of the eye in *The Last Journey*, featuring a man sitting on a giant eye, and René Magritte's *Philosophy in the Boudoir*, featuring a nightdress with breasts attached and a pair of high-heeled shoes with fleshy toes. The body also features as an object, to be costumed and paraded. It provokes and is highly sexual.

The surrealist looked to the methods of mediums and spiritualists, for whom the body provided a channel to a world beyond. But for the surrealists, this world was the unconscious. Whereas mediumism focused on the transcendence of spirits or the dead, which dissociated the personality, the role of automatic writing was to *unify* the personality.[98] Automatic writing was one means of gaining access to the unconscious, and the body was a channel, but could be an impediment, a source of constraint. Breton describes his attempt to obtain from himself "a monologue spoken as rapidly as possible without any intervention on the part of the critical faculties, a monologue consequently unencumbered by the slightest inhibition."[99] This involved a kind of ecstasis: "For this writing to be really automatic, the mind has to succeed in placing itself in a condition of detachment from the solicitations of the outside world as well as from his own individual practical or sentimental preoccupations."[100] Thus surrealism developed its own version of the Cartesian method and its own variation on the Neoplatonic theme. These themes resonate with the claims people make of the technologically mediated ecstatic experience of cyberspace. To appropriate the surreal, as for the "consensual hallucination" of cyberspace, is to transcend the constraints of the body.

The surreal depends on multiplication, the *repetition* of memories and recollections. Rhythm and multiplication is an obvious theme in surrealist painting, as in Paul Delvauz's *The Echo*, which shows three nude women in identical posture within a single point perspective of a street, giving visual expression to the diminishing repetition of the echo. Through collage and machines of reproduction, surrealist art reproduces, mimics, copies, and repeats other works, objects, scenes, and symbols. But writing, rather than painting, provided the surrealists with the greatest access to the unconscious, through the repetitive nature of codes and symbols. Experimentation with automatism demonstrated surrealism's interest in the mechanical. Breton tried to prove the superiority of the verbal phrase over the "visual hallucination," and automatism provided a kind of repetitive experimental procedure. According to Chénieux-Gendron: "The interplay of the auditory and the visual seems to be peculiar to the experimenter, as Breton himself emphasizes, calling on each person to make repeated experiments so as to establish a statistically wider base for an opinion."[101] But repetition is not primarily a means of validating the experience of the surreal, though it is a means of gaining access to it.

For surrealism, the *ineffable* is the unity of the real and the imaginary. According to Chénieux-Gendron, the surrealist quest is for what Dalí called "paranoiac-critical" activity, a will to "systematize the confusion and to contribute to the total discredit of the world of reality."[102] The world sought by surrealism has two properties. It is a world in which there is no such thing as logical contradiction, and it is world in which desire rules over lack. That something is absent from this world "is no more painful than its presence, since although absent, it could become present."[103] For the surrealist, expressions of desire bring worlds into being. But surrealism brought attention to bear on the theme of desire, as much as on what is desired. As we shall see, the theme of desire has been developed by others influenced by surrealism, and desire features in discourses about information technology, which requires us to turn to Freud.

The Surreal and Freud

Turkle, Haraway, and Stone[104] have already established the meeting of Freud, psychoanalysis, and computing, through the notion of distorting identity on the Internet and the computer as a tool for understanding the mind. As we shall see, digital narratives can be woven around Freud's theme of the Oedipus complex.

Breton was inspired by one of Freud's essays published in 1907, "Delusions and Dreams in Jensen's *Gradiva*,"[105] which analyzed a contemporary popular novel by Wilhelm Jensen about a man obsessed with finding the young woman who appears depicted in an antique stone relief. In the story, the man visits Pompei on impulse and meets a young woman, who resembles the stone relief, walking among the ruins. Although the story appears to be a fantasy, it transpires that the woman was a childhood love object of the man, she lives near to his home, and their meeting is impelled by his unconscious desire to renew their love. Freud brings out these features of the novel, and, as explained by Sarup, Freud's essay provided many of the themes of the surrealists: "the mechanism of repression, the dynamism of the repressed, the myth of love and the primacy of desire."[106] For Breton, Freud's essay also served to unite science and art around the issue of the unconscious. It seemed as though science (as psychoanalysis) and art had both arrived at the same understanding of the unconscious. As science "proceeds through conscious observation of abnormal mental processes in others," so the artist "directs his or her attention to his or her own unconscious and gives it artistic expression."[107]

After the manner of Freud, Breton embarks on a painstaking analysis of one of his own dreams in his book *Communicating Vessels*, but in the introduction to the English translation, Caws points out that Breton is careful to avoid what he regards as Freud's weaknesses, particularly in separating our private mental life from the material world and not taking his analysis far enough.[108] Breton relates his own dreams to everyday life, shows the similar structure in each, and examines how dreams and everyday life contribute to the "reconstitution" of himself,[109] claims that resonate with those made of the "consensual hallucination" of cyberspace.

In spite of certain kinships, the surrealists' regard for Freud was not reciprocated. Freud once wrote to Breton: "Although I have received many testimonies of the interest that you and your friends show for my research, I am not able to clarify for myself what Surrealism is and what it wants. Perhaps I am not destined to understand it, I who am so distant from art."[110] Apparently Breton picked up on the early Freud but ignored most of the later developments of Freud's theories. According to Chénieux-Gendron, this was due to some fundamental points of disagreement between the objectives of surrealism and those of psychoanalysis. Surrealist techniques are not capable of producing anything other than a studied copy of an unconscious that is already representable and put into words: "Where Freud is trying to unveil the mechanisms of the unconscious to bring the actions of the analysand [patient] into harmony with the world as it is, Breton solicits the powers of our unconsciousness to reach a surreality."[111] But Freud's language of repression, ego, and id were taken up by others under the influence of surrealism.

The Oedipus Complex and Fantasy

Freud developed the theme of desire and was able to unify diverse theories of mind, and an extensive corpus of clinical practice, under the rubric of the Oedipus complex, the condition derived from a three-way sexual tension between child, mother, and father. Freud's use of the Oedipus story exemplifies the power of myth and narrative, particularly as subsequently developed by structuralists, in advancing theories that range in application from the identification of mental disorders to critiques of capitalism and, as we shall see, insights into computer narratives. That so much could be attributed to the sexual drive, or the libido, confounded Jung and other students and critics of Freud. Jung suggested to Freud that in clinical practice other basic drives such as hunger might be just as manifest in patients, to which Freud acquiesced that a basic drive other than the libido might do, "if only it would assert itself unmistakably in the psychoneurosis."[112] Freud's ability to see the assertion of the libido everywhere is impressive. From a

Freudian perspective, we do not need to examine "phone sex," "cybersex," or pornography on the Internet to link theories of desire with computer technology.[113] The means by which the diverse applications of the Oedipus complex and theories of the libido are accomplished is often in the manner claimed of the psychoanalytic process itself: the reliance on indirect references, a peeling back of layers, and an uncovering of the essential condition in the gaps of the theories rather than in their substance. As it is so central to subsequent developments of Freud's ideas, it is worth recounting the Oedipus myth here, or at least one version of it.

The classic Oedipus myth, as presented by Homer (c. 800 BC) and Sophocles (496–406 BC), is an epic story involving coincidence, murder, rape, incest, dragon slaying, and conflicts among mortals and gods.[114] Sophocles' version focuses on the incestuous family relationship in which Oedipus kills his father Laios (the king of Thebes) and marries his own mother Jocasta. As told by Sophocles, the train of events begins when the oracle prophesies to Laios and Jocasta that their son (Oedipus) will one day murder Laios and marry Jocasta. To avoid this fate, they cruelly rivet the child Oedipus' ankles together and abandon him in the wilderness. Fortunately, Oedipus is discovered there by a shepherd and taken to eventually become the adopted son of the king of Corinth. As a man, Oedipus is haunted by the same prophesy that he will kill his father and, thinking his foster father to be his natural father, flees Corinth so that the prophesy may never come true. But he unwittingly meets his natural father on the road and kills him during a skirmish. Oedipus does not recognize that the man he killed was either his father or the king of Thebes. By demonstrating valor and winning the affections of the people of Thebes, Oedipus later accedes to his natural father's position as king of Thebes and husband of Jocasta, his mother. They have four children together. Sophocles' play involves the successive unfolding of Oedipus' awful realization of his predicament as murderer of his father, that he is husband of his mother and begetter of "brother-sons."

In simple terms, the Oedipus complex according to Freud involves the desire to marry (have sexual union) with one's mother, to wish to kill or overcome one's father in order to do so, to incur the threat of castration from the father, and to deal with the sense of guilt that accompanies the ordeal. As pointed out by Fromm and others,[115] there is no suggestion in Sophocles' account that Oedipus is in love with his mother and therefore kills his father so as to possess her. He falls into his predicament unwittingly, though it is fated that he should do so, and the prospect has preyed on him in his adult life and on the life of his natural parents. Oedipus did not suffer castration but blinded himself after he discovered he had committed incest, which Freud regards as a symbolic form of castration. As understood by Lévi-Strauss, and other theorists of myth, the details of a myth do not matter as much as the relationships between the various elements of the story, and the Oedipus story is one variant of many that tell of the basic conflicts influencing our psychological development.

For Freud, the Oedipus myth pervades the human condition and features in his account of society, civilization, and history. Freud conjectures that the Oedipus complex was enacted time and again within the tribal behavior of "primal man," as brothers rebelled against the strictures imposed by the tribal leader: "We cannot get away from the assumption that man's sense of guilt springs from the Oedipus complex and was acquired at the killing of the father by the brothers banded together."[116] The guilt lingers whether the act is carried out or not. According to Freud, one is bound to feel guilty in either case, as the sense of guilt is an expression of a conflict between the instinct to live or love (Eros) and the instinct of destruction or death. This ambivalence and conflict starts as soon as people try to live together (an account that goes against the myth of a golden age of communal unity). Insofar as community is based around kinship, the conflict expresses itself in the struggle against the father, leading to the development of a conscience and the first sense of guilt. The community that extends beyond the family inherits and intensifies the same conflict and sense of guilt. For Freud, civilization obeys an impulse for people to unite in a closely knit group in keeping with the instinct to life, but it can only achieve this aim through reinforcing the sense of guilt: "What began in relation to the father is completed in relation to the group."[117] Civilization seems to involve the extension of allegiance from the family to humanity as a whole. So civilization is inextricably bound up with an increase in the sense of guilt. Freud's intention is to represent "the sense of guilt as the most important problem in the development of civilization and to show that the price we pay for our advance in civilization is a loss of happiness through the heightening of the sense of guilt."[118]

Freud's development of the Oedipus myth really belongs to the sphere of the individual and the family unit, and his application of the theme to society as a whole actually came after its development in relation to the individual and therapeutic practice.[119] The small boy fears castration by the father as punishment for his exclusive bonding with the mother. The boy knows about castration from observing girls, who are obviously devoid of a penis and have clearly suffered some such fate.[120] So the phallus features prominently in Freud's account of the human psyche, of both men and women.[121]

We should note that the components of the Oedipus myth become more diffused as Freud extends its application. Desire for union with the mother becomes the desire for life (Eros), and resistance to the father, or rather the *law* of the father, reveals itself as an instinct for death. We note also that literal fathers and sons persist in the various struggles, and the desire for the mother becomes an increasingly metaphorical representation: life, love, happiness, wholeness, freedom, a return to childhood innocence, youthful omnipotence, and *jouissance*. Early in his career, Jung shared Freud's conviction that ancient myths "speak quite 'naturally' of the nuclear complex of neurosis,"[122] which is to say their basis in the family is exemplified by the Oedipus myth. In his later work, Jung demonstrates the generality of the mother figure and how she is implicated in notions of rebirth, regression, and the union between matter and spirit.[123] For Fromm, the Oedipus myth is a story of conflict between the matriarchal versus patriarchal order. Matriarchal society emphasizes blood ties, "ties to the soil, and a passive acceptance of all natural phenomena."[124] It also pertains to universality, an order in which all are equal, sharing the unconditional love of the mother, and all are children of Mother Earth. On the other hand, patriarchal society is characterized by inequality (the concept of the favored son), hierarchical order, restriction, and obedience.

We do not have to look far to see the Oedipus myth retold, with various degrees of invention, in science fiction and fantasy narratives that implicate information technology. For example, by one reading, *2001*[125] tells of an errant computer (child) seeking to preserve the omnipotence of its childlike state (union with the mother), by reluctantly murdering an astronaut (its father), only to be rendered impotent through dismantling (castration). The computer regresses through childhood babble to its eventual demise. *Metropolis* is about a young man who seeks union with a priestess (mother) and is deceived by his natural father and a wicked magician (his collective father) into following a mechanical simulacrum of the priestess, who ostensibly seeks to drain him (castration) of his will to liberate the oppressed workers of the underworld.[126] *Star Wars: The Empire Strikes Back* tells of a young man's conflict with his father (the evil emperor), who seeks to frustrate the son's attempts at union with "the force" (the mother) and in a showdown leaves his son dangling semiconscious and helpless from an aerial platform (castration). *Bladerunner* is about an errant son, a short-lived but otherwise perfect replicant/cyborg, Roy, who seeks "life" (union with the mother) realized through his awareness of a lack of family, history, and memories, and in so doing confronts his father (the genius who made him) whom he blinds before murdering. The replicant suffers various self-inflicted tortures, such as driving a nail through his hand (castration), before he eventually dies.[127] Most explicit of all is the conformity of the story in the computer game *Myst* to the Oedipus myth. It tells of two sons who, in their attempt to enhance their hold on the state of youthful omnipotence in the worlds they inhabit (eros, union with the mother), conspire to kill their virtuous but censorious father. As the game unfolds, we see the sons become increasingly guilt-ridden and mad. The various worlds are littered with symbols of castration, weapons, and tools—many of which don't do anything—and the sons' ultimate end (depending on which ending you play) is to be locked away forever in the pages of a book (castration) followed by annihilation at the father's hands. (In the sequel, *Riven*, the literal mother appears, with the attire and features of a pre-Raphaelite femme fatale, and is treated as a goddess by the indigenous inhabitants of the island where she lives.)

Not all narratives easily succumb to interpretation through the father/son version of the Oedipus story, particularly narratives that feature women as the central character: *Alice in Wonderland, The Wizard of Oz, Aliens.* But Freud's earlier theories about human development extend its applicability, albeit in ways more complicated and contradictory. Freud identifies three phases of childhood development, in which the child focuses erotic attention on different parts of the body: initially the mouth, then the anus, then the genitals—characterized as the oral, anal, and genital phases of development.[128] The successful transition through these phases allows the child to develop through puberty and, as an adult, to direct his or her affections to an attainable human love object. Freud applies these phases to the relationship with the mother and father, implicating the phallus, money, and childbirth. At the earliest stage of development, the child seeks approval from the parents. According to Freud, the first gift of the infant by which it gains parental approval is the passing of feces: "Defaecation affords the first occasion on which the child must decide between a narcissistic and an object-loving attitude. He either parts obediently with his feces, 'sacrifices' them to his love, or else retains them for purposes of auto-erotic satisfaction and later as a means of asserting his own will."[129] The retention of feces constitutes the exaggeration and perseverance of the anal phase of development, characterized by self-love, or narcissism, and is associated with the hoarding of wealth in later life. For the little girl, the gift of feces eventually extends to the idea of the gift of a baby, which after all seems to pass through the same bowel: "Thus the interest in feces is continued partly as interest in money, partly as a wish for a baby, in which latter an anal-erotic and a genital impulse ('envy for a penis') converge."[130] The baby and feces are also associated with the phallus: "the fecal 'stick,' represents as it were the first penis."[131] The little girl wishes for a penis, which is transformed to the wish for a baby to offer her father and then the wish for a man as a possessor of a penis. The female version of the Oedipus complex is for the little girl to feel contempt for her mother who has so badly endowed her (without a penis), to desire the father, and eventually to transfer her affection to an obtainable human love object. (Such constructions on the theme of "penis envy" are widely refuted by theorists otherwise supportive of Freud's approach, notably by Irigaray.[132])

In this light, various narratives retell the Oedipus story as success or failure in processes of coming of age or awakening, passing from the anal, narcissistic phase of self-love to maturity, one of many interpretations explored by Armitt in her explanation of the *Alice* stories.[133] So the otherwise confounding meaning of the dark prism from outer space in *2001*, space ships, and slow docking procedures, are symbols of the rejection of self-love and a coming of age through the ultimate gift of a baby in the closing sequence. The genius father figure, the creator of the replicants, in *Bladerunner* is impotent, lives alone surrounded by wealth, has never passed through the narcissistic phase, and is unable to give life. One of the sons in *Myst*, clearly lacks refinement, enjoys filth and animal pleasures, while the other is afflicted with self-love. Both have not moved beyond the anal phase, both seek wealth and power, and both fail to give and come of age. Sons do not feature prominently in the *Alice* stories, but there is disappearing in and out of rabbit holes, the growth and contraction of body parts, and the transformation of the baby into a pig, which speak of the trials of passage of a young girl on the verge of womanhood. There are also many father figure incarnations, to whom Alice provides verbal gifts of one sort or another. Similarly, there are various exchanges of gifts in *The Wizard of Oz*. As Dorothy can hardly offer the wizard (father figure) a baby (Toto will not do), she presents him with the gift of the witch's broomstick (the fecal stick), aided by the phallically charged and potent ruby slippers, which she has to give up.[134] *Aliens* is about a lonely, embittered, and therefore "immature" woman, Ripley, who comes of age by adopting an orphan girl as her own "baby" and defending it against the slimy female alien whose copious fecundity has the appearance of endless defecation, though she will not willingly give any of it up—the ultimate anal retentive, the most vile narcissist. The film presents two models: one of the human mother, able to hand over her baby to the benign father figure of the humanoid robot (though he is eventually destroyed), and the monstrous alien who will not relinquish anything and has nothing to give. By this reading, the final battle between the heroine and the alien is a conflict between two stages on the passage to womanhood, a spectacular allegory of the emergence from childhood to maturity. By a slightly different reading, the story is about the rage of a young woman against her mother (the alien who in her fecundity presents as the mother of all mothers) who denies her the phallus—that is, she tries to take her baby away.[135]

None of these interpretations can hope to cover every aspect and nuance of these narratives or to explain why the narratives are successful or appealing. Nor need we assume that the inventors of such narratives are versed in the Oedipus story or in Freudian psychoanalytic theory. Narratives also commonly present as distortions of the basic plot, or rather the commonality between narratives (the essential myth) is so rarefied that it is barely recognizable in any one of them. So for Sophocles, Oedipus does not suffer at the hands of his father but his own hands. The initial gift in *2001* comes from out of the sky (the mother). In the *Alice* stories, there are no obvious rebukes to Alice's actual mother and father. In *Aliens*, the aggressive alien is a mother not a male, the little girl is not the one who comes of age, but her adopted mother and the alien's (the mother's) encompassing grasp present a living hell rather than a blissful union. Such shocks and reversals do not necessarily diminish the Oedipus myth but show the contradictions and ambivalences resident in it. As we have seen, the surrealists appreciated that access to the real should reside in the workings of contradiction. The ambiguity of which symbols might apply and where, and the deviation of the application of the symbols from our normal expectation, provide much of the appeal of these narratives. This ambiguity also supports the description of these narratives as "surreal" and indicates, as we shall see in subsequent sections, that a more basic version of the Oedipus story is at work.

At most, these narratives indicate the preoccupation of us all with certain themes pertaining to growth, maturity, development, family conflict, and the sexual drive. The science fiction and fantasy stories just outlined are instantiations of broader ranging metanarratives, some of which implicate information technology. Such metanarratives include the myth of progress, the introduction of democratic communities through digital communications, participation in the "consensual hallucination" of virtual environments, the invention and taming of chaos, the development of artificial intelligence and artificial life, and the construction of cyborg identities. Many of these metanarratives focus on the grand narrative of the real, to which we shall return in order to see how much further we can apply Freud.

Freud and Realism

Freud's commitment to science is that of an empirical realist. He was committed to an external world of reality that is accessed through the ego. The ego is the representative of the "external world of reality," while the superego represents the "internal world of the id." Conflicts between the ego and the superego reflect the contrast between what is real and what is psychical, "between the external world and the internal world."[136] He also regarded his theories about the ego and the id as "a demonstration of the theorem that all knowledge has its origin in external perception."[137]

At times, Freud was a biological determinist. On the subject of pacifism, he said: "We are pacifists because we are obliged to be for organic reasons."[138] He was an evolutionist on the subject of culture: "For countless ages man has been passing through a process of evolution of culture";[139] and at times a chauvinist: "Uncultivated races and backward strata of the population are already multiplying more rapidly than highly cultivated ones. The process is perhaps comparable to the domestication of certain species of animals and it is undoubtedly accompanied by physical alterations."[140]

However, certain aspects of Freud's method betray the tenets of empirical realism. Freud's critics include those who doubt his credentials as an empirical scientist.[141] His appeal to myth, as in his use of the Oedipus story, is antagonistic to the methods of empiricism, and his work after 1920, which includes his theories about the structure of the psyche and about civilization, is regarded by some as highly speculative.[142] Freud himself regarded many of his theories as provisional hypotheses,[143] prescientific speculations that may later be borne out by observation. As we will explore subsequently, the branch of psychology known as cognitive science and artificial intelligence research seem to find little use for Freud, preferring the instrumental theories of the positivists and others. Appeal to myth puts Freud's work out of the realm of repeatable experimentation, though it accords with the methods of the "science" of culture, or the human sciences, such as those of Lévi-Strauss. Ricoeur regards Freud's studies as on par with history rather than empirical science. So Freud's psychoanalysis "is not to be compared with the theory of genes or gases, but with a theory of historical motivation."[144] As Foucault argues, within the human sciences Freud has set in train a highly productive discursive practice, which exceeds the original Freud, or, more precisely, "never stops modifying it."[145] Clearly, outside clinical psychoanalysis, Freud's theories have found application in many areas, including culture studies, literature, the study of myths and narratives, philosophy, and aspects of information technology.

Other aspects of Freud appeal to romanticism. For Freud, the resolution of the Oedipus complex is realized in the structure of the mind, or psyche, into the superego, the ego, and the id. The authority of the father, or the parents, forms the nucleus of the superego, which takes over the role of the father and perpetuates his prohibition against incest. The superego keeps the ego in check, the ego being the organized, "realistic" part of the psyche, and the id is the site of uncoordinated and unrestrained instinct.[146] These concepts of the mind, or human psyche, divided resonate with Enlightenment thinking. The ego is the rational, mediating part of the mind, the id is the source of base instincts and passions, the source of creativity, and the superego is the constricting force of parental disapproval. For the romantic, it is a case of casting off the strictures imposed by the superego to let the genius of the id shine through, albeit mediated by reason. Freud says little that would dispel this romantic interpretation, which supports the concept of IT, the mass media, and virtual reality presenting an unfettered world for the imagination, or Turkle's advocacy of the computer as a means of developing models of "psychological well-being."[147] Fromm's interpretations of Freud amplify Freud's romanticism. According to Fromm, the more that a person "gains freedom in the sense of emerging from the original oneness with man and nature and the more he becomes an 'individual,' he has no choice but to unite himself with the world in the spontaneity of love and productive world or else to seek a kind of security by such ties with the world as destroy his freedom and the integrity of his individual self."[148] Such reflections, and the conflicts they set up, resonate with the Enlightenment and romantic sensibility, and the unity theme, and persist in much "popular psychology" literature.

However, irrespective of his explicit views on the real, Freud provides a space for the phenomenology of the real larger than that suggested by empiricism and romanticism. On the matter of *proximity*, Freud is intent on probing the mind, the closest but most ineffable entity. The topology of the psyche trades in the issue of proximity, distancing, and pushing what is proximal into the unconscious through the process of repression. "Repression" is the name Freud gives to the mechanism by which we deal with early sexual impulses, which is a kind of amnesia,[149] the essence of which "lies simply in turning something away, and keeping it at a distance from the conscious,"[150] leaving symptoms in its wake.[151] Such symptoms include phobias such as the fear of wolves,[152] sadism, hostility toward someone who is loved,[153] social and moral anxiety, and unlimited self-reproaches, with the anxiety often directed at something very small or indifferent[154] (and, we may add, phobias about the computer). (Such displays of anxiety pertain to the ontic, in Heidegger's terms, rather than the ontological condition of anxiety, which pertains to nonbeing.) The theory of the unconscious derives from the theory of repression: "The repressed is a part of the unconscious."[155] Freud presents the issue of proximity as an issue of repression in the discourse on the real. The real is therefore something we dig for. It lies beneath the surface. So, from a Freudian perspective, in our analysis of computer narratives we should look beyond the appearance, beyond the surface rhetoric about computers representing space, beyond the observation that new identities are forming through new technologies, or that we can jack into a new cyberreality, to why are we saying these things. What are we repressing? What myths are we retelling?

There are several ways that the real is *shared* through Freud's formulation. We are never an isolated subject but already situated within the triangle of mother, father, and child, and this constitutes our being. For Freud, the instinct to share (and love, Eros) works together and is in tension with the instinct for hate and destruction (the death instinct). Furthermore, our state of guilt is complicitous. The brothers conspire together to murder the father. But more important, psychoanalysis emerges as a complicity between the analysand (patient) and the analyst. According to Ricoeur: "The field of analysis is intersubjective both regarding the analytic situation itself and regarding past dramas recounted in that situation."[156] The determination of what constitutes the real for the analysand, the reality behind her or his condition, emerges in a situation of complicity. The difficulty of the application of Freud to an analysis of texts and other cultural phenomena is that complicity takes place in the broader context of intellectual debate, rather than between analyst and analysand, or critic and author. A Freudian analysis of computer myths has little hope of culminating in a field of consensus analogous to the diagnosis, prognosis, or treatment of a patient's condition.[157]

Freud brings the *body* and bodily functioning into a consideration of the psyche, as in his identification of concepts such as the anal phase, the phallus, and sexuality. No function of the body is sacrosanct. In fact, the apparently unseemly aspects of the body are the more revealing, as our attitudes to these are implicated in repression. This is no more evident than in Freud's account of how defecation passes into commerce, as we have seen. Freud's service in the case of the real is to demonstrate how the mind is imbued with the body, which, as we have also seen, is a theme developed by Lakoff and Johnson (from a different psychological perspective than Freud's), and generalized to the identification of bodily metaphors in how we account for mind and cognition. Freud's themes support Dreyfus' phenomenological arguments about the disembodied nature of supposed computer intelligence and, therefore, its impossibility.[158] So much of being human seems to be caught up in bodily drives and repressions. If we strip them away to reveal the realism of the ego, which is itself inscribed with the body,[159] then we may wonder what is left that can be called reason or that can be simulated in a computer.

Repetition features as a symptom of neurosis, returning to the same fetish, rehearsing the same script, and uncanny encounters. Freud describes the repetitive gone-there (fort-da) game of his nephew,[160] which in its most developed form involved throwing away a cotton reel with a piece of thread attached and declaring that it had gone. When the child recalled the object by pulling it into view by the string the child declared that it was there (da).[161] According to Freud, the key to understanding the game lies in the child managing to allow its mother to leave the room without protesting. The child compensated for this "renunciation of instinctual satisfaction" by "staging the disappearance and return of the objects within his reach."[162] The child may only have been able to repeat his unpleasant experience as a game, which yielded its own pleasure, the pleasure of repetition.[163] According to Freud, the pleasure of repetition also comes from another source as the child repeats the game in the company of his playmates. Children will enact unpleasant experiences, such as a visit to the doctor, in the company of other children: "As the child passes over from the passivity of the experience to the activity of the game, he hands over the disagreeable experience to one of his playmates and in this way revenges himself on a substitute."[164] If this is the paradigmatic repetitive situation, then all forms of repetition, such as repeatable experimentation, have their seeds in a kind of protoneurosis. Furthermore, computers repeat, and they force us into habits of repetition: repeated mouse clicks, typing commands, following sequences of operations, and so on—a property that features in Freud's account of the uncanny to be discussed below.

On the theme of the *ineffability* of the real, Freud's quest was complete knowledge of the mind and, in the clinical situation, the cure of the patient. But Freud makes three major contributions to a psychology of the quest for the real. First, he provides an account of the psychology of doubt—doubt in reality. He ties this in with the punitive role of the father and the superago.[165] Freud recalls the first journey he took abroad with his brother, during which, due to unforeseen circumstances, they had to change their plans and were unexpectedly able to visit Athens as a consequence. Both had longed to visit Athens, but rather than filling them with delight the prospect plunged them both into depression. The following day, when they made it to Athens and stood on the Acropolis, Freud was moved to remark: "So all this really does exist, just as we learnt at school!"[166] Freud links the depression of the previous day to the declaration of the removal of doubt in the reality of the Acropolis: "The actual situation on the Acropolis contained an element of doubt of reality."[167] By his account, as a child, Freud had never doubted that the Acropolis existed, only that he should ever see it, constrained as he was by his parents' modest means.[168] Seeing the Acropolis was equated with running away from home, a common childhood dream that involved establishing a distance from the father. Why was he depressed the day before?: "It must be that a sense of guilt was attached to the satisfaction in having gone such a long way: there was something about it that was wrong, that from earliest times had been forbidden. It was something to do with the child's criticism of his father, with the undervaluation which took the place of the overvaluation of earlier childhood. It seems as though the essence of success was to have got further than one's father, and as though to excel one's father was still something forbidden."[169]

We can extend Freud's psychology of doubt to the problem of the question of reality that is at the heart of the various traditions we discussed in the previous chapters and that seems to drive research into cyberspace: the assertion of the nonreality of matter and the quest for a unity beyond the body. By a Freudian reading, at the heart of such longings is a retreat from the domination of the father and a guilt at fighting his will, or at least the force of authority, and as we will see subsequently, categories, order, and rule. We have already considered how, from a Freudian point of view, various science fiction and fantasy narratives seem to present the desire for the mother as a retreat into the omnipotence of early childhood, with mother and child indistinguishable, a return to a primordial unity, ecstasis, perfect freedom, world conquest, the ideal realm, and the consensual hallucination of cyberspace. Participation in such realms is one way of retreating from the iron rule of the father, which is tantamount to denying the materiality of the world, or at least materiality as presented by a classified and ordered world. But such constructions put too much store in the ideal/real distinction and will have to be ameliorated by the post-Freudian reflections of Lacan and others in the next chapter.

Freud's second contribution to the ineffability of the real is his account of the uncanny. This is the feeling of not being at home. If something is uncanny, then it is *unheimlich*, "unhomely."[170] Computer worlds readily demonstrate this property. Ray-traced imagery, for all its sophistication, can present to us as unhomely and uninhabited. Part of the uncanniness of the worlds of *Myst* and *Riven* is that they are all but deserted, the inhabitants have fled, and there is no one at home.[171] Sarup ties Freud's concept of the uncanny to the surrealist intention to enjoy the "convulsive beauty" of "something that shakes the subject's self possession, bringing exaltation through a kind of shock." In this situation, "shock, mixed with the sudden appearance of fate, engulfs the subject."[172] Futhermore, we often equate a feeling of the uncanny with the prospect that something inanimate may in fact be alive, as in the case of a doll. Freud uses Hoffman's story *Coppelius* to illustrate this point, but we can adapt his case to a consideration of computers. We could say that some people regard computers as uncanny. At times, computers appear to be alive. Freud belittles this construction. Children are used to seeing their dolls as real, and this concept causes no disquiet. Likewise, we could say that the lifelike quality of some computer programs causes us no alarm; neither does the prospect of virtual reality. But the *Coppelius* story also includes the terror of the Sand-Man coming to remove the eyes of naughty children. Freud thinks that such aspects of the story contribute more to the sense of the uncanny, as they implicate the child's father. The uncanny becomes "intelligible as soon as we replace the Sand-Man by the dreaded father at whose hands castration is expected."[173] For Freud, fear of blindness is "often enough a substitute for the dread of being castrated."[174] (It was the fate of Oedipus.) In this light, we need to look for how the Oedipus myth is worked out in narratives of the computer to find the source of the uncanny, rather than its supposed simulation of life. So by this reading the uncanniness of *Myst* resides in the final moments of the loser's version of the ending, where we suffer the fate of the brothers and are locked in the book, unable to see or hear anything, reminded of the terror of breaking the rule of the father and of getting caught.[175]

According to Freud, the uncanny is also encountered in the concept of the double, as encountering someone who looks just like you do, or even in one's attitude to one's own picture. An image can instill a sense of longevity, as in the case of the ancient Egyptians making images of the dead in lasting materials. For Freud, this reflects an early stage in childhood development or of primitive man—namely, that of self-love, or narcissism, but when we surmount the narcissistic phase, in early childhood, the significance of the "double" takes on a reversal. Whereas it was initially an assurance of immortality, the double later becomes "the uncanny harbinger of death."[176] At an earlier stage, the double wore a more friendly face, but later the " 'double' has become a thing of terror, just as, after the collapse of their religion, the gods turned into demons."[177] By this reading, we encounter the computer as uncanny, unsettling, insofar as we see ourselves in the machine—literally, in terms of photographic compositing of images, our representation as "avatars" and other simulations, and figuratively, when we reflect on how the computer reveals ourselves.

A further instance of the uncanny is where seemingly random events exhibit a pattern, as when the number on a ticket for a theater seat turns out to be the same as the seat number on the train. According to Freud, repetition itself is uncanny, as in the child's repetition of the fort-da game: "Whatever reminds us of this inner 'compulsion to repeat' is perceived as uncanny."[178] In this light, the computer presents to us as uncanny in the repetitive nature of its processes and in the repetitive behaviors it requires of us. The reserve some people have against the computer is not just that the computer may challenge our creativity, take our jobs, and scrutinize our actions but that it turns *us* into automata—data entry clerks, tracers, copyists, pieceworkers, and hacks. The computer reminds us of the inner compulsion to repeat by requiring us to undertake repetitive actions.

A further case of the uncanny for Freud is where we take what is imaginary for reality, as when old, discarded beliefs in ghosts, death wishes, and animism present themselves as confirmed after all: "As soon as something *actually happens* in our lives which seems to confirm the old, discarded beliefs we get a feeling of the uncanny."[179] Freud says that it is as though we have had to change our beliefs: " 'So, after all, it is *true* that one can kill a person by the mere wish!' or, 'So the dead do live on and appear on the scene of their former activities!' "[180] An uncanny effect is often produced when the distinction between imagination and reality seems to be removed: "as when something that we have hitherto regarded as imaginary appears before us in reality, or when a symbol takes over the full functions of the thing it symbolizes."[181] For Freud, we encounter the uncanny "when infantile complexes which have been repressed are once more revived by some impression, or when primitive beliefs which have been surmounted seem once more to be confirmed."[182] In this light, insofar as the computer and surrealism participate in this tension between imagination and reality, they present to us as uncanny because of the opportunity they provide for infantile complexes to resurface. Robins, a critic of cyberspace rhetoric, intimates as much: "In the virtual world, it is suggested, we shall receive all the gratifications that we are entitled to, but have been deprived of; in this world, we can reclaim the (infantile) illusion of magical creative power."[183] As we have seen, by the Freudian reading, the quest for the consensual hallucination of cyberspace is a desire to return to the mother, and the magical objects we find on the way there strike us as uncanny as they revive infantile beliefs and complexes.

The feeling of not being at home in the world, looking to a future informational oneness, and ecstasis, therefore, have a psychological basis in the revival of repressed childhood complexes. By this reading, Plotinus' contempt for the rendering of his own image in a painting stems from the emergence of feelings from the narcissistic stage, which also may account for the alienation to materiality and embodiment supposedly experienced in cyberculture. More accurately, the mythic contempt for the material world and the quest for a transcendent reality is but another telling of the Oedipus story.

The third contribution of Freud on the issue of the ineffability of the real is on the subject of the drives, or impulses, which are instinctual. Freud identifies several drives and the principles behind them. The death drive is the tendency of organisms "to return to the inanimate state."[184] The biological tendency for organisms is ultimately to die through internal means. On the other hand, the pleasure principle is a tendency toward activities and states that maximize pleasure. The pleasure principle is in conflict with the reality principle: "This latter principle does not abandon the intention of ultimately obtaining pleasure, but it nevertheless demands and carries into effect the postponement of satisfaction, the abandonment of a number of possibilities of gaining satisfaction and the temporary toleration of unpleasure as a step on the long indirect road to pleasure."[185] The Nirvana principle is the impulse to reduce tensions, which is the dominating tendency of mental life "to reduce, to keep constant or to remove internal tension due to stimuli."[186] Wish fulfillment is the feature of dreams whereby we dream what we want to be the case, albeit in disguised form. We will return to the drives following a discussion of Lacan's radical contribution to the interpretation and application of Freud. In particular, Lacan radicalizes the concept of the drive as *desire*.

The relevance of Freud to digital narratives is already apparent. At first reading, Freud sustains the various empiricist/romantic distinctions that inform computer narratives, but on closer inspection he provides insights into the constitution of the real that present certain challenges, particularly on the issue of the uncanny in machines that behave in some respects as we do, the machine as our double. The Oedipus myth also serves cultural critique, by providing accounts of recurrent themes in digital narratives, such as the presence of the father and rebellion against his laws of prohibition, union with the mother, and repression. To this tormented household, we can add denial of the necessity of parental lineage presented by advocates of artificial intelligence and artificial life (life without parents). To introduce the latter, we shall return to the legacy of surrealism and Lévi-Strauss's version of the Oedipus myth as the myth of autochthony.

Autochthony and Artifice

Freud's use of the Oedipus myth is brought into the ambit of surrealism through Lévi-Strauss, whom we have considered in the context of structuralism. The surrealists shared anthropology's passion for ritual objects, myths, taboos, magic, alchemy, sorcery, and other manifestations of the uncanny. Lévi-Strauss developed the concept of "bricolage," making things out of whatever comes to hand, as the means by which mythic thought develops and perpetuates itself. Taking whatever is to hand clearly resonated with surrealist thinking about collage and the ready-made. Lévi-Strauss often associated with the surrealist group in the 1940s.[187] Structuralist concepts of language and the real, to which Lévi-Strauss subscribed, also resonated with the surrealists. According to Chénieux-Gendron: "The Surrealist philosophy of language, such as we find it—more or less implicit, more or less contradictory—takes cognizance of a privilege of language around which reality redefines itself."[188] For both surrealism and structuralism, meaning is thought to be produced "through the juxtaposition of images and the clash of associations rather than as deriving from some ideal correspondence between sign and referent."[189]

For Lévi-Strauss, Freud's telling of the Oedipus myth is one variant of many. It is helpful to review Lévi-Strauss's explanation of the workings of myth, which provides us with an alternative account of the Oedipus story and which touches the world of computing in a different way. That there are different interpretations of the Oedipus myth, Freud's included, causes Lévi-Strauss no disquiet: "The mythic value of the myth is preserved even through the worst translation. . . . Its substance does not lie in its style, its original music, or its syntax, but in the *story* which it tells."[190] For Lévi-Strauss, meaning resides in the way the elements of a myth are combined (relations and bundles of relations) rather than in the isolated elements. There is more to the Oedipus story than the simple account of the son killing his father to marry his mother. When their incest is discovered, the son (Oedipus) puts his own eyes out and his mother kills herself. There is the brother (Cadmos) seeking his sister (Europa), who has been ravished by the father of the gods (Zeus). Cadmos kills the dragon and Oedipus kills the Sphinx (the monster with a woman's head and lion's body). The myth also involves the naming of Oedipus, meaning "swollen footed," and Laios (Oedipus' father), meaning "left sided." For Lévi-Strauss, these elements can be grouped in terms of: (1) the overemphasis of blood relations (marrying one's mother); (2) the underrating of blood relations (killing one's father); (3) the names of Oedipus and Laios that pertain to difficulty in walking and are characteristic of humans in stories of primal origins, stories in which humans emerge from the earth through a single lineage; and (4) denial of the direct lineage of humankind from the earth, by killing the earth monsters. Lévi-Strauss examines the relationships between these clusters, concluding that: "The myth has to do with the inability, for a culture which holds the belief that mankind is autochthonous [comes from a single parent]. . . , to find a satisfactory transition between this theory and the knowledge that human beings are actually born from the union of man and woman . . . the Oedipus myth provides a kind of logical tool which relates the original problem—born from one or born from two?—to the derivative problem: born from different or born from the same."[191]

Lévi-Strauss says that it is futile to try to discover the original and true version of a myth. The myth consists of all its versions and remains a myth as long as people think it to be so: "Although the Freudian problem has ceased to be that of autochthony versus bisexual reproduction, it is still the problem of understanding how *one* can be born from *two*: How is it that we do not have only one procreator, but a mother plus a father?

Therefore, not only Sophocles, but Freud himself, should be included among recorded versions of the Oedipus myth on a par with earlier or seemingly more 'authentic' versions."[192]

The complex of autochthony is no less pertinent now. Clones, machine intelligence, and autonomous half-human robots, in actuality or in prospect, are produced by means independent of bisexual reproduction. From a Freudian reading, the affront they cause is due to guilt and anxiety at defying the father, authority, order, and rule, but we more commonly express this as "playing god." The concept of "playing god" captures the dilemma presented by the myth of autochthony and implicates the technological imperative in the Oedipus myth, which is about the will to achieve total control and mastery of our world and of life by defying their natural processes, and the consequences of doing so. By this reading, many digital narratives present a reworking of the ancient conflict between being "born from different or born from the same." Alternately, the ancient conflict anticipated, or was an early manifestation of, the modern conflict between life and its simulation, the natural and the artificial.

By this reading, the antagonism between the great artifices that feature prominently in digital narratives, such as artificial intelligence and artificial life, on the one hand, and the human and natural, on the other, tell the myth of autochthony, which is also the myth of the one and the many, unity and fragmentation. The artificial represents an attempt to return to a primal unity, from the divided (natural) world of species boundaries and rigid categories. Such is the contradictory working of myth. As the quintessential reductive machine, the computer is employed as the device that will restore us to the unity. If we are to extract anything useful from such ideas, then we need to see the computer through the filter of Freudian theory as it has been transformed by Lacan.

7

Schizophrenia and Suspicion

The antagonism between unity and fragmentation, the one and the many, born from one and born from two, the ontological and the ontic, union with the mother and the law of the father, also present as the antagonism by which the real resists language. As controversial interpreter of Freud, Lacan opens digital narratives to further provocation, and to further contributions from critical theory and hermeneutics. How do digital narratives participate in the antagonism between the real and the symbolic order, and what does this reveal about the information age?

The Mirror and the Real

Surrealism plays on the theme of the image, including the interreferentiality of images—that one image refers to another, that images are reflections of each other—and the ambiguity of image and reality.[1] The mirror provides a potent metaphor in the surrealist concept of the image, featuring prominently in absurd (*Alice Through the Looking Glass*) and surrealist iconography: the eye, the mirror, the look, the gaze, aperture, window, frame, mimicry, and perspective. Such terms and metaphors resonate with the IT world, which purports to present openings into worlds, windows, and hyperlinks, which can return to themselves, and which suggest the interreflections of a chamber of mirrors. For IT commentators such as Chaplin, Alice's looking glass is a precursor to cyberspace[2] in which the fundamental laws of physics, logic, and language are inverted. Cyberspace functions as a "substitute reality,"[3] as does an image in a mirror. For others, IT subjects us to relentless surveillance, which divides and fragments the self. According to Tabor: "Each form filled, card swiped, key stroked and barcode scanned, replicates us in dataspace—as multiple shadows or shattered reflections."[4] Furthermore, as attested by many commentators, computer images are copies of other images, modified and distorted, and the computer renders the concept of an original uncertain and ambiguous.

Biology is also imbued with concepts of mimicry, which, by one reading, is even more basic than concepts of fitness and survival. Lacan was impressed by Caillois's account of mimicry in nature, particularly the form assumed by the praying mantis, which exemplifies a creature being captivated by the image, that is, assuming the form of a twig.[5] As Sarup summarizes Lacan's position: "the human being, like the praying mantis, is captivated by the image. . . . we are dominated by a structure of images and . . . this has a toxic, poisonous effect on the human subject."[6] For Lacan this concern with the image eventually expresses itself in terms of consciousness, which trades in the "illusion of *seeing itself seeing itself*,"[7] which is based on the "inside-out structure of the gaze."[8]

Lacan was a leading psychoanalyst/philosopher whose controversial interpretations of Freud now inform much cultural critique.[9] Lacan's interpretation of Freud is informed by structuralist language theory and by surrealism. Surrealism influenced Lacan's view of language but also his use of it, as he exploited puns and word games, using language to provoke, shock, and even confuse. According to Bowie, Lacan is in the company of the surrealists "and demonstrates his prowess in a self-conscious parade of puns, pleasantries, conceits, learned allusions and whimsical etymologies."[10] The surrealist influence also emerges in Lacan's adoption of knotted, spatial metaphors to describe the psyche[11] and his attraction to quasi-scientific objects as a focus of his arguments: particularly notional and actual optical machines. Lacan explicitly refers to surrealism in *Écrits*, and in *The Four Fundamental Concepts of Psychoanalysis* he refers to surrealist montage as a way of understanding psychological drives.[12] In turn, some of the surrealists, such as Dalí, were informed by Lacan's ideas,[13] among whom he served as a physician during the early part of his career.[14]

Under the scrutiny of Lacan, the confluence of language, the gaze, and psychoanalysis conspire to form particular conceptions of the self and the real that run counter to the Cartesian and empiricist tradition, to which even Freud was prone. For Descartes, the concept of the subject emerges through an understanding of the process of observation. The subject is whole, unproblematic, and aware. According to Lacanian commentators Casey and Woody, Descartes's subject "achieves certainty by recognizing its being through self-reflection, and can be defined with metaphysical precision as a *res cogitans*, an undivided thinking substance."[15] (In the information age, the Cartesian view provides the impetus for the expectation of automated computer intelligence among other things, and reason as an abstract process that can be automated.) By way of contrast, for Lacan, the best way to grasp the concept of the subject is through an understanding of the psychological moment at which the developing person (the child) starts coming under the influence of language—that is, when they enter the world of signification, the symbolic order. (And here the symbolic is equated with modern conceptions of language use rather than traditional, premodern conceptions of symbol as explored by Eliade and others.) At this moment, the child starts using the concept of reference. Prior to this moment, the child begins life emersed in the whole, absorbed in a world that is undifferentiated, fully participating in a state well elucidated in Heidegger's concept of being-in-the-world, and mythologized in Freud's concept of Eros and Jung's mother archetype.[16] The moment when language intrudes comes even before the child learns to speak. For Lacan, this is best illustrated by a child's thoughtful encounter with its reflection in a mirror, when the child is confronted by an image of itself independent of the parent. This is the start of the "mirror phase."

At that moment, the child is presented with the prospect that it is other than the world, that there are things that are not within its control, that the mother is other than itself. For Lacan, the ensuing confusion, alienation, and sense of loss indicate that the essence of the self/subject resides in rift and division.[17] In interpreting Lacan on this point, Gross summarizes the prognosis for the child: "From this time on, lack, gap, splitting will be its mode of being."[18] This is an effect of language for Lacan, which is essentially a matter of separating, dividing, distancing, and objectifying, where "the subject as such is uncertain because he is divided by the effects of language."[19]

Such accounts of the psyche resonate with the enthusiasm of the surrealists for collage, the juxtaposition of diverse realities, exaltation through a kind of shock, and the revelation of differences that play on our "mental equilibrium." Lacan defines the human psyche as permeated by rift and disequilibrium, with language as the cause. As we have seen, language seems to involve dividing the world into compartments and categories, and pointing something out (the basic act of signification) involves separating out the thing from everything else. This view of language finds support from the basic structuralist thesis that meaning resides in difference and in systems of oppositions. Lacan adds the primal opposition between the state of blissful primordial unity (for the child who is one with the world and its parents prior to the mirror phase) and initiation into the symbolic order. As we shall explore subsequently, this insight forms the basis of the Oedipus story for Lacan.

The mirror phase is a useful starting point for understanding Lacan's concept of the real. For Freud, "reality" is ostensibly what is external to the human mind and places limitations on it, through the "reality principle"—though we also see that Freud introduces other, nonempiricist considerations into the debate on the real: proximity as an issue of repression, the shared as a matter of the intersubjectivity of family relations and the analytic situation, the ego representing the surface of the body, repetition as a manifestation of neurosis, and the link between the ineffable, guilt, the uncanny, and desire. As Lacan subscribes to structuralist concepts of the real, which share no contract with empiricism's concept of "independent reality," the issue of what constitutes the real is open to interpretation and redefinition. The real becomes contingent and cannot be taken for granted.

In simple terms, for Lacan the real is the primal realm enjoyed by the child prior to the mirror phase.[20] Lacan opposes the real to the symbolic realm. The real is complete and in its place, absolute and unified. But the real is not what we represent through symbol (language) as the empiricists would hold. The real is precisely what *resists* symbolization. (The emphasis on *resistance* rather than *representation* is a key point in Lacan's thinking and renders it nonsensical from an empiricist point of view.) According to Lacan: "The real does not wait, and specifically not for the subject, since it expects nothing from the word. But it is there, identical to its existence, a noise in which everything can be heard, and ready to submerge in its outbursts what the 'reality principle' constructs within it under the name of external world."[21] The real is therefore indivisible, without fissure, but noisy and elusive—ideas that resonate with concepts of the surreal. But Lacan's position is contrary to a constructivist, idealist view of the real. Psychoanalysis is contrary to idealism, claiming its substance in the praxis of the analytic experience: "It in no way allows us to accept some such aphorism as *life is a dream.* No praxis is more orientated towards that which, at the heart of experience, is the kernel of the real than psychoanalysis."[22]

Lacan's concept of the real is ambiguous and difficult but cannot be dismissed. Žižek illustrates Lacan's concept of the real through the workings of paradox and joke. Two men are sitting in a train, and one of them asks what is in the package up in the luggage rack. The other man replies that it is a MacGuffin for trapping lions in the Scottish Highlands. The other man remarks that there are no lions in the Highlands, to which comes the reply, "See how efficient it is." According to Žižek, such paradoxes provide "the precise definition of the real object: a cause which in itself does not exist—which can be present only in a series of effects, but always in a distorted, displaced way."[23] Another example is Wittgenstein's "Whereof one cannot speak, thereof one must be silent," which raises the question that if there are things it is impossible to talk about, then why add that we must not talk about them? Prohibition by the law of the father against the return to union with the mother as presented through the Oedipus myth is a further example of the elusive nature of the real, "the prohibition of something which is already in itself impossible."[24] Another example is the fraudulent position of the Wizard of Oz, where people make him "do things that everybody knows cannot be done,"[25] and yet the land of Oz thrives off this hollow core of impotence. To this catalogue of testimonies to the ineffability of the real we can add the concept of cyberspace. There is no technology that enables brain implants linking us to a data matrix, and as phenomenology tells us, experiencing and knowing do not function that way anyway. Cyberspace is an imaginative fiction that provides a stand-in, a substitute, or a wild card, in various digital narratives—for what we are not sure. Cyberspace functions as "a pure void which functions as the object-cause of desire."[26] This is not to privilege cyberspace as the ultimate expression of the real, any more than it is to privilege the Land of Oz, but to show how the real feeds on just such absurdities. In so doing, the Lacanian real draws on early forms of the myth of unity and aspects of the Platonic real discussed in chapter 2. The real is characterized by paradox and "contradictory determinations,"[27] and it can only be apprehended through them.

As we can see in the case of the real, it is difficult to pin Lacan's terms to neat definition.[28] But problems with Lacan's use of language seem to exemplify Lacan's account of the problems of language and the problems with the real. We can review the Lacanian real in the light of our five determinants of the real: proximity, sharing, embodiment, repetition, and the ineffable. For Lacan, the real implicates *proximity*, as the real is that which we enjoy prior to the mirror phase, but it also represents an impossibility and attempts to return to it are implicated in anxiety and trauma.[29] The real resists rupture, breach, and distance, exemplified in the disruptive moment of the mirror phase, where symbolization attempts to break through the real. By this reading, the notion that computers may take us close to reality by allowing us to represent it is less interesting than what passes through the representational schema. Insofar as the computer purports to represent, the real is what resists the computer's advances.

How does the real implicate what is *shared?* Lacan develops the concept (derived from Hegel) of the Other. As explained by Bowie, the term "other" is extremely important to Lacan, but also very pliable. Lacan uses the concept in at least two senses. The first is where the other is precipitated by the encounter in the mirror. What is encountered is other than the subject, part of the subject-other dialectic. But in his second usage, Lacan suggests a "limitless field and overriding condition in which both members find themselves—'alterity,' 'otherness.' "[30] According to Lacan, "the subject as such is uncertain because he [or she] is divided by the effects of language,"[31] and so he or she is pursuing "more than half of himself."[32] So otherness is already there in the language situation, and in a way that precedes concepts of a language community. In keeping with Lacan's terminology of fragmentation and reversal, "otherness" captures the antagonistic nature of participation in the real, and in a way that does not require the presumption of community. When Lacan declares that "the unconscious is the discourse of the other," he is indicating the presence of the alien within.[33] A Lacanian reading suggests that the alien does not reside with the computer, the cyborg, or the robot. The alien is what is within us, which the computer may disclose. So-called digital communities that construct their own realities pertain more to "digital otherness," that process by which we encounter the other on-line, which is never simply what is out there, but within. There is little place for romantic concepts of a united whole, a return to the tribe through electronic networks, in this tortured discourse.

The *body* features prominently in Lacanian discourse. In place of Descartes's moment of reflection on the self, which in turn led to the divide between mind and body, the sight of the body in the mirror provides the crucial moment from which stems the fragmented self, which is also the "fragmented body."[34] Lacan also implicates the body in concepts of desire (to be explored below) in a very basic way: "Even when you stuff the mouth . . . it is not the food that satisfies it, it is, as one says, the pleasure of the mouth."[35] But for Lacan, bodily orifices and organs are transformed into metasignifiers. The notion of the phallus precedes the notion of a bodily organ. For example, the phallus is *the* metasignifier, the symbol of signification itself.[36] Lacan's style of argument reminds us of Heidegger's appeal to concepts of preembodiment. There is a way of talking about the eye, hand, mouth, phallus that precedes the necessity of a material bodily organ. By a Lacanian reading, the primordial roots of the computer and the robot exist in the bodily concept of prosthesis, which they also share with the self. The mirror phase presents the body and the self as something broken off, appended, in keeping with Haraway's development of the cyborg. Insofar as we grasp the real by being able to hold on to it, the agent of that grasping (the body) is already characterized by otherness. It is a foreign body. The disembodiment of cyberspace provides a metaphor for our relationship to the real by exaggerating the ineffability, the ungraspability of the real.

Repetition features overtly in Freudian and Lacanian discourse as the ritual reenactment of original conflicts. Such rituals are compulsive. From an empiricist point of view, the most real is that which we enact and reenact. But for Lacan: "The real is beyond the *automaton*, the return, the coming-back, the insistence of the signs, . . . The real is that which always lies behind the automaton . . ."[37] In other words, the real is what resists the pattern: the wateriness of water is what the regularity of two hydrogen atoms to one oxygen atom leaves out; the human is what is not in the DNA code; the building is what is not in the geometrical description in the computer. This does not mean that the codes, patterns, formulas, and signs are redundant, but they reveal, as symbols by the resistances they expose. Virtual reality does not represent the real, or construct new realities, but exposes the real through resistance.

According to Bowie's commentary on Lacan, the real comes close to meaning "the ineffable" or "the impossible."[38] The *ineffable* features prominently in Lacanian discourse as an issue of desire: "*Desidero* is the Freudian *cogito*."[39] Freud starts with "I desire" rather than Descartes's "I think." But desire is illusive, as is the thing desired. In the mirror phase, the infant first detects the splitting of the subject and detects an imagined unity in the mirror image. But the imagined unity, the object of desire, does not appear in the mirror. Desire "can only refer to what is missing, to the object wanted, or, in this instance, to the very unity and coordination which are still lacking in the infant."[40] The lack is due to the child's entry into the world of language.[41]

Lacan's theory of desire implicates language. Desire is closely linked to need for Freud and Lacan, who take needs as the constant requirements of brute survival: nourishment, shelter, warmth, freedom of movement, and a minimal community.[42] The conventional (empiricist) wisdom is that we desire what we need, and eventually we may demand it (express that desire in language). The child needs love to develop, desires what it needs, and so articulates various demands for affection. But for Lacan, need is followed immediately by articulation (a demand made in language), which, if it is not met, is then inevitably suppressed. Then demand becomes desire; desire is repressed demand. Lacan therefore identifies desire as the difference, gap, or discrepancy separating need from demand. Demand is something spoken, but desire is repressed from articulation. Desire "is structured like a language, but is never spoken as such by the subject."[43] For Lacan, "Desire is neither the appetite for satisfaction, nor the demand for love, but the difference that results from the subtraction of the first from the second."[44] So desire follows from demand. Need is only a short-lived phenomenon for the child as need is soon transformed into demand and then desire. Bowie summarizes Lacan's concept of desire: "It is not a state or a motion but a space, and not a unified space but a split and contorted one. Need and demand are its coordinates, but they cannot be coordinated."[45] If needs belong to the natural order, and demands pertain to language, then desire does not belong to either. Desire is at the intersection between the natural and the signifying, which are both "infected" by it.

We can explain Lacan's inversion of the concept of desire by relating it to information technology. It seems that we have needs: to ease the burden of work, remember large quantities of information, calculate, and communicate. We articulate those needs as demands: statements of need, technical specifications, and social commentary. The outcome is electronic and other technologies, which inevitably do not satisfy us, leading to a rearticulation of needs, further demands, and further technological developments. This technological treadmill is ostensibly driven by the relationship between need and demand, while paying little heed to the role of desire, the discrepancy between what we need and what we say we need. But desire is precisely the space of indeterminacy in the discourse on need. The vocabulary of empiricist or systems approaches to technology see this discrepancy in terms of misfit, mismatch, incongruity, and error.[46] Lacan's discourse places the phenomenon of misfit at center stage and graces it with the appellation "desire," which, after all, touches on our psychic makeup. More tangibly, it implicates the nature of our uses of texts and language, and the social and cultural contexts in which they make sense.

As an issue of desire, technology is brought into the wider picture.[47] In this light, the various digital narratives we have been discussing (of cyberspace utopias, transcendence, and representation) are not *expressions* of deep-seated desires, but generate desires, and not as some arbitrary effect of the mass media, but by virtue of the differences they set up between need and demand. By this Lacanian reading, desire is what is left out of the texts, the ineffable, paradoxical realm between our elusive "computational needs" and what we say about them, but desire is parasitic on them. So digital narratives, and the texts that promote them, are not only reports on what is happening that we scrutinize to see if they conform to best intellectual practice, but they are co-implicated in constituting the topic under discussion. Lacan's is one reading of the concept of intellectual space, which is rendered less enigmatic under the scrutiny of hermeneutical pragmatics, to which we shall return subsequently.

Oedipus and Symbol

Lacan "demythologizes" Freud's concept of the Oedipus complex. That is, he saves it from a literalist interpretation, and brings it back to an issue of language. For Freud, desire is often set against the concept of law, which sets limits on our desire. The prime example of the law is the injunction against incest operative in the Oedipal situation. But Lacan finds something in the Oedipal situation that, as Lévi-Strauss indicates, is beyond cultural specifics. Lacan urges that it is not the actual father, even one's own father, who invokes the prohibition against incest so much as the father's name, which is not to say the name of a specific father, but Name of the Father. The essence of fatherhood as presented by Freud resides in the concept of lineage, or, more tangibly, in the concept of the deceased father. It is actually the dead father who constitutes the law of the signifier, whose prohibition is perpetuated through his name.[48] Casey and Woody explain the situation as one in which the father's prohibition, which is to say, the law, divides the subject, both amplifying our desires and subverting them: "Hence the law of the signifier sets up a bar dividing the subject, and is both constitutive and subversive of desire. It also bars the way to *jouissance*, that primordial union with the Mother whose recovery is prohibited by the paternal 'No' and which signifies that completion of being which is forever inaccessible to the split subject."[49] The Oedipus myth is ostensibly about the father preventing the child from reintegrating with the mother, but it is also about language denying us participation in a prelinguistic, primordial wholeness, frustrating our access to the real.

This retelling of the Oedipus myth sets the unity against the symbolic order, the real against language—language being the instrument that divides and classifies the world. The coming into being of language establishes our identity and keeps us apart from the world. The Oedipus story tells of the father's word frustrating our desire for union with the mother, a conflict for which we should find ample evidence in fictional and digital narratives: in the exaggerated reliance on symbol and the restorative power afforded to verbal nonsense.

Allusions to the symbolic order appear in various guises, including the rantings of tyrannical rule, where the words of a villain are to be obeyed without question, and where there is an excessive dependence on symbols, as in modern fantasies involving magic, where talismans and other objects have powers to change things.[50] In *The Wizard of Oz*, the conflict between unity and the symbolic order appears where the gifts of intelligence, courage, and compassion turn out to reside in trivial symbols (a diploma, a medal, a heart-shaped clock) and the wizard's hollow valedictories, which expose the actual residence of such virtues in the unity that exists between the main characters, who, after all, in total (or Toto), are human, animal, vegetable, and machine. Various forces conspire to break up this unity, including the excessive symbolization of the Witch of the West, who feels she must possess the ruby slippers, and finally the incantation "there's no place like home," which verbal nonsense demonstrates the uncanny (not at home) nature of Dorothy's necessarily transient sojourn in the world of perpetual childhood and androgyny,[51] amid the unity of all things in Oz. This interpretation is heightened when we appreciate that Baum, the author of the book on which the MGM film is based,[52] was a theosophist. Theosophy (mentioned in chapter 2) combines overt Neoplatonism with a tendency to valorize symbols, as in its collapse of the distinction between magic and science. According to Wolstenholme, power in Oz comes from natural forces, like cyclones, but sometimes it resides in manufactured items (symbols) such as slippers, houses, medals, and clocks, which share in the magic that is resident in nature.[53] Such oversymbolization adds potency to this Lacanian reading.

Similarly, in *Star Wars*, it is the excessive use of symbols by the evil emperor that attempts to break up the unity between humans, machines, and animals (the camaraderie between the heroes, robots, and the Wookie), and the unity that is "the force." Such symbol abuse is best expressed through the presumption of an absolute bonding between word and action, the language abuse of the absolute dictator whose word is everyone else's command.

The film *2001* is permeated by the menace of the enigmatic large black prism, which provokes the questions "What does it mean?" and "What is it a symbol of?" This metasymbol (phallus) awakens primitive humanity from its oneness with nature (the state of the primordial apes) and sets it on the path to technological development. In the book of the film, it becomes apparent that the large black prism is a gateway (Star Gate) to the "unfathomable secrets of space and time."[54] That the dimensions of the prism are the first three numbers of a quadratic series that extends beyond three dimensions implicates the prism in the language of metasymbology. On entering it, one encounters "reality, grasped as a whole with senses now more subtle than vision."[55] The language/unity theme recurs throughout the film. It is the computer's facility in language that ultimately destroys its omnipotence, its omnipresence, and its bonding with the humans. It "sees" words uttered in secret. By one reading, as the computer's circuitry is disabled, it attempts to hold more tenaciously to its unity by resorting to increasingly nonsensical and regressive language.

By a similar reading, the film *Lawnmower Man II*, is also about the clash between the unity of cyberspace and the symbolic order. Jobe can pervade the unity of virtual reality, where he can be what he wants to be, or stay in the outside world, where he has to be what other people want him to be. The latter requires conformity to the excessive symbolization of the real world, conformity to categories and expectations. As with each of these narratives, the choice is not a simple one. There is no either/or, but rather an agonistic play where the real and the symbolic find residence in each other, and the intrigues of the plot are taken from mutual resistances.

The myth of autochthony presents a similar reading. To be born of one, out of the earth, is to be one with the cosmos, made of the same stuff. To be born of parents is to be a special species, different from nature, subject to categories, subject to the symbolic order. The narratives we construct around the technologies of artificial intelligence and artificial life rework this conflict. If we can make intelligence and life, then we prove the unity of all things and extricate ourselves from symbolization. In this case, "playing god" is a means of getting back to the unity between humanity and nature, albeit in a way that is highly technological and ostensibly "unnatural."

We can see the conflict between the unity and the symbolic order worked out in other contemporary narratives. As we have seen, McLuhan exposes the two great epochs of the senses: the aural sense and the visual. The sense of hearing is immediate and unitary. It is the sense of preliterate, tribal humanity, for whom there was a unity with nature and a lack of differentiation. The epoch of the visual sense came with the invention of writing and manuscript culture and pertains to distance, objectivity, classification, language, and the symbolic order. The electronic age sees a return to the aural sense. This narrative retells the story of the unity (the aural) in conflict with the symbolic order (the visual), as does Dawkin's presentation of the meme pool, a kind of distributed reductionism that seeks the unity of all things in information, to be contrasted with Descartes's centralized reductionism, with the homunculus, the home of the soul, at the center, made of other stuff than nature and sending instructions to the material world. The homunculus serves as the ultimate instrument of the symbolic order. There are also the narratives of the authenticity of speech, which is close to the mind; which is dynamic and implicates a unity between conversants, as opposed to writing, which is inauthentic, fixed, distant, and deals in symbols. Narratives of virtual communities posit the nonhierarchical, egalitarian world of digital democracy against the hierarchical world of classification and symbol: the cyberspace world where you can be yourself, against a fragmented world in which you have to conform to the expectations of others. The new artificial intelligence presents holism, against the fractured methods of formal systems and hierarchical control: connectionism against classical (symbolic) AI (to be examined further in the next chapter). The Lacanian slant on this identification of the ubiquity of the real against the symbolic order is that it is an agonistic relationship, of mutual dependence and resistance.

No position is immune from this application of the Oedipus myth, as the antagonism between the unity and the symbolic order, which is also evident in pragmatism, that treats words as primarily forms of action, caught in indivisible modes of praxis, as opposed to systems of propositions. The premodern definition of symbolism, as expounded by Eliade and others, as participation in the cosmic order independently of reflections on the meaning of those symbols, suggests there is an authentic unitary view of symbol in opposition to a modern linguistic view, Lacan's "symbolic order."

The Lacanian interpretation of these narratives speak of a confused schizophrenia, the Oedipus myth at its most intrinsic, which yields pure profit for the Lacanian economy, and for those who have developed it further.

Schizophrenia and Cybernetics

The theme of psychical antagonism also finds support from critical theorists and Derrida, who argue against the kind of psychologizing that seeks wholeness, the reintegration of humanity and the self, through IT, or by whatever means. According to Marcuse, conservatives have used Freud to legitimate the identification of deviance and abnormality against a model of wholeness, normality, and social integration. For Fromm, and others of the school of ego psychology popular in the United States at the time of Marcuse's writing, the goal of therapy is "the optimal development of a person's potentialities and the realization of his individuality,"[56] which is not far from Turkle's hope for the development of "multiple yet coherent" identities on the Internet. For Marcuse, such theory revises Freud for the worst, promoting concepts of a healthy and whole society in which people can reach their optimal development. It seeks to integrate individuals into a societal whole. Most important, it is a revisionism that characterizes advocates of subversion or radical unrest as a psychopathic type,[57] thereby marginalizing them.

For Marcuse, as for Lacan, the human condition is fraught with fragmentation and disintegration, and it cannot be otherwise, a position that Marcuse develops to show the endemic nature of repression and the necessity to resist it. The Oedipus myth tells of the rebellion of the sons against the domination and repression of the father. But on killing the father, the sons soon appoint a new father to perpetuate the process. So the sons are implicated in repression. According to Marcuse, the sons' sense of guilt includes "guilt about the betrayal and denial of their deed."[58] They are also guilty of "restoring the repressive father, guilty of self-imposed perpetuation of domination?"[59] So there is no innocent party in the Oedipus myth, which speaks only of conflict.

In *Anti-Oedipus: Capitalism and Schizophrenia,* Deleuze and Guattari[60] criticize and further radicalize Freud's project while building on Lacan.[61]

Psychoanalysis, as a positive project aimed at integration, is inextricably linked with capitalism. In exchange for money, it reduces "flows of desire" to the domain of the family (father, mother, child) so central to capitalism, according to Marx.[62] Oedipus is "the last word of capitalist consumption—sucking away at daddy-mommy, being blocked and triangulated on the couch."[63] According to Deleuze and Guattari, psychoanalysis makes the absurd error of assuming that conflict begins with the child,[64] as though the real actions and passions of the father and mother are the child's fantasies. Freud also neglects that the family is immersed in the social field, and it is only in this field that the priority of the father-child relationship makes sense.[65] Neither does Freud do justice to the social field as the medium of transmission of the unconscious. For Deleuze and Guattari, "the family is never determining, but is always determined."[66]

They propose a new kind of "analysis" based on the exaltation of schizophrenia, the condition of the fragmented, deluded, unstable personality, which they label "schizoanalysis," the task of which is destruction: "a whole scouring of the unconscious, a complete curettage."[67] They urge: "Destroy Oedipus, the illusion of the ego, the puppet of the superego, guilt, the law, castration."[68] The task of the psychotic subversive is to "develop action, thought, and desires by proliferation, juxtaposition, and disjunction," in opposition to the process of building structures: "subdivision and pyramidal hierarchization."[69]

In surrealist mode, and prefiguring Haraway's cyborg, Deleuze and Guattari play on the theme of the machine, as a foil to Marx's capitalist machine, which thrives on the division of labor.[70] They rename the capital of capitalists, or rather the "capitalist being," as the "body without organs," which also produces ironic dissonances against Descartes's concept of mind independent of body.[71] This machine gives "to the sterility of money the form whereby money produces money."[72] They also identify the "paranoiac machine" and the "desiring machine," which attempt to "break into the body without organs," where they are repelled, as capital "experiences them as an over-all persecution apparatus."[73] This radical, postsurrealist rhetoric does not characterize machines in systems terms with input, output, determinacy, and a neat division into parts. The essence of the machine resides in flow, or rather the disruption of flow: "A machine may be defined as *a system of interruptions* or breaks (*coupures*)."[74] For Deleuze and Guattari "every machine functions as a break in the flow in relation to the machine to which it is connected, but at the same time it is also a flow itself, or the production of a flow, in relation to the machine connected to it."[75] The desiring machine defies union as a whole: "In desiring-machines everything functions at the same time, but amidst hiatuses and ruptures, breakdowns and failures, stalling and short circuits, distances and fragmentations, within a sum that never succeeds in bringing its various parts together so as to form a whole."[76] This fragmentation is celebrated as providing a foil to the outdated, restrictive, and oppressive forces of capitalism.

Haraway's cyborg, Gibson's cyberspace, and Virilio's vision machine function in similar fashion as catalysts for schizophrenia. Virilio's "vision machine" is an imaginary product of the emerging technologies of digital imagery and automated vision for use in robots, expert systems, production lines, and intelligent missiles. Such an object brings the issue of reality contested by the physics of Einstein, Bohr, and others into sharp relief. But according to Virilio, it also speaks of the reduction of international conflict to a war of images and sounds rather than objects and things, "in which winning is simply a matter of not losing sight of the opposition."[77] Virilio seems to want to show the primacy of the image, and the menace inherent in its deceptions, and in this he implicates the IT enterprise. Such machines are not unitary phenomena but trade in disruption, deception, and breakdown. For the surrealist-philosopher, even chimerical machines never disappear into the background as part of an equipmental whole. The computer appears as a metaphor of disjunction rather than totalizing and integrating.

Deconstructed Identities

The reflections of deconstruction also testify to the perpetuity of schizophrenia and antagonism, which resonate with Derrida's project to unsettle notions of foundations, fundamentals, and the metaphysical, and to further transform Freud's notions of the Oedipal condition. Ulmer identifies Derrida's writing with verbal collage and montage.[78]

As we have seen, for Derrida the issue of proximity, as a constituent of the real, is taken on as an issue of *presence*. According to structuralism, the various oppositions that make up our understanding of the world entail a privileging of one term over another. So we pit the concept of privacy against the concept of going public, the authentic against the inauthentic, the real against the imaginary, and the human against the merely technological. We tend to define one pole of an opposition in terms of the other. Going public is extending your thoughts outward from the private realm, an inauthentic work of art is a copy of the authentic original, the imaginary is a mental copy of what is real, modified, and distorted, and technologies are extensions of human capabilities. In general terms, there is a central, essential term: the private, the authentic, the real, and the human. Then there is the peripheral, derived term: going public, being inauthentic, the imaginary, and the technological. In Derrida's terminology, the first term engages the language of presence, the second is supplemental. For Derrida, this privileging constitutes the language of metaphysics, the language of presence.

Digital narratives are metaphysical, insofar as they assert foundations, principles, and underlying order, or where they are based on metanarratives of progress, liberty, unity, and representations of the real world. In Derridean terms, such narratives are metaphysical where they present original identities and false or constructed identities on the Internet, distributed computing (personable and empowering) and centralized computing (constraining and inauthentic), "softer epistemologies" (Turkle) and hard AI. Many of the concepts from psychoanalysis expounded by Freud also fall under the rubric of "metaphysics," including the concept of a real world versus an imaginary world, the reality principle versus the pleasure principle, the unconscious versus the conscious, the ego versus the id, and an original trauma versus its repetition.

According to Derrida, the language of metaphysics is ultimately unsustainable. It is possible to demonstrate how the claim of any privileged term to primacy can be usurped by the claims of its opposite. The concept of the private has little meaning unless we first have a concept of something against which we wish to mark out the private. The concept of the public is already inscribed within the concept of the private. Similarly, the concept of authenticity is parasitic on the notion of copying, repetition, and derivation, which we associate with the inauthentic. The real is commonly understood (by empiricism) as that which is not imaginary, idiosyncratic to the individual, but depends on the impartial observation of human subjects. Objectivity is already imbued with subjectivity. Being human is already inscribed with being technological—in defining what is human we might appeal to systems of organs, set apart from but engaged with a context and processes, each of which can be tied back to technological ideas of system, machine, and causal connections between parts. Adopting the cyborg logic, the body is already a prosthesis. If such reversals are possible, then epistemological edifices built from basic premises, with other axioms based on these, are very shaky. The foundation is shown to depend on what it supports, or it is made of foreign material, the center is shown to depend on the periphery, and they can readily change places. The present is shown to depend on its supplement.[79]

To undertake a deconstructive reading of a text is to identify the metaphysical assumptions of the text and to show how they are betrayed by the language of the text. This requires a close reading. It is a productive process and commonly reveals new terms and concepts that illuminate the theme under discussion. A deconstructive reading of the IT literature would require us to demonstrate how the notion of true identities on the Internet is already imbued with notions of constructed identities; talk of distributed computing contains the rhetoric of control and manipulation normally accorded to centralized computing; soft epistemologies embody the assumptions of hard AI and are simply another form of it; and the claims to the unifying power of computer networks already contain the seeds of fragmentation, as suggested by Deleuze and Guattari and others. In the case of Freudian psychoanalysis, we could demonstrate how the reality principle is already tainted by the concept of pleasure (eros), the unconscious depends on the notion of consciousness, the ego depends on the id, and the concept of the trauma is already imbued with the concept of repetition.

In this light, the various discourses of the real are metaphysical insofar as they participate in the language of proximity understood as a matter of foundations and absolutes. Empirical realism is clearly metaphysical in asserting an objectively real world independent of human knowing.[80] Digital narratives that focus on unity without recognition of the antagonisms that sustain them are also metaphysical. The privilege accorded to unity, transcendence, and disembodiment in technoromanticism is metaphysical.

Derrida concurs with Lacan and Deleuze on the antagonistic nature of the self, and its disqualification as foundation. According to Caputo, summarizing Derrida: "There is no prelinguistic stratum, no private sphere of self-consciousness in which the self is in naked contact with itself."[81] Derrida targets the psychoanalytic project as presented by Freud by focusing on the ubiquity of repetition in psychoanalytic theory.[82] So much of metaphysical discourse depends on the language of repetition—copying, mimicry, reproduction, representation—with the repeated version assuming an inferior position in relation to an original. Psychoanalysis also addresses the issue of repetition, as the compulsive, ritual reenactment of original conflicts, as exemplified in Freud's account of his young nephew's fort-da game that rehearsed and repeated the departure and return of his mother. Derrida offers a critique of Freud's account that shows Freud is playing the same game.[83] Freud explains the process by which repetition establishes the pleasure principle but notes: "What obviously repeats itself . . . is the movement of the speculator [Freud] to reject, set aside, make disappear (*fort*), defer everything that seems to call the PP [pleasure principle] into question."[84] Freud is caught in the game of repetition in repressing and deferring any criticism of his own position.

For Freud, occurrences of repetition can be uncanny insofar as they remind us of childhood traumas. For Derrida, it is not repetition that should strike us as strange but it is "the very idea of a *first time* which becomes enigmatic."[85] For Derrida, experience "begins" with repetition: "For repetition does not *happen to* an initial impression; its possibility is already there . . ."[86] How is repetition already there? In the opposition between forces, the resistances implied by the notion of the trauma: "in the *first time* of the contact between *two* forces, repetition has begun."[87] Derrida's arguments about the primacy of repetition are captured by Caputo: "As we have seen, the 're-' in repetition, representation, reproduction does not come second, as a re-presenting of a prior presence, but first, as an enabling condition of possibility, as a code of iterable, repeatable signs, which generates presence (perceptual objects, ideal objects, and indeed every possible unity of meaning). We might call this an 'originary' repetition, inasmuch as repetition functions like a transcendental condition of possibility, except that the point of this analysis has been to undo the notion of something originary, if that means something purely present which then gets copied by more or less inadequate reproductions. Repetition as a nonoriginary origin is the repetition which moves forward."[88]

Derrida provides little support for the view of many commentators from Benjamin on that mechanical repetitions are a recent condition,[89] that the mechanical and electronic reproduction and dissemination of texts and images represents a fundamental (postmodern) shift in how we understand

and use words and images. For Derrida, the concept of reproduction, repetition, endless signification, and the interrefereniality of signs already pervades language, long before writing, printing, computers, and electronic networks. Similarly, the self has not recently become divided, fragmented, manifold, elusive, the product of repetitive forces of writing, print, and electronic communications, but has always been, according to Caputo, "a temporary inscription on the flux."[90]

Repetition leaves something behind—a trace, wake, memory, recollection—which is not a thing that can be grasped. It does not consist of objects. It is not "empirical reality." The essence of things resides in this elusive, ineffable game of repetition, flux, and trace. The foundation on which we build our understanding of the world, the real, is permeated by flux and indeterminacy, even before the computer came along with its confused referents. So Derrida adds to the theme of Lacan, Deleuze, and Guattari of the residency of the real in indeterminacy, rift, division, antagonism, and the ineffable.[91]

Psychoanalysis and Suspicion

Such reflections, which conclude with the primacy of rift and antagonism, emerge primarily from the tenets of structuralist language theory. Having decoupled the signifier from the signified, structuralism understands context as a matter of the play of signifiers. The vagaries and indeterminacies of social practices become the workings of ludic games perpetrated by language, which, according to Lacan, is always antagonistic to the elusive real. On the other hand, pragmatic and phenomenological readings of the nature of language draw us back to praxis. According to Giddens: "Meaning is defined through difference, certainly; not in an endless play of signifiers, but in pragmatic contexts of use."[92] In contrast to Lacan, Deleuze, and Derrida, the pragmatic slant on Freudian theory draws attention to the grounding of psychoanalysis as a practice, which in turn draws on the primacy of interpretation.

The psychoanalytic process involves interpreting what is happening in the life of the analysand (patient). Ricoeur regards the Freudian project and its variations as "a tearing off of masks, an interpretation that reduces disguises,"[93] otherwise known as a "hermeneutics of suspicion." In this, Freud's work resonates with the critical theorists (such as Marcuse) and constitutes much of his appeal to them. Critical theory looks beneath the surface of social manners, the intentions people declare, the symbols they use, and institutional structures and identifies the presence of oppression in one form or another.[94] Freud and others begin with the assumption that we do not know ourselves. There are things hidden from introspection, presented with most potency in the concept of the unconscious. In this vein, certain critical IT dystopian narratives speak of the networks of oppression lurking beneath the illusion of freedom, where we are in fact "controlled by the machine."[95]

Ricoeur contrasts this suspicious hermeneutics with a more trusting kind, the hermeneutics of Gadamer.[96] Presumably, the psychoanalyst is involved in the interpretation of signs, symptoms, accounts, histories, and reports. The psychoanalytic process involves discussion. It takes time and involves false conjectures, revisions, refinement, and new starts. For all its specialized knowledge and language, psychoanalysis involves processes similar to those that take place between a student and teacher, a lawyer and client, a programmer and a user, a detective and a witness, an architect and a client, and within many other professional, diagnostic, consultative, adjudicatory, and designerly relationships. Apparently the Freudian psychoanalyst does not arrive at a diagnosis and a cure, which is eventually presented to the analysand as a solution, in the same way that a medical practitioner might produce a prescription, or an architect might present a building design that is the interpretation of the client's wishes. According to Ricoeur, the analysand cooperates in his or her own analysis, and works toward gaining insight, encouraged to overcome his or her own resistances. This is true of any interpretive situation, though it is perhaps made obvious in the practice of psychoanalysis.

For Gadamerian hermeneutics, interpretation is crucial in the process of understanding: to interpret is to understand. Descartes doubted even his prejudiced standpoint and so sought to develop a state of mind free of prejudice, the so-called objective position. Gadamer demonstrates the impossibility of this position, arguing that we always come to a situation of understanding from a base of prejudice. Through the process of interpretation, we develop understanding, which in turn invokes challenges to our prejudices. This process applies to both the analyst and the analysand, both of whom have to work against resistances, not in order to discover the unassailable truth, but to arrive at pertinent and therapeutically useful interpretations, what Giddens describes as a process of "constructing narratives of self."[97]

This understanding of the psychoanalytic process suggests other metaphors than those of exposing, revealing, and probing beneath the surface. The terminology of the unconscious and repression is supplemented, or replaced, by that of interpretation, which embraces rhetoric, narrative, conversation, difference, negotiation, and resistance. By this reading, psychoanalysis is no more an uncovering of the deep recesses of the psyche than is the process by which a student and teacher quiz each other about what they think, a detective finds the truth from a witness, an architect works out what the client wants, or a programmer ascertains "user needs." By this reading, the therapeutic narratives that emerge from a psychoanalytic situation are not the products of hidden psychical forces and repressed memories but interpretations. To ask where interpretations come from is like asking where legal judgments, medical diagnoses, or ideas for a book come from. There is no single source, but the dialogical situation, fields of practice, discourses, background awareness, history, and, in the case of psychoanalysis, also Freud's evocative and influential language of the unconscious and repression. In fact, Derrida remarks that what psychoanalysis uncovers is psychoanalysis itself.[98] Whatever it scutinizes (neurosis, history, culture, art, literature, film, information technology), it is only capable of recognizing its own (Oedipal) schemes.[99]

The Oedipus myth provides a good example of the workings of the hermeneutical process. In Sophocles' play, Oedipus is presented with petitions from the people of Thebes about the terrible plague that has befallen them. Various interpretations of the situation emerge and are successively elaborated, beginning with the proposition that there is some "unclean thing" in the city. It is a story about the overwhelming prejudice on the part of Oedipus, that someone other than himself is responsible for the plague. The story is of Oedipus's pride, his resistance to the processes by which the interpretive situation challenges his prejudices. For Ricoeur, the Oedipus drama is a drama of self-consciousness and self-recognition, the collapse of Oedipus's assumption of innocence. Freud also points to this alternative reading of the drama in indicating how the play involves the gradual unfolding of the truth of Oedipus's incest and even likens this process to psychoanalysis. For Ricoeur, the play is about the conflict between truth and nontruth evident in poetry and all art. For Ricoeur, the play reveals "the profound unity of disguise and disclosure."[100] In this light, the problem posed by the story is not primarily the problem of desire for union with the mother, against the authority of the father, but of interpretation.[101]

But then even the way Freud and various interpreters tell the story speak of the problem of interpretation. The guilt of sons in wanting to kill the father, or in succeeding, speaks of the frustration of resisting authority. In raw form, the oversymbolization that is the law of the father, whose word is to be obeyed, is what resists hermeneutical inquiry. The realm of the mother invites inquiry, though it offers its own resistances. Desiring the mother against the law of the father also speaks of a conflict between a primordial unity and the symbolic order, which is one version of the conflict between the whole and the parts. One of the metaphors used by Gadamer to problematize the interpretive situation is how to get from an understanding of the whole of a text by means of understanding the parts. You cannot understand the whole unless you understand the parts, and the parts cannot be understood without the whole. The metaphor is that of a circle: the hermeneutical circle, the circularity of understanding. The circle is not a "vicious" one according to Gadamer, and the resolution involves the projection of a fore-understanding derived from our background of prejudices, a horizon of anticipation, a further holistic awareness subject to scrutiny and interpretation. By this reading, myths of maternal unity and fatherly rule, guilt and repression, the real and the symbolic order, all present as the problem of interpretation, the circle of understanding, the problematic of the whole and the parts, as do issues of deep-seated antagonism, rift, and division within the psyche.[102]

As we saw in chapter 5, digital narratives follow the form of the hermeneutical play between the whole and the parts, excursion and return, to which we shall return in the following chapter.

Interpretation and the Real

How does interpretation fit into the phenomenology of the real? *Proximity* plays a part in the Gadamerian model of interpretation as an issue of distanciation. A conservative view of interpretation suggests that to interpret a text is to get close to the "original intention" of its author, to find common agreements between what the author thinks and what the reader thinks in order to progress to new understandings. In establishing common ground, the skillful author is able to lead the reader along new paths of understanding. Plausible as this sounds, Gadamer suggests that at its most basic, the process of interpretation requires establishing distance. To use Whitehead's terms, it involves rendering the strange familiar, and the familiar strange. It is in the space between the strange and the familiar that interpretation can have play. The process is well illustrated in the case of education by Gallagher.[103] Learning involves the exploration of material in ways counter to one's expectations. For an architecture student, the familiar environment of the house is problematized, rendered strange, through all kinds of analyses, ranging from the mathematical calculation of heat flow through walls to the language of phenomenology, from the familiar praxis of living in a house to having to design one. In the converse way, the strange language of heat flow calculations, phenomenological discourses, and the "peculiar" way architects speak about buildings is rendered familiar and becomes part of the habitual praxis of the student. Listening to a lecture and engaging in project work are modes of interpretive practice, and the same processes are at work in reading a text, using on-line course material, and presumably in psychoanalysis. Any therapeutic, diagnostic, professional, consultative, textual mode of practice involves such distanciation.

The importance of distanciation finds exaggerated expression in surrealist art and certain computer imagery and interactions. We expect objects to behave in certain ways and belong in particular contexts. Distance is established by moving objects into strange contexts that provoke new understandings. For the surrealists, this involved putting tubas and chairs together in cloud formations, classical streetscapes, and desert landscapes. In the computer world, it may involve plotting data using unusual dynamic projections, assembling avatars, icons, and Internet links in unusual ways, and people presenting themselves in disguise in on-line chat groups. Of course, newness cannot be sustained indefinitely, and ever new means of establishing distance need to be maintained for the provocation to continue, which is to say, for it to provide ever new understandings. But the importance of distance is also operative in routine interpretive activities, such as the interpretation of an email message. From the hermeneutical position, for any communication to operate as a performative, it has to establish a distance within the fields of practice of the recipient of the message. To be effective, a message such as "the file server is down" needs to establish a distance within our usual expectations, and the interpretation of the message involves working out how it applies in the particular situation. To really understand the message is not to translate it into propositions but to apply it, and application is contingent on the fields of practices and contexts with which we are engaged. So distanciation, as with the interpretive act, does not only reside with the text but extends to the practical field in which the text makes sense.

Similar processes are at work in psychoanalysis. Freud's construction of the topology of the psyche as a mechanism for repressing traumas into the unconscious establishes a particular language of distanciation and suggests that some work has to be accomplished in negotiating this distance. This work is the work of therapy. Ricoeur makes this point in his account of the topography of the psyche: "The justification for the topographic differentiation into [Freud's] systems is to be found in praxis; the 'remoteness' between the systems and their separation by the 'barrier' of repression are the exact pictorial transcription of the 'work' that provides access to the area of the repressed."[104]

How is the issue of distanciation worked out in computer narratives? Narratives operate within backgrounds and horizons of counternarratives that establish various differences: against the cozy rhetoric of utopian digital unity, there are narratives of disunity and rupture; and against the familiar story of the perils of rationalism and technological thinking, there is the admonition against technoromanticism. Artifacts can also establish distance. Computers operate as interventions into practice and as such disclose aspects of our practices. We can regard a computer as a foreign agent, a provocation, a surrealist object planted into our otherwise comfortable work and leisure contexts. So the household that has never had a computer before finds the new object (the computer) prompts new patterns of activity, different members of the family appropriate the machine in their own way, there is a new source of conflict, people find that sending email is nothing like writing letters, solitary games assume prominence, and there is the treadmill of keeping up to date with the latest software. It is not simply that the computer makes people do things differently but that the new object reveals aspects of the family's practices: desire for isolation, separation of parents and children, the ambition for success of the young, patterns of keeping in touch, defining boundaries, schooling practices, and myriad other ways, and this distanciation applies to the introduction of a mobile phone, cable television, and the second car, though some technologies are more disclosive than others. The computer here operates as an interpretive device. We understand something about ourselves when we encounter this strange new object, as when the arrival of a houseguest, winning the lottery, or a member of the family leaving brings out hitherto unseen affections and tensions in the family. How the interventions of these technologies might disclose the Oedipal triangle in family life is something else again.

In a similar way, the experience of total immersion environments, VR—the consensual hallucination of cyberspace—may impress us with the differences it presents. Initially, it is a concept, a half truth, a utopian or dystopian projection that brings into relief certain issues having to do with community, space, politics, and the form of cities. It provokes by virtue of presenting as a possibility—which seems to call on the language of the city, of connections, spaces, gateways, storage, and commerce—but it renders each of these entities strange, as we think of connections that are always changing, spaces that can be traversed in an instant, gateways to multiple places, immediate access to stored data, and information as a good. Here cyberspace serves as a device for interpretation. It features as such in Boyer's provocative interpretation of the city, as the Cybercity,[105] which is similar to Calvino's multiple descriptions of the city of Venice, in *Invisible Cities*.[106] The city is rendered strange and multiple by describing it in terms of its plumbing, its waste system, the way it treats its dead, and its differences.

In a similar vein, wearing a headset and data glove impresses us with the differences it presents. Having the visual sense presented with dynamic stereoscopic images of three-dimensional geometries is nothing like walking down the street, and this is not just because of the current poor resolution of VR images,[107] but because being in places is not a matter of just looking around you but of doing things. Reports of computerized flight simulators, the most successful application of VR to date, indicate how lifelike and persuasive is the experience, yet the instrumentation is as important as the visual and kinesthetic simulation, and it is the engagement in the praxical field of piloting and the examination procedures that provide the immersive experience. Because of their emphasis on practice, computer games probably provide the most accessible form of immersive spatial environments. But even in this praxical world of engagement, it is the nonreality of the VR experience that is the most disclosive: being able to walk through walls, excavating rock samples in remote locations,[108] and performing medical operations at a distance.[109] The VR endeavor reveals something about current practice by emphasizing the impermeability of walls, the intricate texture of rock held by a human hand, the color and elasticity of human tissue, and our dependence on touch, but also by presenting familiar objects of walls, rocks, and human tissue as strange. When these experiences are seen as such, we are able to deal with them in new ways, shape walls into new arrangements, collect rocks on mars, and undertake operations in hospitals we have never visited. The hermeneutical concept of distanciation draws attention to these aspects of IT in other ways than do concepts of uncovering the unconscious.

Whereas Lacanian narratives overlook concepts of community in favor of otherness, hermeneutics starts from the primacy of the *shared*. Gadamerian hermeneutics recognizes the situation of the interpretive act within community, well expressed in Fish's account of interpretive communities: an interpretive community is not "a group of individuals who share a point of view, but a point of view or way of organizing experience that [shares] individuals."[110] The digital narratives that have been set up around the themes of unity and fragmentation, democracy, freedom, and transcendent realities take place within communities and structure those communities dynamically. Establishing and negotiating distance provides the space for interpretive play, but as this space implicates resistance, it also implies work. Interpretive communities are "engaged in doing work, the work of transforming the landscape into material for its own project,"[111] and this project is in turn "transformed by the very work it does."[112] According to Ricoeur, in the case of psychoanalysis, the analysand and the analyst, for the purposes of the therapy, constitute a community of two, and both share in the dynamic work of change within that relationship: "From the side of the analyst the analytic procedure, from start to finish, is a 'work,' to which corresponds, on the part of the analysand, another work, the work of gaining insight whereby he cooperates in his own analysis. This work in turn reveals a third form of work, of which the patient was unaware—the mechanism of his neurosis."[113]

By this reading, digital communities are best understood in terms of the work that they are engaged in, which fits within what Giddens describes as "self help."[114] Giddens identifies the need to construct personal narratives as one of the conditions of late modernity, or post-traditional society. In traditional societies, people know their place. The role of father, mother, and child is worked out over a long period and people assume roles according to their traditions.[115] But the definition and maintenance of identity is today a matter of working things out for yourself: "The self today is for everyone a reflexive project—a more or less continuous interrogation of past, present, and future. It is a project carried on amid a profusion of reflexive resources: therapy and self-help manuals of all kinds, television programmes and magazine articles."[116] We can add the self-help "manuals" of the Internet, cyberculture, computer games, MUDs, and digital narratives that insists we are being redefined through computer technologies, each of which serves as a kind of therapy.[117] Identity is now a matter of negotiation: "Where large areas of a person's life are no longer set by pre-existing patterns and habits, the individual is continually obliged to negotiate life-style options."[118] The task is abetted by self-help groups, such as Alcoholics Anonymous, where criticism and judgment are suspended and people assist one another in rewriting self-narratives. Intimacies also have to be negotiated: "In modern social life, self-identity, including sexual identity, is a reflexive achievement."[119] In this light, digital communities are best understood as interpretive communities engaged in the work of constructing narratives of the self, establishing new spheres of legitimacy and intimacy.

For Ricoeur's hermeneutics, we can work with meanings of which we are not conscious by virtue of having a *body* as our "mode of being." The body is "neither ego nor thing of the world."[120] This is not to say that the Freudian unconscious is the body. The unconscious is "neither representation in me nor thing outside of me"[121] for Ricoeur. This is what it means to be embodied. Embodiment is neither within nor without. This account of "the body as incarnate meaning"[122] indicates that "sexuality is not an isolated function alongside many others; it affects all behavior. . . . Sexuality is a particular manner of living, a total engagement toward reality."[123] Here, a resonance exists with concepts of embodiment developed by Heidegger, Merleau-Ponty, Lacan, and Foucault, which also suggest that there is no transcendence beyond the body. As a further example, the body is already inscribed within the computer, and the body already bears the marks of computation, or, at least, the same forces are at work in the creation of both.[124]

Foucault provides the most eloquent argument for the involvement of the body in the hermeneutical process by demonstrating how meanings can be inscribed within the body, in ways that diminish the need for the concept of the unconscious. For Foucault, history in the industrialized/Enlightenment age involves the transformation of power relations from violence to docility.[125] The body is implicated in transformations of power. Foucault begins *Discipline and Punish* with a lurid description of the public spectacle of drawing and quartering a man for regicide in 1757. The body was used in a violent spectacle, to punish, to ensure that justice is seen to be done, and to warn others. Subsequent developments in the justice system involved transformations from violence to heavily structured relations as in the case of prisons, schools, and hospitals. The objective of these institutions is to form bodies that may be subjected, used, transformed, and improved. The means of rendering the body docile include containment, restricting movement, confining people within prisons and hospitals, surveillance, training (dividing the body into parts, military exercises), stripping the body of its signifying dimensions (imposing uniforms and marching in formation), imposing standardized routines, and inculcating certain postures (standing for long periods of time, sitting up straight, and adopting military bearing). The issue is not that certain people control other people by subjugating their bodies but that we, society, history, and our institutions, have written certain attitudes, beliefs, practices, and values into our bodies, manifested in how, in different contexts, we define, adorn, admire, abuse, and comport our bodies. For Foucault, it is not the mind that is subjugated, but the body. And here the body is not just the container of the individual, but it is our clothing, furniture, machines, and architecture.[126]

That the disposition of the body comes before concepts of a state of mind or way of thinking resonates with Heidegger's concepts of embodiment, in which care, anticipation, and orientation are to be understood in bodily terms. It also accords with Lakoff and Johnson's concepts of "the body in the mind" and with other pragmatic understandings of thought and language.[127]

A history of computing in terms of transformations of disciplinary practice and posture has yet to be written. Such a history would include an account of the discipline of manual card punching (at least in universities), a repetitive, tedious, and error prone process but carried out in the casual environments of university libraries and cafeterias. Card punch machines confined people to dedicated rooms. Along with mechanical typewriter interfaces, there was the posture of the dedicated data entry clerk, pounding and cringing at the impact of noisy keys, with little to arrest the gaze, and only able to withstand short periods of work. With VDUs (video display units) and electronic keyboards, there was the prospect of mesmeric concentration for hours on end in the world of the screen, with the posture of the gawking, arched neck, only ameliorated by disciplined work regimes, scheduled breaks, special furniture, and "correct" posture—the straight back and discipline of the modern, ergonomically wise data entry clerk. With the LCD (liquid crystal display) and portable computers, we have the executive hunched over a laptop on the plane or train. Certain disciplinary micropractices have emerged, such as using the screen as a means of watching behind you through its reflections, lest anyone should invade your space—the self-conscious posture of the clandestine web browser, game player, or voyeur. The ubiquitous computing scenarios of Weiser[128] and others show work environments populated with a range of interactive LCD screens of different sizes and orientations, which you can nurse in your lap, have built into desktops, or erected as markup boards, that reflect varied and casual postures in keeping with the aspirations of the new egalitarian office. Then there is the posture of the cybernaut stooping under the cumbersome load of wires, sensors, headset, and dataglove, grasping at space, and whose hand becomes conspicuous by its inability to do anything useful. Contrast this with the reconstructed heroic iconography of Vitruvian man, popular in cyborg imagery,[129] and the imaginary postures of floating, flying, and digital coitus. By this Foucauldian reading, the computer is implicated in the transformations of power, and not only because now we are all being watched (the Panopticon model) but in the way the computer has been invented and developed to inscribe in our bodies certain postures we deem appropriate to our modes of being in the digital age.

Ricoeur draws attention to the centrality of *repetition* in psychoanalysis, the "automatism of repetition, which is part of a very significant sequence—resistance, transference, repetition."[130] By the Freudian account, repetition promotes a sense of the uncanny as it reminds us of the repetitions of childhood, by which we suppressed certain traumas. For Derrida, repetition is a concept prior to that of the original that is repeated. For Foucault, repetition, the drill, is one of the means by which the body is implicated in transformations of power. So the computer strikes us as uncanny due to the modes of repetition it instills in us, but that repetition is also one of the means by which power is transformed, by which bodies are rendered docile. By this reading, repetition goes further than the concern of individual neuroses and implicates regimes of power.

The hermeneutical account also raises the dichotomy of repetition versus play. Repetition pertains to the way that power is inscribed in the body. Play pertains to productive exploration. Engagement in the hermeneutical circle implies repetition, but it is a repetition in which each iteration between the whole and the parts produces something new. We should replace "repetition" with "free play," an indeterminate exploration, a transformative process of excursion and return, in which we return differently than when we set out. The fort-da game of Freud's nephew is not the archetypal game, or at least not as Freud describes it. The play element of the game is in its variation, in how it becomes different each time, in how it means something new as the context of its enacting changes. By this reading, at its most productive, the repetition of a formula and the empirical observation of regularities (in the laboratory) are never simply repetitions but applications, which implicate context, use, technologies, and practice. The application of observations, rules, and patterns are a kind of play. Certain computer systems developers have woven the idea of play into their methods of software development and into the systems themselves.[131] It is easy to forget how gamelike, and even frivolous, the first graphic interfaces appeared. The first Macintosh computer was more like an executive toy than a tool for the office. Though it instills modes of repetitive practice, effective computer software often does so in terms of variation, discovery, engagement, and the gradual acquisition of demonstrable skills: the hallmarks of play.

This brings us to another of Gadamer's themes, that of representation in art. A representation is not primarily a copy, but a re-presentation, presenting again, best exemplified by the re-presentation of a stage play. A work is never complete, and each performance of a play brings about a new interpretation. But each showing of a film, each viewing of a painting, and even each replay of a ten-second digital movie on a computer screen, is a re-presentation, even if it is to the same audience. We are never in the same situation and the same context, and each showing discloses something new (until we become inured and the work ceases to function as art). For Gadamer, this new disclosure is inseparable from the work of art, as are the performers and the audience, which make up the whole that is the work.[132] In this light, the issue of computer representation goes further than concepts of correspondence between objects and data in a computer. To represent a scene in a three-dimensional modeling program is to present again, to offer an interpretation. At each offering, it is a reinterpretation, as the context has changed, if not the viewpoint. The iterative nature of computer image generation testifies to this concept of representation. The file directories of designers and artists engaged in computer modeling are populated with a host of variants, re-presentations, from which a selection must be made for final presentation. The computer medium brings to light the re-presentational aspect of representation.

What of the *ineffability* of the real, which for Lacan becomes a matter of desire? As we have seen, Heidegger introduces the notion of care, which indicates we are already disposed toward the world as we encounter it, and care comes before the psychological issues of caring for things (looking after my car or my health) and making assertions about wanting something. For Heidegger, care precedes notions of articulation, putting things into words, and other ontic presentations. Care is an ontological condition, which is to say it pertains to our primordial condition. The ontological is contrasted with the ontic, which pertains to everyday contingent notions of caring for particular things (the garden, children, unemployment[133]), reduction, measurement, and making assertions. In Lacanian terms, the ontic is the symbolic order, pertaining to language and categorization. In *Being and Time*, Heidegger indicates a conflict between the ontological and the ontic but does not see the conflict in productive terms, as though the human condition resides in the conflict between them. But Lacan brings precisely this conflict into his various accounts of the human condition. For Lacan, desire is the space of nondeterminacy generated by the conflict between need and demand, the real and the symbolic. For Heidegger, care does not emerge from a conflict but resides on the side of the ontological. But in his later work, Heidegger brings his thinking more in line with the Lacanian position by introducing the productive nature of the conflict between "earth" and "world." Earth is that which resists our attempts at classification, ordering, empirical probings—what Lacan calls "the real"—and "world" is here the "destiny of an historical peoples," or the constructions of history and culture, that we can roughly equate to the symbolic order. So Heidegger later concedes the antagonistic nature of being: "The earth is the spontaneous forthcoming of that which is continually self-secluding and to that extent sheltering and concealing. World and earth are essentially . . . different from one another and yet are never separated. The world grounds itself on the earth, and earth juts through world."[134]

For Gadamer, Heidegger elevates the role of understanding beyond epistemology, the ontic, a concern with methods for gaining knowledge, to an ontological issue: "Understanding is the original character of the being of human life itself."[135] When someone understands something, they don't just possess knowledge, they move into new areas of intellectual freedom. This freedom "implies the general possibility of interpretation, of seeing connections, of drawing conclusions . . . Someone who knows his way around a machine, who understands how to use it, or who knows a trade . . . all this kind of understanding is ultimately a self understanding. . . . a person who understands, understands himself, projecting himself according to his possibilities."[136] Gadamer reinforces the ontological dimension to interpretation: in the interpretive act, "neither the knower nor the known are present-at-hand in an 'ontic' way, but in a 'historical' one."[137] For Gadamer, Heidegger shows that the hermeneutical circle has "an ontologically positive significance."[138] Interpretation involves participation in the ontological game. In this respect, technoromanticism is right in projecting the particular world of computer praxis into a universal narrative of the human condition, where "the cyborg is our ontology."[139] The issues of computer representations, network communications, computer role games, and on-line databases do not only belong in the instrumental realm of the everyday but have ontological consequences. In fact, the everyday world of practice is already caught in the ontological, and the simplest way to grasp interpretation and practice as ontological is to consider how the self is defined through everyday practice. The quest for understanding is not a luxury appended to the essential processes of survival, or an act of the will by an individual in a particular situation, but our condition in history: "Long before we understand ourselves through the process of self-examination, we understand ourselves in a self-evident way in the family, society and state in which we live."[140] We can add that we understand ourselves in relation to the technologies we use and the narratives we construct around them.[141]

As we have seen, Gadamer's ontological account of interpretation implicates application, which is to say action. Gadamer illustrates this in the case of the law (which has a bearing on Lacan's concept of the law of the father): "the recognition of the meaning of a legal text and its application in a particular legal instance are not two separate actions, but one process."[142] We understand a rule or procedure by applying it; application pertains to a particular situation and serves as a test or indication of our understanding. In fact, we demonstrate our understanding of a text not only in what we say a text means but in what we say it means in a particular situation, in what we do with the text or in what actions follow. So to demonstrate that we understand Plato's allegory of the cave in *The Republic* is not simply to render a précis of it but to apply it, which is another way of saying that a précis of a text is a particular application, as is contrasting the text with other texts, showing what it meant to other scholars, and applying it to the world of information technology. And the demonstration is indistinguishable from the interpretation. The law does not presume absolute authority to be contrasted to the pleasures of hermeneutical engagement, but law participates in the hermeneutical process.

It is useful to consider the various components of the Oedipus myth in terms of the hermeneutical process. The rule (law) of the father can be taken as a text that applies in a particular situation. Desire for the mother is a kind of text, a message from the primordial past that gets interpreted and applied in various ways. We can also think of these components as the background awareness from which we make our interpretations, our horizon, history, and prejudices, as though the Oedipus myth is part of our heritage and constitutes our horizon. In this light, the desire for wholeness and unity with the mother is a background assumption that we bring to the interpretation of our relationships within the family, and prohibition against incest is undeniably part of cultural heritage. Each of these components is also an interpretation. Desire for the mother is one interpretation of the human condition, and the rule of the father is another. The most general explanation we can give of the Oedipus story is that it involves a series of interpretations, interpretations of interpretations, and conflicts between interpretations.

In this light, desire is simply an interpretive projection that encounters contrary interpretations. Ricoeur reminds us that for Freud, desire meets resistance in the form of prohibition. Freud always describes desire in an intersubjective context. So the desire for union with the mother meets prohibition from the father. There is no such thing as repression or censorship outside our encounter with other conflicting desires: "that the other and others are primarily bearers of prohibitions is simply another way of saying desire encounters another desire—an opposed desire . . . desire becomes educated to reality through the specific unpleasure inflicted upon it by an opposing desire."[143] The metaphor of desire therefore captures that aspect of the fore-projection, the expectation, within the process of interpretation that is in conflict with the interpretation of others. This is a different construction than Lacan presents, where the essence of desire resides in the space between need and demand (the real and the symbolic), and that in articulating demands we generate desires, which are always elusive. From the hermeneutical point of view, we do not need to say that digital narratives "express" desires, nor need they be seen in terms of antagonisms between the elusive real and language as a play of texts. Digital narratives are in conflict with each other, with other narratives, practices, and other ways of accounting for the world. Desires are recognized as such when narratives encounter resistance from other narratives, when narratives of progress are traded in a situation where narratives of regress hold sway, where optimistic narratives of cybernetic rapture encounter narratives of the limitations of digital technology, and where digital narratives that claim to be breaking new ground are placed in the context of time-worn romances.

But Lacan brings the tyranny of the symbolic order into this hermeneutical conflict. For Lacan, the rule of the father is precisely that which defies interpretation. It attempts to break the hermeneutical circle. The authority of the father is such that his word does not call for interpretation. The symbolic order tells it as it is, or tries to. We approach the primordial wholeness, the ineffability of the real, against the constraints of the symbolic order. By this reading, Lacanian desire is the attempt to return to the hermeneutical circle against the resistance of the imperialism of law. In this light, Gadamer is describing an ideal situation of interpretation, when everything is going well, when there is a unity in play, in conversation, art, reading texts, and applying rules. Lacan's attention lies elsewhere, in the friction that denies this pleasure. This is the realm of desire.

Such is the divide between what Gallagher and others[144] have identified as moderate and radical hermeneutics, which forms a circle of sorts. Gadamer's hermeneutics suggests that the radical discourse of friction and rupture is illuminating but is already subject to the hermeneutical process. It is a mode of discourse already subject to the workings of hermeneutical communities and subject to fields of practice. From the contrary position, radical hermeneutics (Lacan, Deleuze and Guattari, Derrida) is always looking for the ruptures in the universalizing system, including the cozy vocabulary of Gadamerian hermeneutics. From the hermeneutical point of view, the "resolution" of these differences resides in application, the modes of practice that the various discourses uncover. Apparently there are different modes of psychoanalytic practice, each with their own modes of discourse. We could make similar observations about cultural critique and different understandings of the computer. Surrealism seems to reveal and conceal different aspects of computer practice, Freudian discourse another, and Lacan and Deleuze reveal different things again. Technoromanticism comes under scrutiny from two contrasting positions. There is the position of moderate hermeneutics, where digital narratives are to be understood in the context of interpretation, engagement, praxis, and language games. The hermeneutical critique of technoromanticism focuses on its presumptions of correspondence, representational schemas, and subjectivity. The critique from radical hermeneutics focuses on technoromanticism's metaphysical pretensions, its failure to engage the aporias and contradictions of its own schemas.

Summary

It is worth reviewing the account so far of the relationship between unity and multiplicity prior to revisiting major technoromantic narratives in the final chapter. Computer narratives bring the question of the real to the fore. The term "virtual reality" suggests as much. Certain computer narratives also claim to challenge conventional concepts of the real with their promises of new worlds in which anything is possible. Narratives of the real soon invoke an antagonism, between a world of open possibility and a world of limits and constraints. We have characterized the problematic of the real as an issue of the antagonism between unity and multiplicity.

So the issue of the real in digital narratives continues the trajectory of Plato's real, which resides with the unity. Digital narratives participate in this unity in various ways: advocating community, freedom, oneness with nature, and the unity of human and machine. As with all myth, the concept of the real undergoes various transformations. For the pre-Socratics and traditional cultures, the real is a unity best understood through contradiction and paradox, of the one in coexistence with the many, being with becoming. The real is appropriated through logical contradiction, role reversals, the inversion of conventional hierarchies, androgyny, ribaldry, orgy, base laughter, grotesquery, a reintegration of opposites, and a regression to the primordial and homogeneous chaos. To such thinking, the grotesque was not a tragedy but an intimation of primordial unity. Debasement, clowning, and inversion of the sacred order appealed to the enveloping and regenerative character of the earth. Unity and fragmentation coexist and feed off each other.

Under Plotinus the unity of the real is spatialized and literalized. The unity is transcendent. The body does not bring us close to base things but is an impediment, a resistance to access to the unity. The enlightened state is beyond the material. The bodily realm is fragmented. The ideal is unity.

Descartes introduces the subject, the ego, into considerations of the real. The unity stands for the single overview of the observing subject. From here came the divide between the culture of the object (empiricism) and that of the subject (romanticism). As the culture of the subject, romanticism finally strips the unity myth of its celebration of paradox and indeterminacy. Romanticism seeks to counter the fragmentation of the object world through access to the unity provided by art and genius.

On the other hand, empiricism emerged from the culture of the object, ostensibly eschewing the language of transcendent unity by focusing on the object world, though it was never far from Neoplatonism. Rather than countering romanticism, empiricism seems to provide the conditions for technoromantic narratives to thrive, promoting the potential of computers to transcend the material real through their ability to represent. Empiricism also promotes the goal of representing the object world. We may not yet know what it is, and our understanding may at present be only partial, but there exists ultimately one correct description of reality. Empiricism and romanticism seem to collude in digital narratives. The means to unity is data, which the computer has the power to manipulate. The unitary goals of the romantics are met by the reductive, fragmenting tools of empiricism.

Such possibilities are contested by theories of language. Both speech act theory and structuralism argue that language does not work by way of correspondences between words and things and by representations. Speech act theory and structuralism undermine the claim that computers represent reality and can thereby transcend it. Structuralism also introduces the possibility of cultural critique into language study, as all cultural phenomena, including the practices of the empiricists, pertain to formations of signs. At the hands of critical theorists, structuralism is marshaled to expose the hegemony of the fixed language games of empiricism and, therefore, of the claims to the unifying and liberating power of computation.

Phenomenology takes us out of the self-referential language world of structuralism and reveals language as a practice, pertaining to contexts, communities, and the workings of hermeneutics. Phenomenology and hermeneutics offer a potent critique of the subject-object dialectic on which notions of digital transcendence are based. The whole does not emerge from the sum of parts, language is not representation, and developing, using, and understanding computers is a matter of praxis. To apply a geometrical schema, as in three-dimensional computer modeling and virtual reality, is to engage in interpretation, which does not invalidate correspondence and representation but indicates that correspondence is preceded by the more important notion of interpretation. Phenomenology also introduces the distinction between the ontological and ontic. The ontological realm pertains to our being-in-the-world; the ontic pertains to the praxis of everyday systems, categories, and correspondences. This distinction provides a further instantiation of the myth of unity and fragmentation.

The break with the view that language is capable of representation came with Nietzsche at the end of the nineteenth century.[145] If the Enlightenment was characterized by correspondence, metonymy, and metaphor, then the operative trope of Freud, Lévi-Strauss, and the surrealists is irony, the recognition of incongruities between what is and what our narratives lead us to expect, yet to continue anyway, recognizing the ubiquity of antagonism. Freud's use of the Oedipus story demonstrates the centrality of antagonism in the human condition. The paradigmatic situation of antagonism is the family and the conflict between the young child's desire for the mother against the rule of the father. By various readings, through Freud and others, the incestuous desire for the mother is transformed into the desire for *jouissance*, wholeness, and the prohibition of the father is words, law, and order. For Lévi-Strauss, the myth is of the antagonism between born from one (autochthony) and born from two, which also speaks of the perpetual antagonism between unity and multiplicity.

Lacan provides a return to the concept of the real as paradox and "contradictory determinations." The real is an elusive unity that resists intervention by language. The unity is antagonistic to the symbolic order. The antagonism between unity and multiplicity, the one and the many, born from one and born from two, the ontological and the ontic, union with the mother and the law of the father present as the antagonism by which the real resists language. At the pen of radicals such as Deleuze, such reflections speak of the residency of the real in the workings of schizophrenia. The task of the psychotic subversive is to act through proliferation, juxtaposition, and disjunction, in opposition to building structures and subdividing into pyramidal hierarchization.

One can marshal the resources of hermeneutics to show that all subversion, the art of the avant-garde, and other distanciations conform to the workings of the hermeneutical circle, but Lacan brings hermeneutics into the conflict between the real and the tyranny of the symbolic order. The smooth workings of the hermeneutical process pertain to the ontological mode of engagement between the reader and the text, the unitary world of engagement. For Heidegger, the ontic pertains to the fragmentary world of empirical observation, representation, and correspondence. For Lacan, the rule of the father is precisely that which defies interpretation. The authority of the archetypal father is such that his word resists interpretation. The symbolic order tries to tell it as it is. We approach the primordial wholeness, the ineffability of the real, against the constraints of the symbolic order. Gadamerian hermeneutics tells of the ideal situation of interpretation, when everything is going well, when there is a unity in play. Lacan's attention lies elsewhere, in the antagonism between the real and the symbolic order, language as correspondence, that denies this pleasure.

What have we accomplished in this chapter? First, we have shown that claims about the reconfiguration of identities on the Net, and other psychologizing that accompanies IT narratives, are further variations on romantic themes. But we have also identified the radical, Lacanian theme that returns us to the residence of the real in rupture and agonistic relations. Second, I have reiterated and reinforced some of the consequences of adopting a pragmatic take on information technology, which presents IT as disclosing aspects of practice, including narratives of the self, opening up an altogether more subtle and productive line of inquiry than that proposed in the rhetoric of cyberspace as transformative and transcendent. Third, we have identified various escape routes from the enchantment of technoromanticism, more robust than those offered by empiricism. These fall within the purview of surrealism, deconstruction, and phenomenology, and find potent articulation in reflections on hermeneutics. Fourth, and more important, we see that IT narrative, digital utopia, the real transcendent, the expectation that IT returns us to the world of the tribe transformed, is a variant of the perennial theme of unity, the play between the whole and the parts. Digital narrative can stay with the monosemic pronouncements of technoromanticism or venture into the polysemous world of hermeneutical inquiry. If the latter seems to strip IT of much of its enchantment, then it redirects attention to the primacy, efficacy, and complexity of being-in-the-world, of which IT is just a part.

In the chapter that follows, we revisit certain digital narratives from the point of view of hermeneutics and show how each brings out the antagonism between the one and the many, and what this reveals about the information age.

8

Technoromantic Narratives

We now analyze technoromantic narratives in terms of how they deal with the relationship between the particular and the universal, the one and the many, unity and multiplicity, the whole and the parts, and how they attempt to stabilize disparate events by placing them into a sequential whole, with a "beginning, middle and end."[1]

Narratives of Calculation

The narrative enterprise is instantiated even in the enterprise of calculation, following the procedures of logic and algorithms, which implicate a transition from the particular to the general and back to the particular. Hegel describes the basic logical procedure, the form and use of the syllogism, in these terms.[2] From the particular case that Gaius is a man and the universal statement that all men are mortals, we deduce further information on the particular, that Gaius is mortal. The algorithms of a computerized virtual reality system take particulars in terms of the spatial coordinates of an environment, perhaps an actual building, and the location of the viewer. The system processes this data through its geometrical transformation algorithms to produce a further particular, a stereoscopic screen display of the space in perspective. As the input data changes, so does the display output. In classical computational theory, algorithms are generalizations. They describe a class of "machines," that is, systems for processing particular inputs to produce particular outputs.[3] This model of computation includes the possibility that the outputs of the system can return as further inputs, exploiting notions of feedback and recursion.

The particular-universal-particular narrative structure is apparent in so-called classical, or symbolic, AI (artificial intelligence) using logic programming. The progression is from given facts (particulars), through general rules (universals), to inferred facts (particulars). In technical terms, this involves a transition from predicates in terms of "existential quantifiers" (there exists an *x* such that *x* is a door) through rules expressed in terms of "universal quantifiers" (for *all x*, if *x* is a door then *x* is in a wall) to further existential quantifiers (there exists an *x* that is in a wall)—a further example of the classical (syllogistic) deductive process. Formal accounts of computer processes thus implicate the narrative structure of the particular and the universal. How is this also a narrative of unity and multiplicity?

Theories of algorithms often call on the terminology of "many to one" and "one to many." A deductive computer program (such as a system for inferring "facts" about doors or calculating energy flows) may start with a set of facts, such as data about the geometry and materials of a particular building, and particulars about climatic conditions. There will be rules, algorithms, or generalizations (which amount to the same thing), generating outputs from inputs, in this case (to use the technical terminology) translating "descriptions" to "performances." The output of the system will be further data about how much energy the building consumes, internal temperature ranges, and perhaps an evaluation that the building is or is not energy efficient. The algorithms of the system provide mappings of input to output data. A particular set of input data produces a particular set of output data. One consistent set of facts as input leads to another consistent set of facts as output. The mapping can be described as "one to one," but it is also "many to one." There are many differently shaped buildings that can give exactly the same performance. There are many different sets of inputs yielding exactly the same set of outputs.

By this reading, the deductive process of inferring building performance is many to one, but the reverse process is of the "one to many" kind. If we were to create an algorithm that produces a building geometry from a performance specification, then it should generate many different geometries. The same applies to the production of perspective views, as in a VR system. There is a many-to-one mapping from the input data (the model geometry and viewpoint) to the perspective view as there are many geometrical configurations that would produce the same perspective view, as demonstrated by the distorted room illusion, in which a room that appears conventional from one angle is seen to have converging floor, ceiling, and wall planes when viewed from a different angle.[4] One of the major challenges of generalized "computer vision," of the kind that might be used in free-ranging robots, is to get a computer to process a camera image, identify edges and corners, and translate these to a three-dimensional model, so that the robot can navigate around the space. It turns out that there are many three-dimensional models that could be derived from the same camera image, so further information is needed, such as camera images from different angles, or data from other sensors. The problem then becomes one of resolving the complex interactions between subparts of the problem. In formal terms, such problems are described as "underdetermined," at least in their parts. As formulated in classical AI, most problems of any interest are of this "one to many" variety, at least during some stage in their resolution. As a further example, a computer system for automatically generating house plans would need to take account of the fact that there are many ways that bedrooms and bathrooms can be arranged so that they are near to each other, there are many ways that a house can be arranged on a site to take advantage of the sun, and there are many ways of ensuring good access from the street. But some configurations of bedrooms and bathrooms do not work well with certain site orientations. The complex interactions of these and many other constraints (for example, proximity between kitchens and dining rooms, and acoustical privacy between bedrooms and living areas) ensure the complexity of the generative process. Were there a successful computer system to tackle the problem of floor planning, then it would proceed from a set of facts (the design brief) through an explosion of possibilities and subpossibilities, the conflicts between which would have to be resolved before it could produce even one suitable plan. (The process of *designing* a building is something else again, involving so much more than the spatial configuration of known components.[5]) There are algorithmic techniques, such as linear and nonlinear optimization, and forms of logic, such as nonmonotonic logics, that attempt to deal with the generative problem, though there is no escaping the complex combinatorics of problems, the subparts of which are underdetermined. Such problems are commonly couched in terms of "many to one" and "one to many." Simple deductive problems are of the many to one kind. Complex problems are of the one-to-many kind. So the narratives of computation and algorithm already invoke the conflict between the one and the many and recount a version of the unity theme. Such accounts of problem solving also invoke notions of convergence and divergence. To proceed from many to one is to converge on a solution. To proceed from one to many is to diverge. And so develop various theories about creativity. Creative problem solving, designing software, inventing new proofs, inventing new scientific models, involve divergent thinking, whereas analyzing problems, evaluating solutions, proving a hypothesis apparently require convergent thinking.[6]

Certain algorithmic approaches take the concept of underdetermined interactions as their starting point, such as in connectionist AI, complex adaptive systems, artificial life (AL), genetic algorithms, and multi-agent systems. Such approaches assume that the whole system behaves in a way that is not preprogrammed, but depends on the complex interactions of the parts. The behavior of an ant colony provides the classic biological analogue for AL. Each ant seems to be performing very simple tasks, operating locally, but as a whole the colony seems to be following paths to food, and building and maintaining bridges and other structures, as though the colony as a whole had goals.[7] The emergence of such intentional and complex behaviors from apparently simple actions is sometimes termed "morphogenesis."[8] Certain systems are built around the idea of discrete algorithmic components that cooperate and compete at a local level, where each component simply responds to its immediate environment. The Game of Life is a well-known example. It is a simple computer program consisting of a gridded array of "cells" presented graphically as a matrix of colored squares on a computer screen, each of which appears or disappears according to the state of its neighboring cells and some simple rules. Over time, cells are seen to cluster into groups, or clusters are seen to migrate across the screen, depending on the rules used and the initial configuration of active cells. AL researchers conjecture that the processes by which the interesting behavior of these simple systems emerges are similar to the processes at work in complex organisms and societies of organisms.

Connectionism is a branch of AI that similarly capitalizes on the emergent properties of systems made up of many simple components, drawing analogies with configurations of neurons in the brain. In connectionist systems, the components are simulated neurons that are connected to each other in a vast array or network. Each neuron provides outputs that act as inputs to the neurons to which it is connected. The algorithm for converting inputs to outputs for each neuron may be simple, but the dependencies between the neurons are so complex that the total network performs in a way that could not be determined from an understanding of the individual parts. In fact, certain neural network models have randomness built in to the behavior of the individual neurons to enhance performance. Such models operate by means of small-scale incremental change with the whole system moving through different "entropy states."

AL and connectionism implicate narratives of the parts contributing to a whole that is greater than the sum of the parts, in that the behavior of the whole is more complex than what can be understood by considering the parts in isolation. Proponents of connectionism as a model of cognition, such as Clark, also extend this holistic functioning beyond the individual organism, proposing "extended brain-body-world systems as integrated computational and dynamic wholes," where we treat cognitive processes as "extending beyond the narrow confines of skin and skull."[9]

Connectionist systems have been configured so that they store and reproduce patterns, including visual patterns. The patterns are not stored locally in individual neurons, but in the system as a whole. Destroying a few neurons may make the system less accurate, but it will not result in the loss of individual patterns. The analogy is often made with a hologram, of the kind that projects a three-dimensional image under certain lighting conditions. A fragment of a holographic plate has information about the scene as contained in the entire plate, though at lower resolution. You can "peer through" a fragment of a holographic plate as though it were a little window into a bigger picture beyond. With certain connectionist systems, and with holograms, there is a sense in which the whole is contained in the parts and the parts in the whole.

Connectionism also addresses the issue of the universal and the particular in positioning itself against "classical AI" and its emphasis on the importance of symbolic knowledge structures, models, generalizations, rules, and hierarchies of categories.[10] A neural network that has been "trained" on a series of patterns does not seem to contain any structures that could be construed as generalizations. Such systems do not classify patterns, though they can reconstruct and display patterns that they have already encountered. A neural network can also produce new patterns consistent with its inputs but that had never been presented to it. So connectionist systems produce new outputs and operate in ways suggestive of solving problems, beyond recollecting patterns, though connectionist systems are described as dealing with cases rather than classes, particulars rather than universals. Neural networks can be designed so that they have no generalizing structures. The corollary for thought (and the behavior of organisms generally) is that thinking does not rely on forming mental representations, models, or generalized structures. We don't have a generic concept of house in mind against which we compare our encounters to see what objects fit the category. Neither do we think through a series of syllogistic rules (if the dwelling has a front door and a back door, then it must be a house). Connectionism sees itself in terms of particulars rather than universals, and situations rather than generalized contexts.[11]

How are these digital narratives of logic, algorithm, AL, and connectionism also narratives of unity and multiplicity? Unity also comes to the fore in the issue of theories. William of Ockham posited that given a choice between a simple and a complicated theory, we should go for the simple one: "Plurality must not be asserted without necessity."[12] The history of science attests to the importance of this quest for the simple, elegant explanation, theory, formula, algorithm, or rule. The ambition is to produce simple, elegant generalizations to account for a wide range of phenomena, simple algorithms linking diverse inputs and outputs of facts and data. The elegant theory unifies the multiplicity of diverse observations. Hawking and others also anticipate "the complete unified theory" that will subsume the phenomena accounted for by the multiplicity of extant theories pertaining to matter, space, time, energy, field, and gravity, and through which "we shall really know the mind of God."[13]

Explained as such, empiricist narratives that implicate the relationship between the particular and the general appear unproblematic. It seems to be the way science operates. Science pertains to the detection of regularities. Such apparently matter-of-fact formulations of the narrative of unity and multiplicity do not resort ostensibly to notions of disembodiment, ecstasis, or rapturous ascent to unity. From our discussion of the nature of scientific theories, I hope that by now it is evident that the account given above is only one construction that can be placed on the role of theories. The pragmatic and phenomenological view posits theories as fitting within complex constellations of practice. Following Deleuze, one can reconstruct the empiricist narrative in terms of repetition rather than universality, which takes us in different directions in understanding the methods of science. What constitutes a regularity, the difference between the simple and the complex, is not "given" but involves issues of interpretation.

That Hawking should situate himself within a line of scientific apologists by ironically invoking concepts of God ("we shall really know the mind of God") indicates that such accounts of science rarely deal simply with "matters of fact." It is not farfetched to construe, in narrative terms, the empiricist formulation of data to algorithm and back to transformed data as but a technologized version of Plato's transition from the world of particulars to the Intelligible realm of ideas. The ideas are in turn imprinted on the world as instantiations. Furthermore, empiricist concepts of representation provide the first steps in the romantic ascent to cybernetic ecstasis. As we have seen, if the world consists of objects that can be represented in computer systems, then the computer representations can be of other worlds that are as real, if not more so than our own. These formulations of the general and the particular offer little to counteract technoromanticism and even support it. The technoromantic unity narrative is of passage from the physical world to the ideal world of computer representations and back to worlds that do not exist, except in the computer.

Technoromantic narratives are sustained by the empiricist narrative, and elaborate it in certain ways, but also set themselves in opposition to it. According to romanticism, to reduce a phenomenon to the syllogism is in fact to resort to particularization, individuation, and multiplicity, where "the intellect exhausts itself in the study of individualities," as opposed to understanding the world as a whole, the "perfection in unity."[14]

The narrative of unity and multiplicity undergoes romantic transformation in the way it implicates the human subject rather than flows of information, propositions, or patterns. It is the person who is transformed, not just information. This subjectivization also implicates the grand sweep of history. It is not just the individual who enters the universal realm of cyberspace but the individual stands in for, is an exemplar or an archetype of, the whole of humanity. The event under consideration is not just the moment at which the cybernaut enters cyberspace, but the epoch in which the whole of humanity enters the cybernetic age. This is evident in the grand technoromantic narratives pertaining to total immersion environments, electronic communities, and the cyborg.

Total Immersion Environments

As we have seen, the technological narrative of the transformation of data through general rules follows a movement from the world of particulars to the generalized world of computer algorithms and representations. It then returns with new, hitherto unseen particulars. In the case of virtual reality, total immersion environments, and cyberspace, this empiricist narrative is of flows and movement, with data as its object. The romantic variation of the narrative "subjectivizes" the sequence, with the human subject as its center. The transformation operates on the human user of a virtual reality system. He or she proceeds from the physical world of particulars to the virtual world of universals, and back to the world transformed, renewed, informed, and enlightened. In certain variations of the story the new subject is the cybernaut, the cyborg, the human-machine, a new particular object.

In its empiricist form, this narrative sets up the problematic of how to present more complete representations to the user of the VR system, with the solution of more sensory channels for heightened realism, greater bandwidths, and faster and more sophisticated algorithms and computer architectures. The romantic version of the narrative is ahead of the technical problems, conjecturing the melding of minds with machines. Here the problems include: how do you present yourself when you are a part of the data stream? does it make sense to talk of a self? what is it like to change identities? and what happens when you meld with other identities?

In the transition from the physical to the virtual and the return to a transformed physical world, the physical is also rendered problematic, challenged, and even superseded, some claiming that cyberspace problematizes the real. After the putative cyberspace experience, how can we be sure that the world outside the data stream is any more real? On analysis, the physical world seems to be permeated with data flows, recursive algorithmic structures; nature appears to be written in the language of mathematics. The physical world seems to reflect the digital one, not the other way around. The permeability of the boundaries between human and machine, mind and algorithm, seem to vindicate the idealist project of a superior reality beyond the material.

The empiricist narrative takes the representational paradigm to its extreme and is disclosive of it. If there are difficulties with VR, in concept, and practically, then these also disclose problems with the representationalist paradigm. The technical problems that seem to inhibit full realization of the VR experience disclose new research programs, but they also disclose problems with empiricism. The tendency for IT commentary to extend the trajectory of VR beyond what is achievable prompts thought of "the unthinkable," but it also discloses how wedded is the digital age to the projects of idealism and realism, the subsumption of the grand idealist myth of transport to unity in technology. The narrative also discloses how Neoplatonism is already technological. It is not an authentic concept recently sullied by technology. Neoplatonism fixes certain categories and oppositions while masking the antagonism between opposites. In Heidegger's terms, it represents the conquest of techne over poesis, a kind of making and reflecting that seeks instrumental causes rather than a mode of being that lets things disclose themselves. The easy transition to the

terminology of IT shows the limits of the distinctions between sp
matter, the ideal and the real, the permanent and the transient. Neop
ism masks the aporia presented by the pre-Socratics of the prod
antagonism between the one and the many, unity and multiplicity, b
and becoming.

VR (virtual reality) is also disclosive in other ways. On examination, VR shows how the concept of the unity as an ideal entity is already disintegrated. According to Žižek, if VR is all surface, with no actual access to substance and the real, then we find this is also the condition of life outside VR. The real is based on just this exclusion. It is also all surface: "The computer-generated 'virtual reality' is a semblance, it does foreclose the Real; but what we experience as the 'true, hard, external reality' is based upon exactly the same exclusion. The ultimate lesson of 'virtual reality' is the virtualization of the very 'true' reality: by the mirage of 'virtual reality,' the 'true' reality itself is posited as a semblance of itself, as a pure symbolic edifice. The fact that 'a computer doesn't think' means that the price for our access to 'reality' is that *something must remain unthought*."[15] The ambitions of VR remind us that the real is that which resists representation. It is ineffable.

The total immersion environment features in both empiricist and romantic narratives, both set up their own problem regimes, and each is disclosive of the technology and our current condition in different ways.

Digital Communities

We can apply a similar analysis to the narratives of digital communities. Digital narratives also presents on the matter of language, communication, and community through the technology of computer networks. By one empiricist reading, the human agent receives stimuli, inputs, which she processes through general inference mechanisms, to produce outputs, in terms of actions, specific thoughts, or mental images. The stimulus may be a string of words as a sound sequence, in which case the inferential process is one of interpretation, and the output may be a set of actions, including the production of further word sequences. Listeners and speakers have access to general rules of language interpretation and production. Word sequences are analogous to facts in a deductive system and are particulars. The particulars of the word sequences have information content.

According to information theory as developed by Shannon and Weaver, information is coded, transmitted in some medium, and then "unpacked" by a recipient.[16] Narratives of this kind set up a problematic in terms of ensuring that intentions are accurately conveyed through the appropriate use of words, or code, with minimum distortion to the message, and establishing methods for ensuring we understand the correct meaning.

According to this information theory, it is possible to broadcast messages to large numbers of recipients through computer networks. Networks allow information to be circulated widely and for people to share their thoughts. But the network also provides opportunities for distortion. If a message is passed from one person to another, it is bound to be modified as the vagaries of individual interpretations are compounded, as in the putative compounding of errors in the passage of rumors. This narrative grants privilege to the individual and her thought processes, with the group and the network providing an imperfect extension to the individual.[17] The individual represents the site of authenticity, where meanings are whole, and from whom meanings make occasional excursions into the realm of the group, through the individuation of words. Communication between individuals must pass through this perilous territory of multiplicity and ambiguity. So here the narrative is of passage from the coherence of the individual mind to the chaos of the group and back to the mind, or the conceptualizing, generalizing, and universalizing faculties of the human intellect to the world of particulars and back to the intellect.

As we explored in previous chapters, counternarratives present language as *primarily* social, as involving diverse practices, with the function of language as basically performative (it makes things happen) rather than representational or intentional. Language trades in difference and ambiguity, the resolution of which is decided by context, which is the whole social network, and everything else besides. The operative human faculty is that of interpretation, which pertains to application. To understand an utterance is not to form some mental image, or place particulars into general categories, but to apply the utterance in a context, and as contexts change, so does the understanding, which can never be exhausted. By this reading, there are only particulars, which is to say the language of the general and the particular loses its potency.

Romanticism historicizes and subjectivizes the unity narrative, further valorizing the processes of individual thought. The romantic tradition, as exemplified in Rousseau's writing, favors the natural spirit, the genius within each of us, that must be allowed to win through the constraints of custom, tradition, and ordering institutions. Speaking, conversation, dialogue, and aural culture are closer to human thought than writing and the culture of the text, which attempts to fix and rigidify thought. In McLuhan's version of this romantic narrative, humankind under the unifying impetus of thought, speech, and the aural sense is tribal and unified, with no sharp individuation. The introduction of writing and print, the culture of vision, has a distorting and stultifying effect on the unity of aural culture and the spoken word. There is the progression from the age of the tribe, crafts, guilds, and community to an age of individuation and dislocation. Electronic communications bring us together again, though in ways somewhat noisy and contradictory. According to the narratives of electronic communities, we can rediscover the unity that is humankind through the immediacy of electronic communications, which provide access to the essential self, unencumbered by the distorting effects of physical appearance, social status, and the uneven access of social hierarchies. The romantic narrative presents the problematic as how to preserve the freedom of the net, how to maintain the spirit of its grass-roots, decentralized origins, reconstructing the Enlightenment dream of an active and informed citizenry.

But the trajectory of digital communities discloses how wedded we are to the culture of the Enlightenment and the culture of the text, the printed and reproduced word. The abrogation of social status seems to rely on anonymity, not being able to see or hear the other person, hallmarks of the culture of the text, features of textual communication rather than the spoken word. These factors may well dissipate with the increased traffic in digital video and the use of sound, where the way you look and sound are permitted to regain ground.

The democratic ideals implicated in the narratives of digital communities are those of the culture of the text. Counternarratives focus on uneven levels of literacy and network access, how text-based electronic communities foster coteries of the literate against the illiterate, and the potential dystopian aspects of ubiquitous networks: alienation, distance, hegemony, substituting fleeting electronic encounters for being with one another—the full weight of critical theory.

Cyborgs and Postmodernity

The concept of the cyborg finds its empiricist home in the researches of NASA scientists in the 1960s. Rather than creating earthlike environments for people in outer space, Clynes and Kline proposed changing astronauts' physiological functioning by artificial means so that they could survive and thrive in the extreme environments of outer space: zero gravity, lack of oxygen, and high radiation levels.[18] Such artificial physiological controls would include sensors planted into the human body that detect excess radiation levels and regulate injections of protective drugs in appropriate doses. Here the narrative is of systems with inputs and outputs. The human organism becomes part of an "integrated homeostatic system": "The Cyborg deliberately incorporates exogenous components extending the self-regulatory control function of the organism in order to adapt it to new environments."[19] Particular inputs are processed through microcircuits in which are contained algorithms and rules to produce particular responses. In keeping with the tenets of systems theory, the whole functions in a way that is more than the sum of the parts.

But the invention of the cyborg is not a purely technical project and has a romantic agenda. The romantic version of the cyborg narrative takes its cue from the unity and multiplicity narrative, subjectivizing and historicizing it. Clynes and Cline begin their seminal paper on cyborgs and space by stating: "Space travel challenges mankind not only technically but also spiritually, in that it invites man to take an active part in his own biological evolution."[20] There is clearly a goal in mind, the progression of humankind to something greater. With such control systems in place and operating without the conscious awareness of the human, the cyborg is free "to explore, to create, to think, and to feel."[21] They conclude that such devices will mark significant scientific advance but "may well provide a new and larger dimension for man's spirit as well."[22] So the cyborg provides a step toward humankind's greater realization of its true nature, greater progress toward making itself truly free.[23]

The relationship between being in outer space (the new cyborg habitat) and ecstasy does not escape Clynes, who relates weightlessness to the emotion of joy, and even proposes an as yet unnamed emotion, best described as "a feeling of not being totally earthbound, but perhaps part of an infusing process."[24] The project of moving into outer space is also a project of movement to ecstasis, aided and abetted by the technologies of the homeostatic system. The astronaut, the cyborg, the virtual reality user, and the cyberspace inhabitant are conflated in this cyborg technoromanticism. In its full romantic version, the cyborg narrative problematizes the subject itself, whether in space or on earth. For Stone, such problems include: "How are bodies represented through technology? How is desire constructed through representation? What is the relationship of the body to self-awareness?"[25] As the transition into cyberspace apparently results in a questioning of reality, so the emergence into cyborg sensibility is thought to challenge concepts of identity and embodiment.[26]

The romantic conception of the cyborg (and the participant in cyberspace and virtual reality) employs the terminology of union between human, nature, and machine, the realization of humankind's potential, evolutionary progression and participation in a state of weightlessness and bliss. The new relationship between the human, the natural, and the mechanical can be described in terms of a melding into a unity, or, alternatively, in terms of a disruption of conventional boundaries. Haraway's antiromantic cyborg discourse invokes a rhetoric primarily of rupture, violence, and decentering. For Haraway: "Cyborgs are about particular sorts of breached boundaries that confuse a specific historical people's stories about what counts as distinct categories crucial to that culture's natural-technical evolutionary narratives."[27]

However, Haraway's rhetoric of breach and dislocation loses its radical edge at the hands of Turkle's holistic ego psychology. On the one hand Turkle affirms that in playing networked computer games one can assume various identities, "the self is not only decentered but multiplied without limit,"[28] and in such role games "we shall soon encounter slippages." But these slippages are not the sites of radical disjunction, discontinuity, contradiction, incommensurability, and schizophrenia but "places where persona and self merge."[29] She points to the ability of the Internet to "help us develop models of psychological well-being,"[30] new identities as "multiple yet coherent,"[31] the importance of exploring the dilemma of identifying the "real self" on the Internet,[32] the scope for MUDs to provide room "for individuals to express unexplored parts of themselves,"[33] and she indicates that "postmodernism" teaches us of the need for an "openness to multiple viewpoints."[34] On the subject of the newly emerging virtual community, we are apparently presented with a challenge: "Will it be a separate world where people get lost in the surfaces or will we learn to see how the real and the virtual can be made permeable, each having the potential for enriching and expanding the other?"[35] In this romantic cyborgian narrative, the language of rupture is converted into the language of essences, wholeness, recombination, and a reaffirmation of the goal of unity and integration.

Such narrative seems to go hand in hand with the temporalization of the "postmodern condition," according to which, in the modern era, we had certainties, there was a strong sense of the self as a unitary entity, and the control of human affairs proceeded in a hierarchical manner. Now everything is disrupted, unsettled, fragmented, and multiple and presents us with new elements and boundaries that have to be reconfigured to produce new systems, new wholes, a new unity. Under this narrative, the cyborg is a genuinely new phenomenon. According to Gray: "It is no accident that the modern has become postmodern as human changes to cyborg."[36] So too for Boyer: "*Bladerunner* is a metaphor of our postmodern condition, as the film presents a concentration of space, in which distances coalesce into a single point, coupled with a division in time, in which simultaneity and seriality define a temporal order of replication or repetition."[37] This narrative is a variant of McLuhan's myth of the integrating power of electronic communications, co-implicated with the human body and its sensorium. The decentered subject, the multiplied self, and the human in an ambiguous relationship with the machine is presented as a new condition.

By way of contrast, the radical narratives of Žižek and others reflect on such temporal assumptions. They indicate how these narratives turn back on themselves. We see the new condition as one that was there all along. Žižek provides an interesting account of the cyborg and the self that recognizes the nature of the narrative enterprise. For Žižek, the *Bladerunner* movie presents the paradox of the replicant, or humanoid robot, that ordinarily knows it is a fake, except for one replicant, Rachel, who has artificial "memory implants," that cause her to think she was born and raised a human. If the difference between human and replicant is undetectable, except by the increasingly sophisticated testing of memories, what difference does it make? Like the robot, we can ask ourselves: "Where is the *cogito*, the place of my self-consciousness, when everything that I actually am is an artifact—not only my body, my eyes, but even my most intimate memories and fantasies? . . . I am only the void that remains, the empty distance toward every content."[38] Žižek concludes that the fiction of the replicant shows us that "the difference which makes me 'human' and not a replicant is to be discerned nowhere in 'reality.'"[39] Replicants seek to combine their implanted memories into a personal myth, but so do we: "Are not our 'human' memories also 'implanted' in the sense that we all borrow the elements of our individual myths from the treasury of the big Other?"[40] The replicant is a new entity, fictional or not, that causes us to reflect on our condition. It is not that there is an authentic kind of memory that precedes the inauthentic, "implanted" kind, nor that our present condition is one in which we can entertain the prospect of memory implants (a representationalist view of memory), unsettling our cherished beliefs about memory. For the *Bladerunner*/cyborg narrative to qualify as a radical proposition, it is necessary to see that the concept of memory is imbued "from the start" with concepts of implantation.

Žižek's example reminds us of the capacity for technologies, devices, and things to *disclose* (chapter 5), not just simply to give us information but to reveal something about the world in a way that presumes neither uncovering something preexisting nor creating something new. This is also the provocation of the surrealist's *object* and is one reading of Heidegger's concept of truth as disclosure. Specific technologies, such as computers, disclose practices and prompt us to construct narratives around such disclosures. Elsewhere, my colleagues and I have explored how the introduction of computerization, computer-aided design (CAD), multimedia, and network communications into the workplace of the architect discloses aspects of the architectural practice.[41] The introduction of the technologies shows up some firms as custodians of databases, image makers, publicists, and dealers in texts. Even small firms are disclosed as "multinational" as they position themselves in the global network economy. These disclosures appear in the way practitioners now talk about themselves, and the narratives they construct and in which they position themselves. Whether these firms now do business differently than they used to is another question, but they tend to see their business in different ways, and they construct new narratives, not least around issues of computing needs. Disclosure operates in different ways. CAD and multimedia technology disclose something about conventional ways of working, evident in the case of manual drawing. The way we now describe hand drawing takes over metaphors from computing, and vice versa. So we describe hand drawing in terms oppositional to computer drawing, in ways that may not even have occurred to people prior to machine drawing: hand drawing as an embodied activity involving gesturing, a process of transformation, a stage in a production process. We remind ourselves that drawing is a function of the hand, that drawing is connected with thought and the free flow of ideas (because computer drawing is so constrained), that drawing is essentially private (as opposed to the potential "world readable" status of computer graphics), and a hand drawing has the potential to be scanned, modified, reproduced, and made "world readable." Computing discloses a potentiality, a lack, in hand drawing that hitherto went unnoticed. Hand drawing becomes an activity along a chain of production, whether or not we ever use a computer. Having come under the disclosive ambit of the computer, hand drawing will never be the same again. Something similar seems to occur in the case of the technologies of cyberspace, electronic networks, and the cyborg. These technologies, and their fictional extensions, disclose things about the world in which they are situated.[42]

This disclosure can occur in direct ways. So computer game players sometimes begin to imagine their encounters in terms of a game plan, the chess devotee thinks of how people are seated in a room in terms of chessboard positions, the world outside the VR headset starts to impress us with the richness and variety of its "texture maps," a trip through a cave strikes us with its similarity to the passageways in *Myst* and invokes further curiosity, and we hear the chirping of birds in the forest as "sound bytes" as if placed there to give us spatial cues (*Riven*). Computer technologies participate in the workings of metaphor and disclose through their trade in similarity and difference with other objects and media. As a further example, the cyborg is not only the astronaut implanted with a technology not yet realized, or the fictional humanoid replicant: "It's not just Robocop, it is our grandmother with a pacemaker."[43] And for Haraway, the concept of the cyborg discloses that we are all cyborgs: "The cyborg is our ontology; it gives us our politics,"[44] even without implants.

Computer technologies provide spaces in which we can dare to think what has hitherto been unthought. Similar processes are at work in what computer networks disclose about the operations of language and texts, disclosing that texts have always been interconnected, that speech is always understood in terms we use to describe texts, that the author was always an illusive concept, and that authority, authenticity, and originality have always depended on social practices and agreements rather than notions of empirical fact, proof, or truth propositions. Similarly, if the self now appears multiple, uncertain, diffused, fractured, and contingent, then information technology may have provided a space, disclosed something about the world, but there is a sense in which what it discloses is what was there all along. In fact, it is equally valid to say that information technology is the product of the working of language and texts, and successive generations of thinking about identity and the self, which have paved the way for information technology to disclose the character of language and identity, and who knows what will be further revealed.

The fact that information technology is used by some to mark this culmination, or point of transition to a new epoch, is itself disclosive of the nature of technology and technological thinking. It is apt that "technological thinking" that so interested Heidegger should become the "cause" of a transition to a "postmodern age." So much of the linguistic invention of Heidegger's later work seems to be bent on implicating Being in this mission of transformation rather than technology itself. The latest disclosure of Being is this covering over of itself, concealing its agency under the mantle of technology. And what is Being but the ineffable, the incalculable, the void, the nonground, which is beyond cause.

Summary

We have examined several instantiations of the theme of unity and multiplicity. Each narrative establishes two terms in opposition, privileging one term over the other. By a Derridean reading, the one implies center, source, authenticity, and presence. The many implies periphery, derivation, copy, and supplement. The center is privileged over the perimeter, the one over the many, and unity over multiplicity. Insofar as we regard this hierarchy as fixed and unproblematic, we are caught in metaphysical thinking. Romanticism is metaphysical. It builds its narratives on the concept of the center, placing the subject there, and implicating the sweep of history, with the subject standing for humanity. Romanticism presents the narrative enterprise at its simplest, as a transformation from one to many and back to one, a narrative with a beginning, a middle, and an end, concealing the agonistic, recursive, cyclical, and hermeneutical aspects of the narrative venture.

Was the narrative of the one and the many ever different? We have seen that for the pre-Socratics, certain scholastic traditions, and in the spirit of the carnival, the one and the many appeared each to imbue the other in a play of paradox, at least by the post-Hegelian reading of Eliade, Bakhtin, Scholem, and others. This tradition is all but concealed by the legacy of Plato, who starkly delineated the unity that resides with the real, the world of being, from the world of multiplicities, shadows, and becoming, privileging the real. Plotinus spatialized the distinction, elaborating the theme of the soul's release from the body (ecstasis), its transformation, and its flight to unity. Romanticism continues the Neoplatonic trajectory, privileging unity over individuation. For empiricism (romanticism's older sibling) the unity theme is a matter of the general against the particular, and theory against observation, with the focus on representation, truth statements, correspondence, and regularities. The remnants of Neoplatonism persist in empiricism's toleration of the romantic impetus, and in its supporting theories of representation, which legitimate romanticism's trajectory toward transcendence. Romantic idealism is also evident in the search for the unified theory, the quest by some to grasp the world picture.

Continuing the romantic trajectory, McLuhan equates the workings of the tribe and the sense of hearing with unity. Writing, print, and the visual sense introduced multiplicity and individuation. Language implicates the unity theme. For pragmatic language theory (to which McLuhan subscribes), the whole involves the context and the praxical field in which the performative does its work. (Multiplicity pertains to representation, language as a matter of conveying information.) Language works by virtue of the slippage and ambiguity of words, made useful in the contexts of language games and practical situations. Hermeneutics also engages the rhetoric of praxis, implicating the whole and the parts as components of a cyclical play, the cycle of understanding. For Heidegger's phenomenology, the hermeneutical project also implicates ontology. Interpretation pertains to being before knowing. For phenomenology, the ontological is contrasted with the ontic, the world of scientific or commonsense individuation and knowledge. The later Heidegger introduces the concept of "earth," which is the impenetrable whole, the ground that resists individuation.

Structuralism construes the unity and multiplicity theme in terms of the arbitrary nature of the sign-signified relationship, dwelling on the matter of difference. For Lévi-Strauss, the narrative of autochthony contrasts being born of one, from the earth, in union with nature, against the inadequate process of birth through human parentage (the many of human lineage), an opposition dealt with through various mythic resolutions.

If we needed any reminder of the potency of the unity myth, Freud makes clear the psychic antagonism between wholeness in the mother, *jouissance*, against the individuation of the father's rule, with attention directed to the conflict between them, the processes of repression and resolution. The Oedipus myth becomes the archetypal myth of unity and multiplicity. Surrealism trades on similar oppositions, positing the surreal against the world of order and rule. For the surrealist intellectual, Lacan, building on Freud, the burden and function of the concept of unity is exacted on the real, which brings the grand narrative back to Plato. But the real is rehabilitated. The real is what resists symbolization, the workings of language, and attempts at explanation. We do not use language to describe reality, as empiricism suggests, but the real is what resists language. Individuation and multiplicity pertain to the symbolic order, with which the real is in conflict. Lacanian and deconstructive narrative returns to the agonistic, paralogical understanding of the pre-Socratics. The hermeneutics of Gadamer and Ricoeur provides a framework for examining the Oedipal myth and the workings of narrative, but even hermeneutics is prone to interpretation in terms of the unity theme. A Lacanian reading construes hermeneutics' problematic of the whole and the parts, and its appeal to the unity of play, as concerned with an ideal situation of interpretation. Lacan's attention lies in the friction that denies this satisfaction.

There is clearly no escaping the unity theme, but postmetaphysical narratives are highly reflexive on the instabilities of their own structures. But there are digital narratives for which this is not the case. The narratives of technoromanticism trade in a dull symmetry between the one and the many rather than the ruptures and resistances they expose. There are counternarratives that displace romanticism and empiricism—for example, narratives that treat algorithms as particulars rather than universals. Algorithms are sequences of code, or instructions, that when run on computers produce effects that by certain interpretive practices we construe as conforming to some or other prediction. Algorithms are tools that operate within particular contexts. The arguments presented in this book are attempts to celebrate the particular over the universal, to favor repetition rather than generalization, and even to move beyond these categories altogether. Deleuze contrasts the dichotomy of the particular versus the general with the dichotomy of singularity versus repetition.[45] Generality pertains to law, but: "In every respect, repetition is a transgression. It puts law into question, it denounces its nominal or general character in favor of a more profound and more artistic reality."[46] In that case, the transition from the particular to the general and back to particularities is not a given. Neither is technoromanticism's trajectory toward a transcendent disembodied reality.

Notes

Introduction

1. See M. McLuhan, *The Gutenberg Galaxy: The Making of Typographic Man;* and M. McLuhan, *Understanding Media: The Extensions of Man.* For a recent account of the condition of primal bliss, see S. Plant, *Zeros and Ones: Digital Women and the New Technoculture:* "Those were the days, when we were all at sea. It seems like yesterday to me. Species, sex, race, class: in those days none of this meant anything at all. No parents, no children, just ourselves, strings of inseparable sisters, warm and wet, indistinguishable one from the other, gloriously indiscriminate, promiscuous and fused" (3).

2. F. von Schlegel, "On the limits of the beautiful," 414.

3. Ibid.

4. Ibid., 415.

5. Ibid., 419.

6. Ibid., 421.

7. Ibid., 221.

8. "The second, to divide each of the difficulties that I was examining into as many parts as might be possible and necessary in order best to solve it" (R. Descartes, *Discourse on Method and the Meditations*, 41).

9. Von Schlegel, "On the limits of the beautiful," xviii.

10. Ibid., xvii.

11. In some cases, this means implicating rationalist notions of the individual, objectivity, the abolition of prejudice, and perception as the communication of sense data.

12. A notable exception is the exposé of IT romanticism by Stallabrass. See J. Stallabrass, "Empowering technology: The exploration of cyberspace."

13. Here I derive the link between unity, narrative, and hermeneutics from P. Ricoeur in *Time and Narrative.*

14. The identification of *facts* is party to the processes of narrative, the various conventions through which particular parts of a story are corroborated and legitimated. See S. Fish, *Doing What Comes Naturally: Change, Rhetoric, and the Practice of Theory in Literary and Legal Studies*, 56. See also H. White, *Tropics of Discourse*, 43.

15. See H.-G. Gadamer, *Truth and Method,* for an explanation of the workings of the hermeneutical circle. I am indebted to Adrian Snodgrass for elucidating the theme of hermeneutics. See A. Snodgrass and R. Coyne, "Is designing hermeneutical?"; A. B. Snodgrass and R. D. Coyne, "Models, metaphors and the hermeneutics of designing"; and A. B. Snodgrass, "Can design assessment be objective?"

16. For postmodernity, the narrative becomes "plastic and manipulable . . . heterogeneous, ambiguous, pluralised" (A. Gibson, *Towards a Postmodern Theory of Narrative*, 12). Here I begin with "teleological" digital narratives, the ambiguities and ironies of which have yet to be revealed.

17. Notably, H.-G. Gadamer, *Truth and Method.*

18. See E. Barrett, *Sociomedia: Multimedia, Hypermedia, and the Social Construction of Knowledge.*

19. This note and subsequent notes in this chapter will elaborate the chapter contents in greater detail.

Virtual communities are an ideal rather than an actuality, the total immersion environment is a technical impossibility now, there are no demonstrations of artificial intelligence meeting the expectations of science fiction, and there are currently no convincing digital life forms. I summarize critiques of the various utopian narratives that motivate much IT development and polemic, preparing ground for a later exposition of how narrativity and hermeneutics are implicated in phenomenological concepts of time.

20. The real is the world of completeness, with its source in the sunlike radiance of the Good. More specifically, the unity bears the contradictory property of being both one and many. As summarized by Snodgrass, the unity "is and is not, changes and does not change; it is all things and nothing at all" (A. Snodgrass, *Architecture, Time and Eternity,* I, 73).

21. This tension between the spiritual and the material informed Augustine, who was intent on uncovering similarities between Greek scholarship and superior Christian doctrine. His work influenced Descartes several centuries later. For Descartes, Platonic idealism (one name for the unity theme) was further transformed by the notion of the autonomous, reflexive, thinking self, or the subject, who could reason about the world independently of prejudice, tradition, and experience, a concept that has provided substantial impetus to the quest for artificial intelligence. The empiricists set themselves in opposition to Descartes' rationalism and asserted the importance of sense experience for acquiring knowledge about the world, rather than pure reflection or contemplation. But the thinking subject was at the center of their epistemological systems as well. Empirical realism asserts the commonsense notion that there is a world that we can know objectively, that it exists independently of us thinking about it, which promotes the common view that the primary function of language is to describe reality, and that the potency of the computer resides in its ability to represent the world. But the empiricist Berkeley read Plotinus, whose apparent disregard for matter prompted Berkeley to assert that we do not need the concept of matter to support the concept of reality, and so began the empiricist form of idealism, the implication that there is only thought and that thought constitutes reality—ideas that resonate with some of the narratives of virtual reality and cyberspace. The inclination of rationalism to represent the real and the impulse of idealism to transcend it conspire toward the current technoromanticism.

22. Each of these positions touches on the theme of unity and multiplicity in different ways. Romanticism construes empiricism as dividing and categorizing the world. Structuralist language theory commonly implicates language in narratives of multiplicity, and phenomenology presents multiplicity as an issue of the "ontic" and the "ontological," or knowing and being, or scientific knowing and phenomenological understanding. These positions connect with IT in many ways but particularly in the way they deal with space and with language. The computer is often thought of as a device for representing or transcending space. It is also a

device for transmitting code, enabling communication, and simulating language. So the chapters in this section examine the themes of space, language, and phenomenology.

23. Pragmatic concepts of language, particularly speech act theory, assert that even science does not operate through mathematical, logical, and linguistic propositions, their truth conditions, and their correspondence with the world. Structuralism also presents theories of language that militate against theories of correspondence and regards the construction of the real as a maneuver in language.

24. Phenomenology constructs narratives around the theme of unity and multiplicity (specifically the ontological and the ontic). Phenomenology directly addresses the issue of the real through concepts of engagement with the world, being-with, corporality, practice, and anxiety.

25. For example, the spatial fundamentals of point line and plane, considered by some to be foundational in the modeling of space by computer, can be explained as components in an indeterminate field of metaphors rather than as the essential basis of space. Behind these supposed fundamentals are various fields of practice, implicating metaphors. Phenomenology points to the primacy of an indeterminate whole, without which it is not possible to construct our categories and other devices of language.

26. The phenomenology of Heidegger suggests that IT narratives are particular, technologically oriented manifestations of our predisposition to care. The digital utopia is one manifestation of our embeddedness in time or, more accurately, our embeddedness in temporality, of which time is but an indication. This chapter also elaborates on the workings of hermeneutics, the cyclical process of resolving the tension between the whole and the parts in an interpretive situation, that forms the book's thesis.

27. Phenomenology introduces the distinction between (i) the unitary realm of the ontological, which is itself pervaded by indeterminacy, and (ii) the ontic, which is the world of representation, science, order, and multiplicity. Calling on structuralist theory, Lacan associates the ontic with the symbolic order, the domain of language, to which we may add language *realized in its most instrumental form*, as a technology of correspondence. For Lacan, the symbolic order interacts with the unitary world of the real, which is best described as that which offers resistance to symbolization.

28. Surrealism strikes a chord with many IT themes, such as the use of montage, the juxtaposition of objects and ideas from diverse sources, and the creation of new contexts. There are allusions in both the cyberspace literature and surrealist

art to dreams and madness. Both cyberspace discourse and surrealism seem to provoke conflicts between the real and the imaginary, they celebrate the labyrinthine, and they valorize the decontextualised object as a provocation. Surrealists characterized the surreal as the realm of a fragmented whole that subsumes the real. Surrealism is caught up in various conversations that involve phenomenology, structuralism, critical theory and psychoanalysis, (particularly Freud's contribution to dream theory and more lately the Oedipus myth).

29. This conflict has to be resolved in some way, generally by the repression of such desires, which may resurface in later life as various neuroses.

30. We have come full circle. The myth of unity persists, but the real is neither beyond in the realm of Plato's real nor is it the material realm of the empiricist's reality. The unity is the real, which resists symbols. The fragmented world is the world of symbolization, of language. Lacan's version is also of the myth of the process of interpretation, the conflict between the whole and the parts. We cannot understand the whole of a text unless we understand the parts, and the parts only make sense in the context of the whole. The unity and the fragmented are locked in the circle of understanding. But this cycle between part and whole is the hermeneutics of engagement, when everything is going well. It suggests an engaged, pre-Oedipal state. Contrasted with this is the fragmented state of language as correspondence, the empirical or ontic order. This is the radical edge of Lacan's writing. He points to the antagonism between the ontological and the ontic. The antagonism between hermeneutical accounts of understanding and rationalist/empiricist accounts also retell the unity myth.

31. R. Coyne, *Designing Information Technology in the Postmodern Age: From Method to Metaphor.*

Chapter 1

1. See M. Punt, "Accidental machines: The impact of popular participation in computer technology," for further critique of the teleological aspects of digital narrative.

2. M. D. Lemonick, "Future tech is now," 65.

3. H. Rheingold, *The Virtual Community: Homesteading on the Electronic Frontier,* 17.

4. W. J. Mitchell, *City of Bits: Space, Place, and the Infobahn,* 20.

5. The propensity to shift attention to the future when describing information technology applies to technical descriptions but also to social consequences, particularly the social consequences of cyberspace. According to Tomas, cyberspace is "a

powerful, collective, mnemonic technology that promises to have an important, if not revolutionary, impact on the future compositions of human identities and cultures" (D. Tomas, "Old rituals for new space," 31–32). According to Rheingold, "Most people who have not yet used these new media remain unaware of how profoundly the social, political, and scientific experiments under way today via computer networks could change all our lives in the near future" (Rheingold, *The Virtual Community,* 17).

We would expect commentary on a rapidly changing technology to focus on prediction. Mitchell offers a tacit (and perhaps tongue-in-cheek) prediction of a feature of the interactive television of the future: "If you receive three-dimensional models of a sporting event rather than a stream of two-dimensional video images, you could take some control of directorial functions by selecting viewpoints and operating a virtual camera" (Mitchell, *City of Bits*, 63). But in some commentary, it is difficult to distinguish description from prediction. For example, there are computer programs designed to perform specific tasks, independently of and invisible to a computer operator. These programs are often termed "agents" or "demons." Researchers are looking at creating agents that process electronic network messages. According to Mitchell: "My software surrogates [agents] can potentially do much more than provide origins and destinations for messages; when appropriately programmed, they can serve as my semiautonomous agents by tirelessly performing standard tasks that I have delegated to them and even by making simple decisions on my behalf" (Mitchell, *City of Bits*, 13). That Mitchell is discussing a *potential* and not an actuality readily passes unnoticed.

6. M. Sullivan-Trainor, *Detour: The Truth About the Information Superhighway,* xi.

7. M. Heim, *The Metaphysics of Virtual Reality,* 89. Commentators commonly assume that the future is already upon us, suggesting that we now have to work out the philosophical and social implications. They present questions and suggest philosophies for situations that do not yet exist, though they clearly believe they will exist one day. In his article "The erotic ontology of cyberspace," Heim lists some of these philosophical problems, beginning with: "When designing virtual worlds, we face a series of reality questions. How, for instance, should users appear to themselves in a virtual world? . . ." (Heim, *The Metaphysics of Virtual Reality,* 83). Similarly, in his consideration of intelligent computer agents, Mitchell poses the questions: "How do you know who or what stands behind the aliases and masks that present themselves? Can you always tell whether you are dealing directly with real human beings or with their cleverly programmed agents?" (Mitchell, *City of Bits*, 14).

8. Ambiguity as to present and future is commonplace in artificial intelligence (AI) literature. For example, *The Handbook of Artificial Intelligence* introduces AI by stating: "There is every indication that useful AI programs will play an important

part in the evolving role of computers in our lives" (A. Barr and E. A. Feigenbaum, eds., *The Handbook of Artificial Intelligence: Volume I*, 3). In defining the AI field, Schank turns to its goals: "First and foremost the goal is to build an intelligent machine" (R. Schank, "What is AI anyway?" 4). Artificial intelligence is concerned with machines that do not yet exist, but their potential is there in the computer and the theories and methods that surround its development. Artificial intelligence is not about the study of the intelligence of existing machines but about machines that do not yet exist, with the expectation that the field will produce such machines. AI is unusual in that most fields are not so defined. For example, the more prosaic field of computer-aided design is about an existing class of computer systems that assist designers in some way. The field endeavors perhaps to make systems that do this better.

AI researchers are reluctant to describe their systems as intelligent, and it is common lore that systems that may have been described as intelligent before they were realized are now regarded as standard. So, according to this lore, intelligence is chimerical. AI is always chasing the future. In our own early work, I and colleagues also described knowledge-based systems with reference to goals: "Knowledge-based systems are computer systems in which an attempt is made to capture and render operable human knowledge about some domain. The goal is to represent knowledge in such a way that it is comprehensible to both human being and machine" (R. Coyne et al., *Knowledge-Based Design Systems,* 35). The goal was not to study computer machines that accomplish comprehensible and machine-readable knowledge representation, or even to produce systems that do it better, but to produce systems that do it—a future promise, and one that has proved the most evasive of all.

9. J. W. Carey, *Communication as Culture: Essays on Media and Society.*

10. T. More, *Utopia.*

11. See R. Levitas, *The Concept of Utopia*, 8.

12. W. Morris, "The society of the future."

13. E. Bellamy, *Looking Backward 2000–1887.*

14. M. G. Plattel, *Utopian and Critical Thinking.*

15. G. Orwell, *1984.*

16. Gibson, *Neuromancer.*

17. Plattel, *Utopian and Critical Thinking.*

18. More, *Utopia*, 84.

19. For example, see R. Owen's utopian vision in his *Report to the County of Lanark and A New View of Society,* published in 1813–21. According to Owen, the errors of society "must be overcome solely by the force of reason" (196), and "were all men trained to be rational, the art of war would be rendered useless" (156).

20. J. R. R. Tolkien, *Lord of the Rings.*

21. According to Plattel, the Enlightenment utopia also differs from the modern phenomenon of the social myth, such as that paraded by fascism and fundamentalism. Fundamentalism is the translation of myth and tradition into contemporary pluralistic society in such a way that options beyond those myths and traditions are censored or otherwise excluded. (See A. Giddens, *Beyond Left and Right: The Future of Radical Politics.*) Social myths are misplaced myths that belong to different historical times. They are often engineered to promote nationalistic fervor in a climate of insecurity. Such anachronisms can only prosper in a regime of strict authoritarian control. (Note that Plattel portrays the social myth differently as a regression, absolutizing, and objectifying. See Plattel, *Utopian and Critical Thinking*, 56. We address this aspect of myth further in chapter 4 in relation to Barthes' critical theory of language. See R. Barthes, *Mythologies.*) For Plattel, the social myths of Nazism (and we may add neo-Nazism) "demand a blind belief in, and a total surrender to their prophetic expectations of salvation. Society thus lives under the tyranny of its own portrayal of the future" (Plattel, *Utopian and Critical Thinking*, 56).

22. Bellamy, *Looking Backward 2000–1887,* 93.

23. W. Morris, *News From Nowhere*, 53.

24. H. G. Wells, *World Brain: H. G. Wells on the Future of World Education*, 106.

25. Ibid., 108.

26. Benedikt, *Cyberspace: First Steps,* 121.

27. S. R. Hiltz and M. Turoff, *The Network Nation: Human Communication via Computer*, 27.

28. Rheingold, *The Virtual Community,* 26.

29. Mitchell, *City of Bits*, 12.

30. Ibid., 41. See N. K. Hayle, *How We Became Posthuman: Virtual Bodies in Cybernetics, Literature, and Informatics* for a recent critique of the trend toward disembodied reason.

31. B. Schneiderman, "Education by engagement and construction: A strategic education initiative for a multimedia renewal of American education," 13.

32. Rheingold, *The Virtual Community,* 275.

33. A. Tofler, *The Third Wave,* 24.

34. McLuhan, *The Gutenberg Galaxy: The Making of Typographic Man*. This is also the gist of S. Plant's polemic in *Zeros and Ones: Digital Women and the New Technoculture,* which seeks to reclaim the craft origins of the computer, particularly in the history of weaving, and celebrate the role of women in this.

35. Benedikt, *Cyberspace: First Steps,* 121.

36. Ibid., 121–122. Even the president of Microsoft celebrates this return to nature with the "Gateses' future home," set in a forest, by a trout-laden lake, with otters, and space for bicycles, under the care of obsequious robot servants. See B. Gates, *The Road Ahead*, 236–258. We discuss this capability of digital technology to generate new attitudes to existing technologies under the rubric of "disclosure" in chapters 5 and 8.

37. Timothy Leary, until his recent death, was one vocal survivor. See T. Leary, "The cyberpunk: The individual as reality pilot."

38. Rheingold, *The Virtual Community,* 48.

39. See Punt, "Accidental machines," 61.

40. Ibid., 73.

41. See W. Straw, "Characterizing rock music culture: The case of heavy metal," 371.

42. See W. Sacks, "Artificial human nature," 59–60, on the theme of the "romantic reaction" in AI.

43. L. McCaffery, "An interview with William Gibson," 269.

44. According to Gibson: "It wasn't until I could finally afford a computer of my own that I found out there's a drive mechanism inside—this little thing that spins around. I'd been expecting an exotic crystalline thing, a cyber-space deck or something, and what I got was a little piece of Victorian engine that made noises like a scratchy old record player. That noise took away some of the mystique for me; it made computers less *sexy*. My ignorance had allowed me to romanticize them" (Ibid., 270).

45. See R. Bernstein, *Beyond Objectivism and Relativism*; and R. Rorty, *Philosophy and the Mirror of Nature.*

46. A. Cunningham and N. Jardine, eds., *Romanticism and the Sciences.*

47. See K. Milton, *Environmentalism and Cultural Theory: Exploring the role of anthropology in Environmental Discourse*; and D. Pepper, *The Roots of Modern Environmentalism*, on the romantic legacy of environmentalism.

48. But there is a critical edge to contemporary countercultural movements that indicates that they are derivative from rather than at the center of romantic culture.

49. Morris, *News From Nowhere*, 87.

50. Compare Max Weber's approval of charismatic leadership. See M. Weber, *The Protestant Ethic and the Spirit of Capitalism*, 178.

51. ". . . the cowardice of the last century had given place to the eager, restless heroism of a declared revolutionary period" (Morris, *News From Nowhere,* 110).

52. H. S. Commager, *The Empire of Reason: How Europe Imagined and America Realized the Enlightenment.*

53. U. Eco, *Travels in Hyperreality*. McLuhan writes of the amenability of different media to different personality traits, particularly in relation to the "cool" medium of television.

54. Leary, "The cyberpunk: The individual as reality pilot," 253. See also K. Hafner and J. Markoff, *Cyberpunk: Outlaws and Hackers on the Computer Frontier.*

55. Ibid.

56. See R. Schmitt, "Mythology and technology: The novels of William Gibson."

57. Levitas explores the relationship between early French and British communism, including the ideas of Henri de Saint-Simon, Charles Fourier, and Robert Owen. In particular, the French strand assumed the form of esoteric sects and a concern with the cosmological order of the universe. See R. Levitas, *The Concept of Utopia*, 38–39; H. de Saint-Simon, *Social Organization, The Science of Man and Other Writings*; C. Fourier, *The Theory of the Four Movements*; and R. Owen, *Report to the County of Lanark and A New View of Society.*

58. J. J. Rousseau, *Emile.*

59. *Emile* is pervaded with notions of genius, the adulation of nature, and an appeal to the natural spirit. Rousseau's theme of "back to nature" has influenced the liberal strand to education. The return to nature also pervades Morris's thinking. In his utopia, children rear themselves in a context of perpetual play, discovery, and good example. See also J. J. Rousseau, *The Social Contract.* Rousseau's predecessors thought it legitimate to allow the population to pass its sovereignty over to the few who rule, through various democratic processes. But Rousseau thought that

sovereignty should always stay with the people, a view that later became the cornerstone of Marx's thinking. In place of the ruler in whom the people invest their trust, Rousseau also prepared the way for the concept of the revolutionary hero. Rulers are not needed, but lawgivers are. We are truly free if we are able to give ourselves over to the person who has our interests at heart. This lawgiver is analogous to the tutor in education. It is someone who can save us from ourselves and our excesses, the archetypal revolutionary leader.

60. See R. Owen, *Report to the County of Lanark and A New View of Society.*

61. K. Marx and F. Engels, "Manifesto of the Communist Party," 69.

62. Note that Marx and Engels distanced themselves from the utopian strand of socialism. They posited a scientific socialism rather than a utopian one, though many commentators regard this as a moot distinction.

63. Information is one of the few commodities that still resides with its originator when it is given away, though its value may be diminished.

64. K. Popper, *The Poverty of Historicism.*

65. For example, Dennett develops an imaginative scenario of a fully operative brain placed in a vat. The brain is incrementally replaced by artificial electronic components to prove a materialist thesis. See D. C. Dennett, "Where am I?"

66. M. Minsky, *The Society of Mind*, 323.

67. D. H. Meadows, H. Donella, N. L. Meadows, J. Randers, and W. W. Behrens, *The Limits of Growth, a Report for the Club of Rome's Project on the Predicament of Mankind.*

68. E. A. Feigenbaum and P. McCorduck, *Fifth Generation: Artificial Intelligence and Japan's Computer Challenge to the World.*

69. The result is an ambiguous discipline involving a dialectic between mathematical and logical rigor on the one hand and a universalizing rhetoric on the other. The rhetoric does not come under the scrutiny of the mathematical rigor. The field is in danger of becoming a mixture of hard-nosed rationalism and covert romanticism.

70. S. Inayatulla, "Deconstructing and reconstructing the future: Predictive, cultural and critical epistemologies."

71. As exemplified in Jencks's recent flirtations with chaos theory. See C. Jencks, *The Architecture of the Jumping Universe: A Polemic: How Complexity Is Changing Architecture and Culture.*

72. G. Deleuze, *The Deleuze Reader.*

73. R. Bonnel, "Medieval nostalgia in France, 1750–1789: The Gothic imaginary at the end of the old regime."

74. Ibid., 145.

75. M. Baer, "The memory of the Middle Ages: From history of culture to cultural history," 291.

76. In his essay, "The return of the Middle Ages," Eco draws attention to the unflagging interest in medieval themes in literature, film, art, and popular culture. See Eco, *Travels in Hyperreality*. He identifies many conceptions of the medieval period, ranging from the superficial backdrop to a drama or opera, to the portrayal of heroic knights in armor, to the work of the serious contemporary medieval scholar. See also P. Domenico, *Eco on Medievalism*.

77. T. Malory, *Le Morte d'Arthur*.

78. P. E. Tucker, "Chivalry in the Morte," 67.

79. The linking of chivalry with individualism was parodied by Louis de Bonald (1754–1840), an early critic of the Enlightenment: "In the days of old, in the century of strength, a gallant knight, mounted on a palfrey, helmet on his head and lance in his hand, persuaded himself in his chivalric dreams that a beautiful princess locked in a tower under guard by a wizard, was going to offer him her hand and lands for freeing her from captivity. Today, in the century of enlightenment, the young literary man, still coated with the chalk dust of school, pen in his hand and the *Contrat social* in his head, imagines in philosophical reveries that a people groaning under despotism will in its primary assemblies confer upon him at least legislative power if he can with speeches and writings break their chains. We have in each case the same passions" (quoted in W. J. Reedy, "Ideology and utopia in the medievalism of Louis de Bonald," 165). But Bonald was contemptuous of the Enlightenment vision. "But the knight was a generous and brave visionary. The literary man is a threatening lunatic" (165–166). So the myth of chivalry appealed both to the supporters of the Enlightenment and those who sought to break away from it and return to a more noble model of human conduct.

80. Jung's influential account of mythic archetypes (Mother, Old Man, Trickster) that recur in fantasy literature does not place such store on the transformations that have occurred to these myths in the modern period. See C. G. Jung, *Four Archetypes: Mother, Rebirth, Spirit, Trickster*.

81. The world of feudal authority is a world in which everyone has their allotted place. It is an unequal world of unquestioned authority, undeserved privilege and rank. The hero comes in to challenge that authority and to claim a position within the hierarchy. Often that privilege is won through magic, but it is a reasonable

kind of magic, perhaps abetted by a mentor such as a Merlin, a Gandalf, or a good witch who is still caught up in the old unpredictable superstitious world. The hero can benefit from their powers without fully being a part of them and can carry on the good work of the mentor, though without the need of magic.

82. This medievalism also shifts into surrealism, to be explored in chapter 6. See R. Schmitt, "Mythology and technology: The novels of William Gibson," on the theme of magic and unreason in cyberpunk culture.

83. M. Bakhtin, *Rabelais and His World*, 19. Eliade also discusses the role of orgiastic ritual and carnival as a means of recreating the cosmic unity in traditional cultures. See M. Eliade, *The Two and the One*, 113–114.

84. Bakhtin, *Rabelais and His World*, 21.

85. Stallybrass and White develop the themes of fragmentation, marginalization, sublimation, and repression in relation to the carnivalesque. See P. Stallybrass and A. White, "Bourgeois hysteria and the carnivalesque," 290.

86. See Schmitt, "Mythology and technology," on the theme of primitivism and unreason in punk culture.

87. See J. A. Dator, "The futures of culture or cultures of the future."

88. For example, in describing a program for animating flocks of birds, Turkle sees it as important to recount the significance of the name of the program: "Reynolds called the digital birds 'boids,' an extension of high-tech jargon that refers to generalized objects by adding the suffix 'oid.' A boid could be any flocking creature" (S. Turkle, *Life on the Screen: Identity in the Age of the Internet*, 163). Often the word is the one constant factor, as the program undergoes constant revisions and metamorphoses. Sometimes the word outlives the program, ensuring that an idea lives on. A sense of the power of words, not to mention a fetish for words, imbues computer narratives.

89. This theme is developed by C. Chesher, "Computers and the power of invocation." See also J. Dibbell, "A rape in cyberspace," 312.

90. See C. Hale, *Wired Style: Principles of English Usage in the Digital Age*. There is a nonserious punning element to the invention and use of such words. The verbal alchemy invokes a mock medievalism.

91. See S. D. O'Leary, "Cyberspace as sacred space: Communicating religion on computer networks."

92. M. Foucault, *The Order of Things: An Archaeology of the Human Sciences*, 38.

93. See C. Gere, *The Computer as an Irrational Cabinet*, for an elaboration of the themes of the art of memory, museums, and multimedia.

94. The art of memory includes the use of places and objects that are well known and accessible as a means of recalling the key points of a public address, reciting a long poem, or recalling a long list. Such techniques include the use of schoolroom mnemonic chants to recall the order of the planets from the sun, the periodic table, or rules of spelling. A common mnemonic technique was to picture the thing to be remembered in a familiar place, which would no doubt conjure up absurd images, but the art of memory seemed to trade in the absurd for its power. See F. A. Yates, *The Art of Memory*, 108–112.

95. Ibid., 128.

96. Ibid., 154.

97. Ibid., 155.

98. E. Hooper-Greenhill, *Museums and the Shaping of Knowledge*, 104.

99. M. Heim, *Virtual Realism*, 67.

100. M. C. Boyer, *CyberCities: Visual Perception in the Age of Electronic Communication*, 31.

101. Ibid., 243.

102. As further support for dystopian narratives, at the time of writing the supposed millennium bug has disclosed the fact that the interconnection of computer systems is "out of control." No one knows what will happen when old hardware and software designed with the two-digit year field in its clock time function reaches the end of 1999 and what systems will be affected, from video recorders to air navigation.

103. P. Virilio, *Open Sky*, 19.

104. Ibid., 21.

105. Ibid., 61.

106. See J.-F. Lyotard, *The Postmodern Condition: A Report on Knowledge;* and J. Habermas, *The Philosophical Discourse of Modernity: Twelve Lectures.*

107. J. Stallabrass, "Empowering technology: the exploration of cyberspace," 10. See also P. Virilio, "Red alert in cyberspace."

108. See M. Castells, *The Informational City: Information Technology, Economic Restructuring, and the Urban-Regional Process*; C. Dunlop and R. Kling, *Computeriza-*

tion and Controversy: Value Conflicts and Social Choices; T. Forester, *Computers in the Human Context: Information, Technology, Productivity and People*; C. C. Gotlieb and A. Borodin, *Social Issues in Computing*; R. J. Long, *New Office Automation Technology: Human and Managerial Implications*; H. J. Otway and M. Peltu, *New Office Technology: Human and Organisational Aspects*; J. Wainwright and A. Francis, *Office Automation, Organisation and the Nature of Work*; Punt, "Accidental machines."

109. Giddens, *Beyond Left and Right.*

110. There is so-called free ware, which people produce and give away. This is often produced by small operators and hobbyists who have to make little initial financial outlay. Sometimes they rely on the good will of the few to pay for the service; there is no liability for the product, and it can be tested in use with feedback from users. The software is updated every year or so. As confidence in the product increases, the inventor of the software adds more and more features and turns it into a commercial product. The payoff for giving away the software initially is that now there is a ready market with a user group looking forward to the latest version of the software with the latest features. (Perennial changes in hardware and operating systems ensure that software updates are required to take advantage of the new systems.) The risk for the developer is that the product is on full view during its development and others are able to produce something better, though the competition is likely to come from other hobbyists rather than commercial developers. The commercial product may be undercut by the free ware. At the time of writing, modes of software development are emerging that rely on virtual machine architectures. Software can be modularized and the same software will run on different computer platforms.

111. According to one critique, the rationalist imperative is nothing more dramatic than the latest realization of the human propensity for record keeping and control. Ancient civilizations kept records of transactions as marks on parchment or stone. We do the same with ever more sophisticated equipment. According to Ong, just prior to Descartes double-entry bookkeeping extended to understandings of knowledge. The Encyclopedists took this to greater sophistication. The current IT age is a realization of the Encyclopedic tendency to capture all knowledge, to categorize and sort it. See W. Ong, *Ramus: Method, and the Decay of Dialogue from the Art of Discourse to the Art of Reason*; and Foucault, *The Order of Things*.

Chapter 2

1. J. Stallabrass, "Empowering technology: The exploration of cyberspace," 21.

2. W. J. Mitchell, *City of Bits: Space, Place and the Infobahn,* 8.

3. Walser, quoted in H. Rheingold, *Virtual Reality,* 191.

4. D. Rushkoff, *Cyberia: Life in the Trenches of Hyperspace*, 15.

5. Ibid., 19.

6. M. Heim, *The Metaphysics of Virtual Reality*, 83.

7. Lacan has developed the Platonic concept of the real as the realm of being beyond appearances, that which is permanent and beyond the contingency of the symbolic and the imaginary order. For Lacan, the real is also elusive and mysterious, the site of radical indeterminacy. In contrast, the concept of "reality" suggests subjective representations. See D. Evans, *An Introductory Dictionary of Lacanian Psychoanalysis*, 159–161. We examine Lacan's analysis of the real in more detail in chapter 7.

8. L. Carroll, *Alice's Adventures in Wonderland and Through the Looking Glass*, 294.

9. See Maffesoli for an examination of the priority of proximity in social theory: "Things, people and representations relate through a mechanism of proximity. Thus, it is by successive associations that what we call the social given is created" (M. Maffesoli, *The Time of the Tribes: The Decline of Individualism in Mass Society*, 147).

10. In a letter to the editor by Liam Murphy in *The Times Higher Education Supplement,* 10 October, 1997, 17.

11. These themes have received explicit attention from contemporary thinkers from Nietzsche onward. See, for example, Nietzsche, on the theme of the body: "I am body entirely, and nothing beside" (F. Nietzsche, *Thus Spoke Zarathustra*, 61) and on repetition: "the complex of causes in which I am entangled will recur—it will create me again" (237). The theme of repetition is explicated by G. Deleuze in *Difference and Repetition* and in numerous other places to be examined in the rest of this book. For Lacan, the real becomes the site of radical indeterminacy, resistance and conflict with the symbolic order, as will be explained in chapter 7.

12. As examined by Snodgrass, under the Platonic regime, one of the ways we are to gain access to the Intelligible is through the workings of symbol, particularly through art, building, and ritual. Symbols "satisfy both a physical and a metaphysical indigence; they act as supports for the contemplation of supra-empirical principles, leading to a state of intellectual identity with the real" (A. Snodgrass, *Architecture, Time and Eternity,* 25).

13. See Plato, *The Republic of Plato*, 227–235.

14. According to some commentators, there may be a hint of irony in Plato's characterization that the intelligible realm is somehow "more real" than the earthly. Platonic idealism is a reaction against the prosaic, embodied realism of Greek

mythology. According to Arendt, Plato effected a radical reversal from the Homeric narrative of a ghostly afterlife in favor of one in which "ordinary life on earth . . . is located in a 'cave,' in an underworld; the soul is not the shadow of the body, but the body the shadow of the soul" (H. Arendt, *The Human Condition,* 292).

15. See R. E. Allen, Greek *Philosophy: Thales to Aristotle,* 21.

16. Concepts such as time are merely an image, or a derivative, of an *eternal* unity. According to Plato, when the Living Being "ordered the heavens he made in that which we call time an eternal moving image of the eternity which remains forever at one" (Plato, *Timaeus and Critias*, 51). See Snodgrass, *Architecture, Time and Eternity,* 72–73, for a summary of issues of time and eternity.

17. Heraclitus, "Fr. 10," in Allen, *Greek Philosophy*, 41.

18. Aristotle, *Metaphysics*, in Allen, *Greek Philosophy*, 330. Skepticism toward myths of the paradox of unity continues to this day. Stokes sees the issue as a lack of subtlety in the Presocratics' use of language. When Heraclitus says that day and night are "one," we are entitled to ask "one what?" A similar response is appropriate to the more abstract "unity." See M. Stokes, *One and Many in Presocratic Philosophy*, 10.

19. Eliade, *The Two and the One*, 82.

20. Ibid., 82.

21. G. G. Scholem, *Major Trends in Jewish Mysticism*, 225.

22. M. Eckhart, *Meister Eckhart: Selected Treatises and Sermons*, 182.

23. Eliade, *The Two and the One*, 114.

24. Ibid. The creation of the world came about as a breach in the primal unity, as in the separation of light from darkness. According to Indian mythology, in our natural state we are made up of pairs of opposites, but in order to attain metaphysical knowledge we must transcend these opposites through ritual sacrifice and other rites of integration (Eliade, *The Two and the One*, 95). Eliade recounts many instances of the unity theme in various myths, as does Jung in *Four Archetypes: Mother, Rebirth, Spirit, Trickster.*

25. Hegel also devotes a number of pages to "the unity of the One and the Many." See G. W. F. Hegel, *Hegel's Science of Logic*, 172.

26. M. Esslin, *The Theatre of the Absurd*, 245.

27. Ibid., 248.

28. Ibid.

29. Carroll, *Alice's Adventures in Wonderland and Through the Looking Glass*, 230–232.

30. Plato, *Phaedrus*, in Allen, *Greek Philosophy*, 247.

31. Ibid.

32. Ibid.

33. J. Marías, *History of Philosophy,* 99. See also Scholem's account of how, according to early Kabbalistic writing, the soul of the devotee must perform magical incantations of increasing complexity to pass through the closed entrance gates of the seven heavens (G. G. Scholem, *Major Trends in Jewish Mysticism*, 50).

34. Plotinus, *The Essence of Plotinus: Extracts from the Six Enneads and Porphyry's Life of Plotinus,* 208.

35. Ibid., 23.

36. Ibid., 152.

37. Ibid., 1.

38. Ibid., 36.

39. Ibid., 220.

40. Ibid., 23.

41. Ibid., 171.

42. Ibid., 80.

43. Ibid., 62.

44. Ibid., 56. Augustine affirms that it is not any body that weighs the soul down, but the corruptible body: "only bodies which are corruptible, burdensome, oppressive, and in a dying state," the body beset with sin (Augustine, *City of God*, 528). The soul's quest for union takes on an Aristotelian cast for Aquinas. He contests Augustine's assertion that the state of bliss can only be attained in God, maintaining that it can be acquired by contemplation but also "practical understanding" (Aquinas, *Selected Philosophical Writings*, 324). This grounded, pragmatic approach to bliss, in which religious sacraments are tools of God, reputedly later informed McLuhan's pragmatic philosophy of the media.

45. See A. O. Lovejoy, *The Great Chain of Being: A Study of the History of an Idea*, for an account of the continuity of the Neoplatonic tradition, particularly the hierarchical ordering of beings beneath and above humankind.

46. Dante, *The Divine Comedy, Vol. II: Purgatory*, 50. Dante was influenced by Neoplatonism through Augustine and read Plato's *Timaeus*.

47. G. Camillo, quoted in F. Yates, *The Art of Memory*, 138.

48. J. Milton, *Paradise Lost*, 55. See also editor's notes on "the scale of nature," 393. Milton did not acknowledge the existence of the soul independently of the body. What made humans different than animals for Milton was that humans can stand, and so look to heaven, and also that they can smile.

49. Ibid., 116.

50. The use of cyberspace for religious expression raises other issues. See S. D. O'Leary, "Cyberspace as sacred space." O'Leary relates the concept of religious invocation as practiced in pre-Reformation liturgy to the practice of using words in on-line chat groups to invoke "virtual objects."

51. According to Wallis, Augustine's concept of consciousness particularly influenced Descartes. See R. T. Wallis, *Neoplatonism*, 172.

52. Prior to Descartes, things (res) had a being of their own, and the ego was merely one thing among others. See Marías, *History of Philosophy,* 222.

53. See Snodgrass, *Architecture, Time and Eternity.*

54. Wallis, *Neoplatonism*. 173.

55. Descartes suggests that truth is less a matter of reality than of reliability. See Arendt, The *Human Condition,* 279.

56. The being of things is based on the ego. According to Marías, modern idealism is the belief that the ego has no sure knowledge of anything other than itself, that I know about things only while I see, touch, think about, and desire them. I know the things only while I have dealings with them and while they are within my gaze, through whatever mediation. There is no way of knowing what things are like when they are apart from me, since I cannot know anything about them without being present with them. This egocentric, idealist position is that things appear as existing or being for me. They are provisionally ideas of mine, and the reality that corresponds to them is an ideal reality. See Marías, *History of Philosophy,* 222–223.

57. See R. Bernstein, *Beyond Objectivism and Relativism.*

58. Wallis, *Neoplatonism*, 173. Note that Locke was also influenced by Plotinus, particularly in his account of hierarchies of being. See J. Locke, *An Essay Concerning Human Understanding*, 288–289.

59. G. Berkeley, "Siris: A chain of philosophical reflexions and inquiries," 476.

60. Ibid., 481–482.

61. G. Berkeley, *The Principles of Human Knowledge*, 81.

62. Ibid., 74.

63. Ibid., 74.

64. For an explanation of Locke on perception, see J. Hospers, *An Introduction to Philosophical Analysis,* 507.

65. Locke, *An Essay Concerning Human Understanding*, 90.

66. Berkeley, *The Principles of Human Knowledge*, 73. Antagonists of Berkeleyan idealism have attempted to counter this view by declaring that there are certain uniformities in our experience that can only be explained if we posit an entity independent of ourselves. According Hospers, Berkeley suggests that the uniformities we observe, as when we declare the objective existence of a tree, are in fact "names for *recurring patterns*, or complexes, of sense-experiences, and nothing else" (Hospers, *An Introduction to Philosophical Analysis*, 508). Hospers explains Berkeley's position. As we look around the world we see things from different angles and under different conditions. They are always changing. How do we establish that the object that appears as a trapezoid one moment and a diamond the next is in fact the same tabletop? Changes are gradual, and whereas the first view may appear very different to the view a few seconds later, a continuity of appearance exists. There is a family of shapes that constitutes the experience of the tabletop. Moving around the lamp stand involves a different sequence of shapes, and so a different family of patterns. Physical objects are nothing more than families of sense experiences (Hospers, *An Introduction to Philosophical Analysis,* 509). Where we have hallucinations, then there is no such continuity. To test whether something is a hallucination, we relate our sense experiences to one another: "We do not do what Locke's view would seem to require: relate our sense-experiences to a reality outside our sense-experience to see whether they correspond" (Hospers, *An Introduction to Philosophical Analysis,* 509).

67. Berkeley, "Siris," 468.

68. Berkeley, *The Principles of Human Knowledge*, 86.

69. Berkeley, "Three dialogues between Hylas and Philonous," in *The Principles of Human Knowledge*, 220.

70. Kant sided with the rationalists yet in doing so attempted to reconcile rationalism with empiricism. Kant was intent on reconciling the real and the ideal, but the idealist tendency is there nonetheless. Kant regarded himself as an empirical realist and a transcendental idealist. See A. R. Lacey, *A Dictionary of Philosophy*, 86.

71. "Doubtless, indeed, there are intelligible entities corresponding to the sensible entities; there may also be intelligible entities to which our sensible faculty of intuition has no relation whatsoever; but our concepts of understanding, being mere forms of thought for our sensible intuition, could not in the least apply to them" (I. Kant, *Immanuel Kant's Critique of Pure Reason,* 270).

72. The thing behind the appearance is a *noumenon*, "an object independent of sensibility" (Ibid.).

73. Heim, *The Metaphysics of Virtual Reality*. 130.

74. According to Heim: "After Kant, philosophers whittled away at the monistic unity until quantum theory in the twentieth century withdrew support for the kind of coherence Kant thought essential to science" (Heim, *The Metaphysics of Virtual Reality*, 130).

75. See Marías, *History of Philosophy,* 299.

76. See R. Scarce, *Eco-Warriors: Understanding the Radical Environmental Movement*, for evidence of the unity theme in modern environmentalism, where there is an appreciation of the "endless connections between everything" and an appropriation of the "experience of unity with all else" (35). Aspects of environmentalism also take on board chaos and complexity theory. See P. Madge, "Ecological design: A new critique," 50.

77. See A. N. Whitehead, *Science and the Modern World*, 118.

78. A. Cunningham and N. Jardine, *Romanticism and the Sciences*, 8.

79. See I. Kant, *Observations on the Feeling of the Beautiful and Sublime*; and E. Burke, *A Philosophical Enquiry into the Origin of Our Ideas of the Sublime and Beautiful.* See also I. B. Whyte, "The expressionist sublime," for an account of these aspects of romanticism in art and architecture, particularly in twentieth-century German expressionism.

80. J. G. Voller, "Cyberspace and the sublime," 18.

81. Plotinus, *The Essence of Plotinus*, 49.

82. Ibid.

83. J. G. Fichte, "Theory of science," quoted in L. R. Furst, *Romanticism in Perspective: A Comparative Study of Aspects of the Romantic Movements in England, France and Germany,* 58.

84. See G. H. Turnbull, "Appendix."

85. J. C. F. von Schiller, "Philosophical Letters," quoted in Turnbull, "Appendix," 257.

86. S. Coleridge, "Religious Musings," quoted in Turnbull, "Appendix," 263.

87. R. W. Emerson, "The Over-Soul," quoted in Turnbull, "Appendix," 265.

88. Arthur Symons, *Poems I*, 204, quoted in L. Hönnighausen, *The Symbolist Tradition in English Literature: A Study of Pre-Raphaelitism and* Fin de Siécle, 216.

89. Hönnighausen, *The Symbolist Tradition in English Literature*, 209.

90. See also A. Crawford, "Ideas and objects: The Arts and Crafts Movement in Britain," for an account of the "Unity of Art" theme within the arts and crafts movement.

91. Introduction to C. Fourier, *The Theory of the Four Movements*, xxvi. For Saint-Simon, "The idea of God lacks unity" (H. de Saint-Simon, *Social Organization*, 20).

92. Fourier, *The Theory of the Four Movements*, 16.

93. Ibid.

94. Ibid., 84.

95. R. Steiner, *Theosophy: An Introduction to the Spiritual Processes in Human Life and in the Cosmos*, 120.

96. Ibid., 121.

97. C. Jencks, *The Architecture of the Jumping Universe: A Polemic: How Complexity Science Is Changing Architecture and Culture*, 9. Jencks rediscovers romanticism, including Coleridge's appeal to soul, imagination, and unity (39–40), as well as teleology (and Gaia), and the metaphysics of aesthetics: "Love, beauty and aesthetics—they pre-date us and exist independently of us" (106), ideas that are apparently radical because they offend the Modernists. Thanks to the reflections of cosmic scientists, we now "know truths that have been revealed to no other generation, and they can give us great hope and strength" (125). Jencks's break with progressive philosophy is complete as he declares that "the new cosmic metanarrative could provide a spiritual and cultural grounding . . . this idea is bound to be resisted. . . . Many will prefer a fragmented 'war of language games,' as do most Deconstructionists, because they feel a unified culture is inherently totalitarian for individuals. Nevertheless, I believe there are compelling arguments for regarding the new emergent story of cosmogenesis as a common narrative with spiritual implications" (128).

98. Maffesoli, *The Time of the Tribes*, 35.

99. Snodgrass, *Architecture, Time and Eternity*.

100. Cunningham and Jardine, *Romanticism and the Sciences*, 8.

101. Voller, "Cyberspace and the sublime," 27.

102. Wallis, *Neoplatonism*, 178.

103. McLuhan, *Understanding Media*, 61. Kroker traces the development of McLuhan's ideas through Catholicism and his indebtedness to the methods of inquiry of Thomas Aquinas (A. Kroker, "Processed world: Technology and culture in the thought of Marshall McLuhan"). According to Wasson:

> As some medieval scholars read the book of nature as symbols revealing God's grace and law, so McLuhan reads the new technological environment as a book of symbols which reveals the Incarnation. Because everything in the world is a symbol, McLuhan can offer endless symbolic interpretation. ("Marshall McLuhan and the politics of modernism," 577)

The theme of information technology's implication in a Reality: divine teleology is taken up by M. Heim, *Virtual Reality: Theory, Practice and Promise*: "Today we are influenced by the technology of the future. We are drawn forward by it as our internal telos or underlying goal. We are evolving, and our evolution inscribes in us a technological destiny, an intimate relationship to information systems" (190).

104. M. Benedikt, "Cyberspace: Some proposals," 131.

105. Which is not to say Berkeley would have endorsed an idealism grounded in electronic technology.

106. The idealist position is reminiscent of those advanced in relation to certain movements in art. At the beginning of the twentieth century, cubism revealed space as multifaceted, heterogeneous, and changing (S. Kern, *The Culture of Time and Space: 1880–1918,* 152–158). Virtual reality allows us to be everywhere and nowhere, subject to different laws. But there is a major difference between the claims made of cubism and those made of virtual reality. In the case of virtual reality, it is not just that something new about reality is revealed, but it is claimed that we participate radically in the medium that provides the new conceptions. That is, by the aid of the technology, we immerse ourselves perceptually in the virtual reality world in ways that are not possible in a painting or a sculpture in a museum.

107. M. Sullivan-Trainor, *Detour: The Truth About the Information Superhighway*, 228. After describing various theme park rides (into outer space, or back in time)

involving computerized video projections, Sullivan-Trainor states, "Believe it. These trips are very real, challenging the concept of reality" (270). From the idealist position, the information technology experience is as real as can be and in some cases more real than the material.

108. Heim, *The Metaphysics of Virtual Reality*, 83–108.

109. Ibid., 88.

110. Ibid., 89.

111. Ibid.

112. Ibid.

113. R. Kling and S. Iacono, "The mobilization of support for computerization: The role of computerization movements."

114. I. Csicsery-Ronay, "Cyberpunk and neuromanticism," 190.

115. Scholem, *Major Trends in Jewish Mysticism*, 223.

116. V. Bush, "As we may think."

117. N. Wiener, *The Human Use of Human Beings: Cybernetics and Society*, 109. These ideas seem to inform research into "telepresence," where you experience the world though the sensors of a mobile robot. See E. Paulos and J. Canny, "Ubiquitous tele-embodiment: Applications and implications."

118. M. Dery, *Escape Velocity: Cyberculture at the End of the Century.*

119. Ibid., 234.

120. Ibid., 235.

121. H. Moravec, *Mind Children: The Future of Robot and Human Intelligence*, 117.

122. Ibid.

123. Ibid., 123.

124. Ibid.

125. Ibid., 116.

126. Ibid.

127. Rushkoff, *Cyberia*, 16. The extension of the brain to the global sphere resonates with Bush's conjecture of the associative nature of thought, the development of hypertext (the "memex") to replicate it, Wells's concept of the "world brain," and Lévy's concept of the new emerging collective intelligence. See P. Lévy, *Becoming Virtual: Reality in the Digital Age.*

128. R. Grusin, "What is an author? Theory and the technological fallacy," 51.

129. R. Markley, "Boundaries: Mathematics, alienation, and the metaphysics of cyberspace," 77.

130. Ibid. See also M. Kendrick, "Cyberspace and the technological real."

Chapter 3

1. We shall discuss how romanticism and empiricism work out the play between the universal and the particular in chapter 8.

2. I. Csicsery-Ronay, "Cyberpunk and neuromanticism," 189. By way of contrast, according to critics of cyberspace, such as Markley, "the mythic structure of cyberspace is based on the identification of mathematics and metaphysical order that has persisted from Pythagoras to quantum physics" (R. Markley, "Boundaries: Mathematics, alienation, and the metaphysics of cyberspace," 59), a narrative with limited scope. See also Porter's account of the depiction of space in architecture, an account that progresses seamlessly from concepts of representing an external world using conventional media to the transcendent world of computer possibilities where "the user and the machine coalesce into a single entity" (T. Porter, *The Architect's Eye: Visualisation and Depiction of Space in Architecture*, 139).

3. See L. March and P. Steadman, *The Geometry of Environment: An Introduction to Spatial Organisation in Design*, 13. Mappings are not always one-to-one mappings: "In a mapping mathematicaly understood, this is not always so. Nor does a mapping necessarily preserve spatial characteristics such as length, area, angle, sense (left-handedness, right-handedness) and space" (13). The idea of the one-to-many mapping is another version of the unity myth, and the behaviors of systems that contain many such mappings soon become difficult to predict.

4. See J. Lansdown, *Teach Yourself Computer Graphics*, for an account of the process. On issues of rendering see P. Richens and S. Schofield, "Interactive computer rendering"; and J. Lansdown and S. Schofield, "Expressive rendering: A review of nonphotorealistic techniques."

5. See N. Goodman, *Languages of Art*, for a refutation of various conventional attitudes to pictures and representation, including the privilege accorded to perspective.

6. M. Benedikt, "Cyberspace: Some proposals," 128.

7. W. J. Mitchell, *City of Bits: Space, Place and the Infobahn*, 37.

8. M. Sullivan-Trainor, *Detour: The Truth About the Information Superhighway*, 264.

9. D. Rushkoff, *Cyberia: Life in the Trenches of Hyperspace*, 13.

10. N. Negroponte, *Being Digital*. 7.

11. Ibid., 6.

12. M. Heim, *The Metaphysics of Virtual Reality*, 33.

13. T. Leary, "The cyberpunk: The individual as reality pilot," 252.

14. Benedikt, "Cyberspace: Some proposals," 128.

15. See W. J. Mitchell, *The Logic of Architecture: Design, Computation, and Cognition*; and Benedikt, "Cyberspace: Some proposals," for relevant accounts of coordinate geometry in CAD and in design for cyberspace.

16. See J. Wann and M. Mon-Williams, "What does virtual reality NEED?: Human factors issues in the design of three-dimensional computer environments," for a catalogue of the major technical issues.

17. H. Dreyfus, *Being-in-the-World: A Commentary on Heidegger's Being and Time Division I*, 119.

18. See C. Norberg-Schulz, *Genius Loci: Towards a Phenomenology of Architecture.*

19. See E. Relph, "Geographical experiences and being-in-the-world: The phenomenological origins of geography."

20. That we are beholden to narratives is unappealing to empiricism, since it implies contingency on the matter of the real.

21. So qualities such as the shape and size of an object are primary, whereas color is secondary, as the color of an object changes depending on the light conditions under which we view it. The empiricists did not necessarily agree on where the boundary between the real and the apparent lay. For example, Berkeley thought that shape and size are also different, depending upon one's point of view.

22. C. Jencks, *The Architecture of the Jumping Universe: A Polemic: How Complexity Science Is Changing Architecture and Culture*, 164. He also says: "For the first time in the West since the twelfth century we have an all-encompassing story that unites all people of the globe, a metanarrative of the universe and its creation. It is quite the most pretentious story ever told because of its explanatory power and because it is true (although parts are in dispute)" (7). This embarrassing metaphysical declaration by the champion of postmodernism in architecture indicates how wedded cultural commentators can be to empiricism and romanticism.

23. See G. E. Moore, "The refutation of idealism," 84. He concluded: "I am as directly aware of the existence of material things in space as of my own sensations;

and *what* I am aware of with regard to each is exactly the same—namely that in one case the material thing, and in the other case my sensation does really exist" (84).

24. B. Russell, *The Problems of Philosophy*, 11.

25. K. Popper, *Objective Knowledge: An Evolutionary Approach*, 37.

26. Ibid., 38.

27. T. Nagel, *The View From Nowhere*, 95.

28. Russell, *The Problems of Philosophy*, 15. See also A. Sokal and J. Bricmont, *Intellectual Impostures*: "The best way to account for the coherence of our experience is to suppose that the outside world corresponds, at least approximately, to the image of it provided by our senses" (53–54).

29. R. Rorty, *Philosophy and the Mirror of Nature,* 12. He adds: "Without the notion of the mind as mirror, the notion of knowledge as accuracy of representation would not have suggested itself" (12).

30. The concept of the *mental map* would be one such instance of representational realism. See K. Lynch, *The Image of the City*; and S. Kaplan and R. Kaplan, *Cognition and Environment: Functioning in an Uncertain World.*

31. P. A. Schilpp, ed., *Albert Einstein: Philosopher Scientist,* 81. Quoted in A. Fine, *The Shaky Game: Einstein, Realism and the Quantum Theory,* 94.

32. Ibid.

33. S. W. Hawking, *Black Holes and Baby Universes and Other Essays*, 41. For Hawking, a good scientific theory or model describes a broad class of observations and does not have to correspond with some notion of reality. Nevertheless, a consistent model of everything is possible.

34. As described by A. Clark, *Being There: Putting Brain, Body, and World Together Again,* 2, 3.

35. P. K. Feyerabend, *Realism, Rationalism and Scientific Method: Philosophical Papers Volume 1*, 8. He adds: "These theories speak about reality. The task of science is to discover laws and phenomena and to reduce them to those theories" (8).

36. M. Johnson, *The Body in the Mind: The Bodily Basis of Meaning, Imagination, and Reason*, 204.

37. See J. Marías, *History of Philosophy,* 17; and R. E. Allen, *Greek Philosophy: Thales to Aristotle,* 35–39.

38. Plato, *The Republic of Plato*, 242.

39. Ibid.

40. R. Penrose, *The Emperor's New Mind: Concerning Computers, Minds, and the Laws of Physics*, 557.

41. Ibid.

42. See Allen, *Greek Philosophy*, 15.

43. Plotinus, *The Essence of Plotinus: Extracts from the Six Enneads and Porphyry's Life of Plotinus*, 69.

44. G. Ryle, *The Concept of Mind.*

45. D. C. Dennett, *Consciousness Explained*, 33.

46. Ibid.

47. Ibid., 202–206. See also R. Dawkins, *The Blind Watchmaker.*

48. See M. Hesse, *Models and Analogies in Science*; and A. Snodgrass and R. D. Coyne, "Models, metaphors and the hermeneutics of designing."

49. Johnson, *The Body in the Mind*, 203.

50. Ibid., 204.

51. According to Arendt, philosophy has always been "dominated by the never-ending reversals of idealism and materialism, of transcendentalism and imanentism, of realism and nominalism, of hedonism and asceticism and so on." She asserts that these systems are reversible and "can be turned 'upside down' or 'downside up' at any moment in history," and the "concepts themselves remain the same no matter where they are placed in the various systematic orders" (H. Arendt, *The Human Condition*, 292–293).

52. Dennett, *Consciousness Explained*, 71.

53. Ibid., 66. The exercise of objectivity is not limited to science. As a skeptic of objectivist notions in history, White, writing in the 1960s, points out that the study of history is beset with outmoded conceptions of objectivity: "Many historians continue to treat their 'facts' as though they were 'given' and refuse to recognize, unlike most scientists, that they are not so much found as constructed by the kinds of questions which the investigator asks of the phenomena before him." See H. White, *Tropics of Discourse: Essays in Cultural Criticism*, 43.

54. Popper, *Objective Knowledge*, 74.

55. Ibid., 73.

56. Ibid., 74.

57. Nagel, *The View From Nowhere*, 27.

58. Ibid.

59. Ibid., 4.

60. Ibid., 5.

61. Ibid., 6.

62. M. Benedikt, *Cyberspace: First Steps.* 4.

63. M. Novak, "Liquid architectures in cyberspace," 226.

64. Ibid., 227.

65. R. Descartes, "Space and matter," 73.

66. J. Locke, *An Essay Concerning Human Understanding,* 89.

67. Ibid., 135.

68. See L. A. Hickman, *John Dewey's Pragmatic Technology*, 83.

69. See A. Comte, *The Essential Comte.*

70. Ibid., 21.

71. In this, he also followed the lead of the socialist Saint-Simon, who founded the sect that was known as Saint-Simonism.

72. Comte, *The Essential Comte*, 45.

73. Mill and others who were otherwise supportive of positivism thought these latter views aberrant products of a short lived but passionate love affair, "which changed his whole conception of life" and Comte's anxiety to defend positivism against the charge of atheism. See J. S. Mill, *August Comte and Positivism,* 131.

74. M. Maffesoli, *The Time of the Tribes: The Decline of Individualism in Mass Society*, 25.

75. See J. Joergensen, "The development of logical empiricism," 852, for evidence of the link between positivists, such as Comte, and the logical positivism of the Vienna Circle, though the latter had greater affinity with British empiricism than the French school.

76. See R. Carnap, *The Logical Structure of the World: Pseudoproblems in Philosophy*, xvii.

77. Ayer, *Language, Truth and Logic*, 16.

78. L. Wittgenstein, *Tractatus Logico Philosophicus*, 63.

79. Ibid., 189. This statement can also be taken as an affirmation of the ineffable.

80. Ayer, *Language, Truth and Logic*, 150.

81. L. Wittgenstein, *Philosophical Investigations*.

82. Nagel, *The View From Nowhere*, 9.

83. According to the early (positivist) Wittgenstein, a statement is true if it corresponds to a state of affairs in the world, for "[t]he world is everything that is the case" (Wittgenstein, *Tractatus Logico Philosophicus*, 31). Any state of affairs in the world can be reduced to independent, simple, atomic facts. Facts, assertions, or propositions provide privileged access to the world, and to all intents and purposes they are the world: "The existence and non-existence of atomic facts is the reality"(37). Logic is the measure against which we compare our thinking, for according to Wittgenstein we "cannot think anything unlogical" (43). Logic is all important to an effective understanding of the world, as it "is not a theory but a reflection of the world" (169). Logic is in fact "transcendental," and the truth and falsity of propositions works by way of correspondence. A proposition is true if it corresponds to reality: "Propositions can be true or false only by being pictures of the reality" (71). We cannot consider anything outside of what our language presents to us: "*The limits of my language* mean the limits of my world. Logic fills the world: the limits of the world are also its limits" (149).

84. A. M. Turing, "Computing machinery and intelligence."

85. See J. R. Searle, "Minds, brains and programs."

86. H. Simon, *The Sciences of the Artificial*.

87. Mitchell, *The Logic of Architecture*.

88. C. Shannon and W. Weaver, *The Mathematical Theory of Communication*.

89. Such narratives also participate in unity and multiplicity, as a story of the universal and the particular, formalized by the extrication of content from word sequences. "Habitable rooms have windows and doors" is a word sequence, the information content of which is a general proposition about certain objects and relationships. It involves "universal quantification." But there are also word sequences, such as "the door is next to the window," the information content of which says something about the relationship between a particular door and a particular window.

90. Alternatively, the proposition provides the essence of the sentence. The question of essences is a metaphysical question and was not normally part of the language game of logical positivism. See R. D. Coyne, "Language, space and information."

91. Though for Aristotle formal logic, or the syllogism, was primarily a tool to counter the confounding rhetorical strategies of the sophists and was no substitute for judgment and other human virtues.

92. J. Meyrowitz, *No Sense of Place: The Impact of Electronic Media on Social Behavior.*

93. See M. Barnsley, *Fractals Everywhere.*

94. M. Gell-Mann, *The Quark and the Jaguar: Adventures in the Simple and the Complex,* ix. See also Jencks, *The Architecture of the Jumping Universe*, for an example of the incursion of such theories into art criticism.

95. G. Leibniz, "The relational theory of space and time," 96. Leibniz also asserts that space and time are ideal. They are constructed from our understanding of the relationship between things: "Nothing of time does ever exist, but instants; and an instant is not even itself a part of time. Whoever considers these observations, will easily apprehend that time can only be an ideal thing" (96).

96. Leibniz was not a philosophical relativist, and in canonizing the proposition "nothing is without ground" he affirmed absolutes. The epistemological distinction between absolute and relative knowledge is a further affiliate of the realism/idealism distinction. Absolute knowledge is knowledge that can be pinned down to a fixed point, a fixed reference. Relative knowledge is knowledge that depends on comparisons that are variable. So the sanctity of human life is an absolute, that we should drive on the right hand side of the road is based on convention or idiosyncratic preference, though for a relativist even rules about the sanctity of human life are conventional, relative to a frame of reference that may be different for different people and in different situations. Insofar as Descartes sought after a first principle or an "Archimedian point" ("I think therefore I am"), he was an absolutist, as was Leibniz, with his principle that "nothing is without ground." For these thinkers, and for others who have followed, the prospect of relativism offends, thought to lead to despair and nihilism—where nothing is certain, then we may as well do as we please.

We would normally associate the absolute with the objective, but certain philosophers, such as Hegel, have asserted that there are absolutes governing the subjective. Even if someone asserts that there is knowing that is subjective, they may still assert that subjectivity is governed by absolute laws.

A relativist on the subject of space would declare that there is a multiplicity of views of what space is and that the matter cannot be decided; there is no fundamental notion to which everyone can appeal. There are few philosophers who were self-declared relativists prior to this century. Many, such as Nietzsche, vehemently denied the possibility of absolutes, but in so doing did not advocate a doctrine of relativism. Denying absolutism does not automatically make one a relativist. From the point of view of late-twentieth-century philosophy, both absolutism and relativism are party to the same dichotomy. As pointed out by Fish, no one is a relativist in practice, since there are things about which we are all prepared to be adamant, bestowing upon them all the conviction of absolutes. Furthermore, as explored by Bernstein, the absolute/relative distinction is an intellectual trap. Why must we see things as either fixed or unstable, permanent or changing? According to Bernstein, there are ways of accounting for the world that avoid drawing us into the "Cartesian anxiety" of the absolute and the relative.

97. G. Berkeley, "Siris: A chain of philosophical reflexions and inquiries, etc.," 468. He adds: "and was accordingly treated by the greatest of the ancients as a thing merely visionary."

98. See S. Körner, *Kant*, 34. According to Kant, left and right gloves are obviously different in that they cannot be superimposed into the same space—that is, one glove does not fit into the other, and you cannot put your right hand into a left-hand glove. But the difference between the gloves cannot be described in terms of the relationships between parts of the gloves, as the relationships are the same for each. The index finger is adjacent to the middle finger in each case; the thumb and little finger are separated by four fingers in each case, and so on. The difference between the gloves must be in terms of their relationship to something else, and Kant decided that this was in terms of absolute space. The gloves are different by virtue of their relationship to space, which is absolute. Mathematicians and philosophers have been intrigued with Kant's formulation of the problematic of handedness and have advanced various explanations to account for the differences in terms of topology, rotational transformations in four dimensional "space," and so on. They have also taken Kant's logic to task.

99. See Körner, *Kant*, 37–38.

100. Space and time are also "transcendentally ideal." Independent of perception, they disappear. Kant's idealism differs from Berkeley's. For Berkeley, things only exist that are perceived. Kant's transcendental idealism allows for things that are experienced but not perceived. Kant's intricate philosophy depends upon many careful distinctions, which cannot be adequately covered here. Whereas philosophers such as Leibniz emphasized the importance and ubiquity of the logical proposition, Kant focused on judgments. Judgments are made by someone. They constitute a personal event.

For Kant, there are two axes of judgment. On one axis are judgments that are true by definition (analytic) and empirical judgments (synthetic). Judgments that are true by definition are facts inferred without appeal to experience or observation. They are tautologous. They elucidate meaning but are not otherwise informative;

for example, a rainy day is a wet day, objects with shape also have size, binary numbers have hexadecimal equivalents, if you are on the Web you are on the Internet. According to Körner, "A judgment is true by definition if, and only if, its denial would be a contradiction in terms or, what amounts to the same, if it is *logically* necessary or, again in other words, if its negation is *logically* impossible" (23). So to say a rainy day is not a wet day would be a contradiction.

Then there are empirical judgments (called "synthetic" by Kant). These are judgments derived from observation; for example, a rainy day is a cold day, some shapes are large, binary arithmetic is easy, the Internet is growing.

There is the axis of judgments according to how we come to know them. These are self-evident (a priori) and contingent (a posteriori) judgments. Judgments that are self-evident require no observation or confirmation. They are independent of the senses and are true at all times and in all places. The obvious examples come from mathematics: 2 + 2 = 4 is a self-evident judgment as are simple logical statements that are self-evident from the way the terms are defined. For example, every father is a male, red object are colored, the World Wide Web is the Internet. All judgments that are true by definition are in fact also self-evident (or rather all analytic judgments are a priori). See Hospers, *Philosophical Analysis*, 182.

Then there are contingent (a posteriori) judgments. These are judgments that are contingent on circumstances. The judgment is based on observations. That my car is in Edinburgh is not a self-evident but a contingent judgment. It just happens to be true at the moment, but the truth or falsity of the statement depends on circumstance and is determined by observation. Scientific judgments may also be of the contingent kind; for example, all bodies deprived of support fall downward. We know this generally to be the case, but it is contingent on a number of factors, qualifications, and refinements. It may be true as long as we are not in outer space, for example.

According to rationalism, all knowledge is true by definition (it is analytic), meaning that it is ultimately derivable through long chains of causal argument from self-evident (a priori) principles. For empiricism knowledge is of the synthetic, empirical variety. It is experience that furnishes us with knowledge. Kant sought to indicate that neither logic nor experience, individually or together, exhaust knowledge. See A. Winterbourne, *The Ideal and the Real: An Outline of Kant's Theory of Space, Time and Mathematical Construction*, 40. He did this by showing that whereas all judgments that are true by definition (analytic) are also self-evident (a priori), there are also judgments that are self-evident and empirical (a priori synthetic), and judgments that are contingent and empirical (a posteriori synthetic).

101. The knowledge we have of the world through our minds is empirical, which is to say it is appropriated through observation and sense experience. Kant

used the term "synthetic" to label this kind of knowledge, or rather the judgments we make derived from experience. But he also described such judgments so derived as self-evident or given, which he labeled "a priori." Empirical self-evident judgments describe particulars that are yet not perceived by the senses, though they are there to be experienced. Particulars are what we apprehend through the senses, as instances. Under this schema, space and time are self-evident particular: "Their structure is described by synthetic *a priori* judgments, in which we apply the *a priori* concepts of geometry and arithmetic" (Körner, *Kant*, 29). That is to say, their structure is described by empirical self evident judgments, in which we apply the self-evident concepts of geometry and arithmetic: "The notions of space and time are for Kant not abstractions from perception but *a priori* [self evident] particulars or 'pure forms of perception.' " In other words, space is an experiential thing, the kind of thing that we can only know about through experience, but we already apprehend it prior to experiencing it (Ibid., 30).

102. I. Kant, *Immanuel Kant's Critique of Pure Reason*, 38. The possibility of external perceptions presupposes the concept of space and does not create it. As we can represent to ourselves space without objects but not objects without space, space is a necessary representation that is self-evident. But the concept of space is not innate. Concepts such as beauty or virtue can be instantiated in different ways, but there is only one space that we can represent to ourselves. Space is not therefore a discursive concept, but rather an intuition, like the color red. Concepts contain their instances under themselves. Intuitions contain their concepts within themselves. The concept of red is not itself red. The intuition of space is itself space. See Winterbourne, *The Ideal and the Real*, 45.

103. R. Descartes, *Discourse on Method and the Meditations*, 102.

104. I. Newton, "Absolute space and time," 81.

105. See R. Barthes' short essay "The brain of Einstein," in *Mythologies*, for an account of Einstein's mythological status.

106. A. Einstein, *Relativity: The Special and General Theory*, 9.

107. Ordinarily, if we observe the motion of an object, such as a ball being thrown between two people on a train traveling at a constant velocity, then the motion of the ball for the passengers would be the same as if the train were at rest, at least for them. For someone at a stationary position outside the train, the velocity of a ball thrown in the direction of travel of the train would be the sum of the velocity of the train and the velocity of the ball as measured by the passengers. The summing of velocities is accurate for most projectiles we deal with, but the radical point of relativity theory is that it does not hold for very high speeds.

Einstein uses the example of a train, but others explain the principles of relativity theory with reference to high-speed rockets and spacecraft. If I am positioned on the Earth and, using a super telescope or radio signal, I observe the clock on a fast moving rocket orbiting the earth, then the clock will appear to be moving more slowly than my own watch. As the rocket gets closer to the speed of light, the on-board clock becomes almost stationary. If there is anyone on the rocket, then they make the same observations about the clocks on the earth. Those clocks appear to be running slowly. According to this theory, fast moving rockets, and all the on-board spatial measuring devices, such as scale rulers, appear shorter than before they left the earth—to us, not to the people on the rocket. By this theory space is therefore affected by velocity. The limit to any velocity, of course, is the speed of light. Technically, the speed of light is an asymptote in Einstein's equations. The rocket could never actually reach that speed without an infinite amount of energy. The velocity of light is also constant, whichever way you look at it. If one could measure the velocity of light as it traveled from one end of the rocket to the other and could make the measurements from our position on the earth, we would find it to be the same as the velocity as we measure it on earth. Velocity is measured as the distance covered over a certain period of time. Both scale (for measuring distance) and time are variable in Einstein's equations. As distances shrink, so time moves more slowly. To the people on the rocket making similar measurements, the speed of light would also appear the same, whether they were observing it on earth from their rocket, or measuring it in their rocket.

108. Einstein extended his theories to a consideration of what happens when objects accelerate and decelerate, that is, when their velocity changes, which is analogous to saying when they are under the influence of gravity. Under Einstein's formulation, a gravitational field is explicable in terms of a coordinate system in which the scales for measuring distance and time vary. A material object, such as the Earth, distorts what is known as the "space time continuum." A satellite orbiting around the earth is traveling in a straight line as far as it is concerned, but the measuring frame within which it is moving is distorted by the presence of matter—namely, the huge mass of the Earth. By this means, Einstein was able to dispense with the notion of force as a way of describing gravitational fields (the "force of gravity"). Einstein's theories indicate a co-dependence between space and time. According to Gregory, Einstein's special theory "requires giving up the firm conviction that space and time are two distinct entities" (B. Gregory, *Inventing Reality: Physics as Language*, 64). In order for the speed of light to be the same for all observers, it is necessary to see space and time as co-dependent aspects of the same phenomenon—space-time.

Einstein's theories also appropriate concepts developed within geometry, such as the notion of non-Euclidean space developed by Gauss, Riemann, and others. The geometry of Euclid was based on a few simple postulates considered to be self evident. Euclid's so-called fifth postulate is that parallel lines do not intersect, or, more precisely, through a given point there is only one line that is parallel to a given straight line (H. Reichenbach, "Non-Euclidean spaces," 216). Riemann and other mathematicians discovered that it is possible to construct alternative geometrical systems that are internally consistent, but that deny the Euclidean axioms. These geometries are not easy to imagine, let alone draw on a sheet of paper, other than by analogy. The circumference of a large circle drawn on the surface of a sphere, such as the earth, measures slightly less than π times its diameter when the diameter is measured on the surface of the sphere. This is because the center point of the circle, as located on the surface of the sphere, is not actually on the same plane as the large circle. As is well understood by navigators, and the designers of computerized flight simulators, the practical system of measurement on the surface of the Earth results in a slightly different geometrical system than that of Euclid. Riemann extends the curved surface phenomenon to three dimensions, which results in a much more complicated geometrical system. In other words, he assumes the "curvature of space." Furthermore, there are different kinds of curvature: convex, concave, uniform and nonuniform. These abstract formulations would simply be mathematical oddities, were it not that they prove extremely useful as the basis of predictive equations in astronomy and physics, including Einstein's concepts of the curvature of space and time under the influence of matter.

109. Einstein's reflections on his theories are positivist, though commentators such as Fine see a transformation from an empirical stance that would not speak beyond what could be empirically verified, to a later realist stance, but one that focuses on the viability of theories: "When asked whether such-and-such is the case, one responds by shifting the question to asking whether a theory in which such-and-such is the case is a viable theory" (Fine, *The Shaky Game*, 87). Einstein's realism did not rest on correspondence between theory and world but rather on the notion that something is true if it appears in a theory that can be empirically verified. Einstein's realism comes through in his objections to quantum theory. According to quantum theory, the decay of a single radioactive atom cannot be decided. The uncertainty formulas for energy and time rule out the possibility of determining the exact time of a specific decay. The quantum theory only provides probabilities for what one will find if we look for the decay. To ask whether there is a definite decay time is to ask about reality, and according to quantum theory, one cannot simply ask whether a definite moment for the decay of a single atom exists. We should rather ask if it is reasonable to assume the existence of a definite

moment for the decay of a single atom within the framework of the total theoretical construction. As Fine explains, we cannot even ask what the question about the reality of the decay time means. We can only ask whether such a supposition is reasonable in the context of the chosen conceptual system, with a view to its ability to grasp theoretically what is empirically given. According to Fine, this is the same as saying: "We don't know what it means to say that the atom 'really' had a definite decay time, nor do we know whether that is 'true.' We can only ask whether a theory that incorporates such decay times is a good theory, from an empirical point of view" (Fine, *The Shaky Game*, 92). According to Fine, Einstein's realism involves deflecting questions of reality to the question of the empirical adequacy of a whole theory.

110. Quoted in Fine, *The Shaky Game*, 2.

111. At the level of very small particles, such as in measuring the behavior of electrons and photons, it is impossible to measure both location and momentum. According to Gregory, Heisenberg's uncertainty principle means that physicists can determine an electron's position to any degree of precision, but the more precisely they determine this position, the less precisely can they determine how the electron is moving (its momentum) at the same moment in time: "They can find out exactly where the electron is, but only at the price of being able to say nothing about where it is going" (Gregory, *Inventing Reality*, 92). On the other hand, physicists can measure an electron's momentum to any degree of precision, but the more precisely they know the momentum, the less precisely they can tell the electron's position: "They can tell exactly where the electron is going, but only at the price of being able to say anything about where it is" (92). Once you have measured one factor, then the other cannot be decided. So the decision of the observer seems to fix aspects of the behavior of the particles observed. Quantum theory also presents the problem of action at a distance. The example commonly given is where a particle with no spin splits into two new particles (an electron and a positron) as it decays. A scientist can measure the spin of the derived electron, in whatever direction she chooses, but when she looks at the positron, it will always be spinning in the opposite direction. Making a measurement of the direction of spin of one of the derived particles seems to fix the spin of the other. This applies even if the particles have moved far apart. (See Penrose, *The Emperor's New Mind,* 365.) Explanations are offered in terms of quantum probability and fields of "quantum potential." (See D. Bohm and F. D. Peat, *Science, Order and Creativity,* 93; Gell-Mann, *The Quark and the Jaguar,* 171.) The other anomaly often cited is the case of the interference pattern produced by photons passing through parallel slits. One photon seems to be influenced by what is happening to another photon, even though they are distant from one another. In terms of

the wave model of light propagation, the waves cancel each other out to produce dark banding where the waves interfere. But in terms of particle theory, photons cannot cancel each other out. They seem to concentrate, as though a photon passing through one slit is "aware" of the presence of the adjacent slit. (See N. Bohr, "The Bohr-Einstein dialogue"; N. Bohr, "Light and life"; and Penrose, *The Emperor's New Mind,* 299–305.)

112. Gregory, *Inventing Reality*, 85.

113. A. N. Whitehead, *Science and the Modern World*, 167.

114. Fine, *The Shaky Game*, 151.

115. Gregory, *Inventing Reality*, 186.

116. P. Davies, Introduction to *Physics and Philosophy: The Revolution in Modern Science*, 4.

117. W. Heisenberg, *Physics and Philosophy: The Revolution in Modern Science,* 40.

118. Whitehead, *Science and the Modern World*, 16.

119. Arendt, The *Human Condition,* 289.

120. See Einstein, *Relativity*, 26.

121. According to Virilio, "Relativity would not exist without the relative optics (physical optics) of the observer" (P. Virilio, *The Vision Machine*, 75), and "It is speed more than light which allows us to see, to measure and thereby conceive reality" (74).

122. See B. Mandelbrot, *The Fractal Geometry of Nature.*

123. M. Gell-Mann, *The Quark and the Jaguar*, x.

124. Jencks, *The Architecture of the Jumping Universe,* 23.

125. J. Gleick, *Chaos: Making of a New Science*, 5.

126. See A. Petersen, "The philosophy of Niels Bohr."

Chapter 4

1. See R. Carnap, *The Logical Structure of the World: Pseudoproblems in Philosophy*, 5–9. According to formal logic, propositions are made up of objects and relations. So if we are dealing with simple two-dimensional spatial geometries then we may define an object in terms of sets of x and y coordinates as objects that are joined by the relation called "line," or if we are dealing with a sentence in natural language such as "All entrance doors should swing inward" then we may translate the sentence into the proposition $\exists x$ Type(x, Door) $\cap$ Location(x, Entrance) $\supseteq$

Swing(x, Inward), or something similar, depending on the nomenclature used for the different relationships. In this case x, Door, Entrance, and Inward are objects, and Type, Location, and Swing are relationships between pairs of objects. $\cap$ and $\supseteq$ are logical relationships, and $\exists$ is the "existential quantifier" for defining variables. The proposition can be read slightly less formally as "If there exists an object x such that the type of x is door and the location of x is entrance then the swing of x is inward."

2. See C. J. Fox, *Information and Misinformation: An Investigation of the Notions of Information, Misinformation, Informing, and Misinforming.*

3. These simple observations weaken the force of certain information technology narratives, particularly from the point of view of symbolic artificial intelligence and the idea that computer representations provide access to the world, as in modeling, virtual reality, and so on. As pointed out by Hubert Dreyfus, Terry Winograd, Fernando Flores, and others, artificial intelligence is so far inadequate to the task of replicating human intelligence, in spite of its access to the manipulation of propositions. So perhaps the proposition, propositional logic, and information are not central to human cognition and language after all, or perhaps logic is not well enough understood. See H. Dreyfus, *Being-in-the-World: A Commentary on Heidegger's* Being and Time, *Division I*; and T. Winograd and F. Flores, *Understanding Computers and Cognition: A New Foundation for Design.* Similarly, the inadequacies of virtual reality to the task of providing the totally immersive spatial environment suggest that perhaps information does not after all provide access to the nature of the real.

4. See H.-G. Gadamer, *Truth and Method.*

5. J. Austin, *How to Do Things with Words.*

6. Ibid., 5.

7. Ibid., 5.

8. B. Gregory, *Inventing Reality: Physics as Language*, 2.

9. Austin, *How to do Things with Words*, 148–149.

10. See M. Reddy, "The conduit metaphor—a case of frame conflict in our language about language."

11. Dennett says that like genes, memes are invisible, "and are carried by meme vehicles—pictures, books, sayings (in particular languages, oral or written, on paper or magnetically encoded, etc.)" (D. Dennett, *Consciousness Explained*, 204). See also Popper's concept of the world of information discussed in chapter 3, and K. Popper, *Objective Knowledge: An Evolutionary Approach*, 74.

12. Such as $\exists x$ Type(x, Door) $\cap$ Location(x, Entrance) $\supseteq$ Swing(x, Inward).

13. See R. D. Coyne, *Logic Models of Design*, for an insider's view of the application of logical rules. Many of the limitations of logic come down to Gödel's theorem, which states that there are always propositions that are consistent with the axioms of a system but which cannot be proved by the system. The "proof" will be undecided. Operationally, the theorem proving system gets into an infinite loop. See K. Gödel, *On Formally Unprovable Propositions*.

14. P. Winch, *The Idea of a Social Science: and Its Relation to Philosophy,* 9. Moore was an early analytic philosopher who supported logical positivism and is mentioned in that context in chapter 3.

15. Winch, *The Idea of a Social Science*, 10.

16. L. Wittgenstein, *Philosophical Investigations*, 15.

17. See P. Davies, *Superforce: The Search for the Grand Unified Theory of Nature*; S. W. Hawking, *A Brief History of Time: From the Big Bang to Black Holes*; R. Penrose, *The Emperor's New Mind: Concerning Computers, Minds, and the Laws of Physics*. D. Bohm and D. Peat, *Science, Order and Creativity;* and M. Gell-Mann, *The Quark and the Jaguar: Adventures in the Simple and the Complex*.

18. B. Gregory, *Inventing Reality*, 184. This echoes the view of the physicist Niels Bohr: "We are suspended in language in such a way that we cannot say what is up and what is down. The word 'reality' is also a word, a word which we must learn to use correctly," as reported in A. Petersen, *Niels Bohr: A Centenary Volume*, 302. Apparently Bohr was an "avid reader" of the Danish theologian Søren Kierkegaard (303). Heidegger also gained much from Kierkegaard (see Dreyfus, *Being-in-the-World*). So Bohr was in touch with aspects of mainstream phenomenological thought. It is also worth noting that it is in their neglect of the pragmatic aspect of language that A. Sokal and J. Bricmont's supposed critique of continental philosophy is such an embarrassing failure. See A. Sokal and J. Bricmont, *Intellectual Impostures: Postmodern Philosophers' Abuse of Science*.

19. See Gregory, *Inventing Reality*, 188–189.

20. Ibid., 192.

21. Ibid., 3.

22. Ibid., 38.

23. Ibid., 70.

24. Ibid., 155.

25. See L. A. Hickman, *John Dewey's Pragmatic Technology*.; T. Kuhn, *The Structure*

of Scientific Revolutions; and B. Latour and S. Woolgar, *Laboratory Life: The Construction of Scientific Facts*.

26. W. Heisenberg, *Physics and Philosophy: The Revolution in Modern Science,* 46.

27. N. Bohr, "The Bohr-Einstein dialogue," 139.

28. Reported in A. Petersen, "The philosophy of Niels Bohr," 305.

29. Gregory, *Inventing Reality*, 95.

30. Ibid., 145.

31. Ibid., 115–116.

32. Ibid., 96.

33. Ibid.

34. Ibid.

35. See Gregory, *Inventing Reality*, 163; and Gell-Mann, *The Quark and the Jaguar*, 203–204.

36. See H. Bergson, *Time and Free Will*, 96.

37. Ibid., 97.

38. Ibid.

39. We may be tempted to ascribe this capacity to see space as homogeneous to our ability to form abstractions, to remove objects from their context, and to consider them mathematically. But Bergson regards our ability to homogenize space as prior to our ability to abstract. Abstraction assumes clean-cut distinctions, the identification and isolation of objects from one another against a neutral background. According to Bergson, "Abstraction already implies the intuition of a homogeneous medium." He identifies two different kinds of "reality," one of which is heterogeneous, the differentiated world of sense experience, the world of sensible qualities; and the other is homogeneous, namely, the homogeneity of space. This uniform conception of space "enables us to use clean-cut distinctions, to count, to abstract, and perhaps also to speak" (H. Bergson, *Time and Free Will*, 97). For Bergson, insofar as we regard time as homogeneous, we also regard it as spatial: "From the moment when you attribute the least homogeneity to duration, you surreptitiously introduce space" (104).

Of time he also says: "It is true that, when we make time a homogeneous medium in which conscious states unfold themselves, we take it to be given all at once, which amounts to saying that we abstract it from duration. . . . We may therefore surmise that time, conceived under the form of a homogeneous medium, is some spurious concept, due to the trespassing of the idea of space upon the field of pure consciousness" (98). He adds that "time, conceived under the form of an unbounded and homogeneous medium, is nothing but the ghost of space haunting the reflective consciousness" (99).

On the subject of time he also says: "In a word, pure duration might well be nothing but a succession of qualitative changes, which melt into and permeate one another, without precise outlines, without any tendency to externalize themselves in relation to one another, without any affiliation with number: it would be pure heterogeneity. But for the present we shall not insist upon this point; it is enough for us to have shown that, from the moment when you attribute the least homogeneity to duration, you surreptitiously introduce space" (104).

40. See A. B. Snodgrass, *Architecture, Time and Eternity: Studies in the Stellar and Temporal Symbolism of Traditional Buildings*; and A. K. Coomaraswamy, "The symbolism of the dome."

41. Snodgrass, *Architecture, Time and Eternity*, 57.

42. M. Foucault, "Of other spaces," 23.

43. See G. Lakoff and M. Johnson, *Metaphors We Live By*; G. Lakoff, *Women, Fire and Dangerous Things: What Categories Reveal about the Mind*; and M. Johnson, *The Body in the Mind: The Bodily Basis of Meaning.*

44. See T. Hawkes, *Structuralism and Semiotics*, 19.

45. F. de Saussure, *Course in General Linguistics.*

46. Ibid., 105.

47. F. Jameson, *The Prison-House of Language: A Critical Account of Structuralism and Russian Formalism*, 14. He adds: "Saussure's concept of the 'system' implies that in this new trackless unphysical reality content is form; that you can see only as much as your model permits you to see; that the methodological starting point does more than simply reveal, it actually creates, the object of study."

48. Ibid., 30.

49. Ibid., 66.

50. C. K. Ogden and I. A. Richards, *The Meaning of Meaning: A Study of the Influence of Language upon Thought and the Science of Symbolism.*

51. Jameson, *The Prison-House of Language,* 32.

52. Ibid., 33.

53. Ibid.

54. Ibid.

55. Ibid.

56. J. Piaget, *Structuralism*, 40.

57. Ibid., 40–41.

58. Ibid., 8–9.

59. Saussure, *Course in General Linguistics,* 116.

60. The significance of the phoneme is discussed at length in R. Jakobson and M. Halle, *Fundamentals of Language.*

61. The same applies to comparisons between any two languages, such as English and French, as in the difference between the French words *pas* and *par.*

62. For Hegel, the dialectic comes before logic. See G. W. F. Hegel, *Hegel's Science of Logic;* and S. Houlgate, *Freedom, Truth and History: An Introduction to Hegel's Philosophy.*

63. Saussure, *Course in General Linguistics*, 149.

64. Jameson, *The Prison-House of Language,* 15.

65. The French word *la parole* is literally "speech" or "word" in English, and *la langue* is "tongue" or "language."

66. Jameson, *The Prison-House of Language,* 26–27.

67. According to Jameson, Chomsky attempted to rehabilitate the issue of sentence formation, the structure of sentences, or syntax, as an issue of *langue* and not merely of *parole.* There are syntactic categories such as articles, nouns, verbs, noun phrases, verb phrases, and so on, and there are transformation rules for determining that a sentence is legal in the language. The syntax is the surface structure of language. Any rules we might devise to determine meaning would be the semantic structure.

68. See R. Barthes, *Mythologies*, 118.

69. C. Lévi-Strauss, *Structural Anthropology.*

70. See Lévi-Strauss, *Structural Anthropology*, 40–42; and T. Hawkes, *Structuralism and Semiotics*, 54–55.

71. See Hawkes, *Structuralism and Semiotics*, 54–55.

72. Lévi-Strauss, *Structural Anthropology,* 224.

73. See Hawkes, *Structuralism and Semiotics*, 48.

74. Barthes, *Mythologies*, 149.

75. This description is a précis from a description of Saint Marie-Magdeleine at Vezelay by Snodgrass, *Architecture, Time and Eternity*, 313. But note that his interpretation does not rely so much on the tenets of structuralism as on the theory of symbols.

76. See Snodgrass, *Architecture, Time and Eternity.*

77. The structuralist program has been productive in architecture for commentators and architects such as Norberg-Schültz, Jencks, Eisenman, Tsumi, and others. For example, see P. Eisenman and R. Krauss, *Peter Eisenman: House of Cards.*

78. B. Latour and S. Woolgar, *Laboratory Life.*

79. See ibid., 81.

80. See Kuhn, *The Structure of Scientific Revolutions.*

81. Johnson, *The Body in the Mind*, 44.

82. Lakoff, *Women, Fire and Dangerous Things*, 363.

83. Barthes, *Mythologies*, 18.

84. See ibid.

85. Snodgrass, *Architecture, Time and Eternity,* 25. See also Coomaraswamy, "Literary symbolism"; and Coomaraswamy, "Imitation, expression, and participation."

86. M. Eliade, *The Two and the One*, 201.

87. By this reading, structuralism is not therefore eloquent on the subject of myths of unity and fragmentation, and it is certainly less reverential than theorists such as Eliade and Coomaraswamy. However, in chapter 7, we will see how structuralism paves the way for considering how the antagonistic relationship between language and the real tells the unity myth in a particular way.

88. Jameson, *The Prison-House of Language,* 106.

89. Ibid., 109. Jameson outlines two options for extracting structuralism from this difficulty: "The entire sign-system would somehow correspond to all of reality, without there being any one-to-one correspondence in the individual elements at any point" (110). A more positivistic point of view is that there is some preestablished harmony between the structures of the mind, or the brain, and the order of the external world.

90. A. Giddens, *The Constitution of Society: Outline of the Theory of Structuration*, 32.

91. The poststructuralist project still calls on the vocabulary of structuralism and, according to critics such as Giddens, does not answer the charge of idealism. Giddens states the case thus: "The foundation of a theory of meaning in 'difference' in which, following Saussure, there are no 'positive values' leads almost inevitably to a view accentuating the primacy of the semiotic. The field of signs, the grids of meaning, are created by the ordered nature of differences which comprise codes. The 'retreat into the code'—whence it is difficult or impossible to re-emerge into the world of activity and event—is a characteristic tactic adopted by structuralist and post-structuralist authors. Such a retreat, however, is not necessary at all if we understand the relational character of the codes that generate meaning to be located in the ordering of social practices, in the very capacity to 'go on' in the multiplicity of contexts of social activity." See Giddens, *The Constitution of Society*, 32.

92. J. Derrida, *Of Grammatology*.

93. One of Derrida's most revealing metaphors is of the postal service, the implications of which to concepts of meaning he exhausts in the book *The Postcard: From Socrates to Freud and Beyond*.

94. R. Barthes, *Mythologies*, 143.

95. I discuss Derrida's refutation of concepts of foundations and the metaphysical in R. D. Coyne, *Designing Information Technology in the Postmodern Age: From Method to Metaphor.*

96. J. Derrida, *Limited Inc.*

97. See M. Poster, *The Mode of Information*; G. Ulmer, *Teletheory: Grammatology in the Age of Video;* and R. Luckhurst, "(Touching On) Tele-Technology."

98. See Coyne, *Designing Information Technology in the Postmodern Age.*

99. Ibid., 108–114. For a similar treatment see R. Grusin, "What is an electronic author? Theory and the technological fallacy," 45.

100. See Coyne, *Designing Information Technology in the Postmodern Age.* Derrida deals specifically with space in an argument about writing. We commonly attribute writing to the development of concepts of the spatiality of thought. That is, as elaborated by Ong, writing is a matter of laying ideas out in space, and we have taken this over into notions about how we think (see W. J. Ong, *Orality and Literacy: The Technologizing of the Word*). Derrida deals with thought by way of speech. Many writers, from Plato to Marshall McLuhan, have regarded speaking as more immediate than writing, since speech is more closely related to thought. Many philosophers have assumed that whereas speech pertains to the nonspatial,

writing takes the internal nonspatial realm of thought into space, which is outside ourselves. Derrida refutes this commonsense view: "For the voice is already invested, undone. . . , required, and marked in its essence by a certain spatiality" (J. Derrida, *Of Grammatology*, 290).

Derrida sets out to show the spatiality in speech, that is, how that we understand speech depends on notions of space, that is, it is imbued with notions of spatiality. In so doing, he shows that notions of speech are dependent on notions of writing. He achieves this by looking at the components of writing and shows how they already inhere within speech. I have explained these arguments elsewhere in relation to information technology and will only summarize them here (see Coyne, *Designing Information Technology in the Postmodern Age*, 116–117). Derrida identifies several components commonly attributed to writing. There are signs or marks on paper that occupy space and that point to a referent. Signs are useful because they can be repeated and copied. Sign sequences (texts, drawings, etc.) can be *disseminated.* Sequences of signs can be recognized as the same even in different circumstances. It is therefore possible to pass sign sequences on from one situation to another. Signs therefore can be distributed throughout space and remain recognized. Sign sequences also operate in the *absence* of an originator. It is possible to present a sign sequence to a third party without the presence of the originator and without knowing that person's situation. The sign sequence is capable of signifying even when stored or presented independently of intention.

These phenomena—of signs organized spatially, able to be copied, transmitted across space, and having meaning even though they end up spatially removed from their source—are commonly taken as attributes of writing, print or electronic communications, but Derrida shows how exactly the same characteristics can be ascribed to speaking. Spoken words function as signs, a spoken utterance can be repeated by another person, it is possible to *disseminate* a speech to a crowd, the original speaker can be *absent*; they do not need to be present for the utterance to be received, and the conveyor of the utterance does not have to understand the utterance in order for a third party to receive and make sense of the sign sequence.

These spatial features of sign, repetition, distribution, and absence are not commonly regarded simply as part of language but as features of writing, hence Derrida's formulation of the study of protowriting or arche-writing, of which writing and speech are but instances. This line of argument from Derrida, that speech and writing are instances of protowriting suggests that speech, and by implication thought, is already a spatial phenomenon. Derrida echoes Bergson's assertion of the priority of space over mathematical concepts of abstraction. Derrida's arguments also disarm complaints about the corruption of the spoken word due to electronic communications technology. The features of sign, repetition, dissemination, and absence already belong to thought, speech, manuscript culture, print, and electronic communications. As we shall see in the following chapter, there is a great deal to be gained by distinguishing between space and spatiality. To say that speech, thought, language, and communication presuppose spatiality is not to accord priority to empirical space, but a more basic, though less determinate phenomenon.

101. Barthes, *Mythologies*, 117.

102. Ibid., 140.

103. Ibid.

104. Ibid., 142.

105. Ibid., 150.

106. Ibid., 155.

107. Ibid., 155.

108. Ibid., 46.

109. J. Stallabrass, "Empowering technology: The exploration of cyberspace," 24.

110. Ibid., 26.

111. J. Baudrillard, "Simulacra and simulations."

112. Barthes, *Mythologies*, 168.

113. Ibid., 160.

114. Ibid., 162.

115. Ibid., 160.

116. H. Marcuse, *One Dimensional Man Studies in the Ideology of Advanced Industrial Society*.

117. T. W. Adorno and M. Horkheimer, *Dialectic of Enlightenment,* 22.

118. J. Habermas, *Theory of Communicative Action*, 95.

119. H. Lefebvre, *The Production of Space*, 89.

120. Ibid.

121. Ibid., 90.

122. Derrida, *Of Grammatology,* 290.

123. Ibid.

124. See Poster, *The Mode of Information.*

125. See H. White, *Tropics of Discourse*, for an examination of the role of transformation in Foucault's writing.

126. See M. Foucault, *Discipline and Punish: The Birth of the Prison.*

127. See A. Borgman, *Technology and the Character of Contemporary Life: A Philosophical Inquiry.*

128. See Coyne, *Designing Information Technology in the Postmodern Age.*

Chapter 5

1. M. Heidegger, *Being and Time*, 226.

2. What follows is a necessarily brief account, glossing over many of Heidegger's careful distinctions. My interpretation of Heidegger emphasizes his pragmatism and is informed by H. L. Dreyfus's analysis in *Being-in-the-World: A Commentary on Heidegger's* Being and Time, *Division I*, as well as numerous other sources, including Ryle's review of *Being and Time* published in 1929. See G. Ryle, "Review of Martin Heidegger's *Sein und Zeit*." I discuss some of the complexities of Heidegger's varied philosophy as it pertains to information technology in *Designing Information Technology in the Postmodern Age: From Method to Metaphor*. See also C. Spinosa, F. Flores, and H. L. Dreyfus, *Disclosing New Worlds: Entrepreneurship, Democratic Action, and the Cultivation of Solidarity*, for an examination of the application of Heidegger's thought to management studies.

3. According to Arendt, we can thank Plato's philosophy for the privileging of contemplation (*theoria*) over work, which reflected the ideal of a leisured and free class of philosophers at the pinnacle of the *Polis*. Before we are thinking beings, we are, according to Arendt, *homo faber*, the fabricator of the world. See H. Arendt, *The Human Condition*, 14.

4. Heidegger, *Being and Time*, 27.

5. Heidegger develops the theme of space and regions at length. See Heidegger, *Being and Time*, 134–148.

6. B. Gregory, *Inventing Reality: Physics as Language*, 197.

7. See M. Heidegger, *The Question Concerning Technology and Other Essays.*

8. I discuss Heidegger's view of technology in greater detail in *Designing Information Technology in the Postmodern Age: From Method to Metaphor*, 53–98.

9. See Dreyfus, *Being-in-the-World: A Commentary on Heidegger's* Being and Time, *Division I.*

10. See Coyne, *Designing Information Technology in the Postmodern Age*, 158–177.

11. R. Barthes, *Mythologies*, 172.

12. Heidegger, *Being and Time*, 255.

13. Ibid., 227.

14. Ibid., 143.

15. Merleau-Ponty also develops the provocative theme of the ubiquity of embodiment as "flesh": "We have to reject the age-old assumptions that put the body in the world and the seer in the body, or, conversely, the world and the body in the seer as in a box. Where are we to put the limit between the body and the world, since the world is flesh?" (M. Merleau-Ponty, *The Visible and the Invisible*, 138).

16. Heidegger addresses the issue of repetition in relation to history: "Dasein does not first become historical in repetition; but because it is historical as temporal, it can take itself over in its history by repeating" (Heidegger, *Being and Time*, 438).

17. Heidegger, *Being and Time*, 66–67.

18. M. Johnson, *The Body in the Mind: The Bodily Basis of Meaning, Imagination, and Reason*, 203.

19. Dasein is the being for whom Being is an issue: "Being is that which is an issue for every such entity" (Heidegger, *Being and Time*, 67).

20. Heidegger, *Being and Time*, 255.

21. Ibid.

22. Ibid., 232. Dasein counters anxiety by "fleeing in the face of itself" (229), "to turn thither towards entities within-the-world by absorbing itself in them" (230). As Dreyfus puts it: "The everyday world is organized precisely to provide Dasein with ways to cover-up its unsettledness" (Dreyfus, *Being-in-the-World*, 182).

23. A notable exception is G. Ryle, who wrote a detailed review of *Being and Time* soon after its publication, and whose book *The Concept of Mind* seems to have been influenced by it. See G. Ryle, "Review of Martin Heidegger's *Sein und Zeit*."

24. See T. Adorno, *The Jargon of Authenticity*.

25. This is one of Gadamer's characterizations of the superficial tourist inspecting buildings as monuments, as objects to be photographed. See H.-G. Gadamer, *Truth and Method*, 139.

26. See Coyne, *Designing Information Technology in the Postmodern Age*.

27. See G. Lakoff and M. Johnson, *Metaphors We Live By;* G. Lakoff, Women, *Fire, and Dangerous Things: What Categories Reveal about the Mind;* and M. Johnson, *The Body in the Mind.*

28. See M. Jay, *Downcast Eyes: The Denigration of Vision in Twentieth Century Thought.*

29. See W. Ong, *Ramus: Method, and the Decay of Dialogue from the Art of Discourse to the Art of Reason.*

30. See W. Ong, *Ramus.*

31. Computer games such as *Myst* and *Riven* seem to rely as much on ambient sound for distinguishing spaces as on visual cues.

32. See Coyne, *Designing Information Technology in the Postmodern Age*, 270–276.

33. P. Ricoeur, *The Rule of Metaphor.*

34. See Coyne, *Designing Information Technology in the Postmodern Age*, 249–301.

35. See R. Rorty, "Unfamiliar noises: Hesse and Davidson on Metaphor"; and M. Hesse, "Unfamiliar noises—tropical talk: The myth of the literal."

36. T. Kuhn, *The Structure of Scientific Revolutions.*

37. Dovey presents an alternative reading of the application of phenomenology to geometrical drawing practice. See K. Dovey, "Putting geometry in its place: Towards a phenomenology of the design process." For Dovey, drawings are to be understood in terms of communication flows (254), and geometrical (working) drawings are concerned "with 'facts' and not 'values' " (253).

38. See G. Deleuze on the subject of the line, in *The Deleuze Reader*, 225–234.

39. Infinite regress is anathema to the deterministic language of empiricism, but grist to the mill of the open discourses of poststructuralism.

40. L. March and P. Steadman, *The Geometry of Environment: An Introduction to Spatial Organisation in Design*, as discussed in chapter 3.

41. March and Steadman present an empiricist interpretation of this openness to difference: "Our argument is that a new pattern will be recognized only by an observer who has available, or develops, an appropriate range of mental sets, abstract or otherwise, upon which to map the data, and who actively seeks not to corroborate the habitual but to conjecture potentiality" (March and Steadman, *The Geometry of Environment*, 30).

42. White explains the succession of concepts of the real in terms of difference, in discussing Foucault's work *The Order of Things: An Archaeology of the Human*

Sciences. Foucault identifies four epochs of Western cultural history that sought to treat the real in different ways. (See M. Foucault, *The Order of Things*.) According to Foucault, there are four major epochs, which he claims are disconnected, though White considers the transition from one epoch to the next to be accountable in terms of a succession of tropes—that is, a succession in language from concepts of a literal view of metaphor, to metonymy, to synechdoche, to irony. By this reading, the sixteenth-century, pre-Enlightenment period was dominated by the desire to find the same in the different. This extended beyond the desire to identify resemblances between things to identifying the relationship between things and the words signifying them. This engendered a kind of word magic, on the one hand, and a belief that mastery of words might enable mastery of the things they represented, on the other. White translates this concern to the issue of metaphor, which asserts both similarity and difference. A science committed to cataloguing all the similarities encountered in nature is eventually overcome by the plenitude of difference. The longer the list of similarities, the more apparent becomes the fact of difference: "The multiplication of data in such sciences would inevitably increase the number of things appearing to be different from one another, and thereby strain the capacities of observers to discern the similarities presumed to exist among them" (H. White, *Tropics of Discourse*, 253).

With the issue of resemblance exhausted, the seventeenth and eighteenth centuries, the early Enlightenment, were concerned with the problem of difference, through the disciplines of order and measurement conceived in essence spatially. How could a sign be linked to what it signified? This period is characterized by tables of essential relationships and the search for the "original language." The focus of its studies was language, organism, and wealth. For White, this is where science set itself the task of cataloguing the relationships presumed to exist among *different* things. The way to do this was in terms of their spatial relationships. The operative trope here is metonymy. To use metonymy is to identify a thing by one of its parts. The part serves as a metaphor of the whole, as in the use of "the crown," to indicate the monarch, or "fifty sail" to indicate fifty ships. For White, this epoch is characterized by the pursuit of the parts as giving access to the whole. So the study of language was a matter of finding the "universal grammar," the true basis of wealth was thought to reside in land or gold, and the essence of organic life resided in external attributes: "Hence the endless constructions of those tables of attributes . . . which are meant to reveal finally the 'web of relationships' that bind the entities together into an 'order of things' " (Ibid.). This epoch eventually hit up against the limits of endless classification, which failed to yield the web of relationships that constitutes life. Closer inspection of things only revealed a greater number of parts that could be used to represent the whole.

The next epoch, the nineteenth century, was concerned with the evolution of the different out of the same. According to White, this period recognized not only do entities differ from one another but the parts and the relationships between parts within the things themselves, by which we classify things, also change over time. Instead of the analysis of wealth, political economists turned to the analysis of modes of production. Measurement and order were displaced by analogy and succession, with prominence given to the issue of time. For White, the operative trope in this epoch is synecdoche, which pertains to the microcosm-macrocosm relationship.

The break with this epoch came with Nietzsche at the end of the nineteenth century, who indicated that language is incapable of representation, of the kind foisted upon it. With Freud and Lévi-Strauss, the categories of succession and analogy are displaced by those of finitude and infinity. For White, the operative trope in this epoch is irony, the recognition of incongruities between what is and what our discourses lead us to expect, and yet to act in spite of these incongruities. By this reading, latter-day empiricism, positivism, and romanticism are caught up in the trope of irony in their discourses of the real, though they have not yet institutionalized this recognition in their discourses.

The history of the real is a series of mythic transformations, transformations through different metaphorical schemas. It is also the transformation of the myth of metaphor.

43. R. Dyer, "Entertainment and utopia," 277–278.

44. See S. Turkle, *Life on the Screen: Identity in the Age of the Internet.*

45. M. G. Plattel, *Utopian and Critical Thinking*, 45.

46. A. J. Ayer, *Language, Truth and Logic*, 32.

47. Plattel points to the tension within utopian writing between a sensuous, pleasure-filled world and a world of order. This is a theme taken up by Freud and, more recently, by poststructuralists. This reflects the Enlightenment tension, which set the body against the mind, etc. See Plattel, *Utopian and Critical Thinking*, 48.

48. See J. C. Weinsheimer, *Gadamer's Hermeneutics: A Reading of Truth and Method*, for an explanation of interpretation and the myth of excursion and return.

49. The earliest systematic treatment of time is contained in Augustine's *Confessions*, to which Heidegger makes reference in *Being and Time*. See also P. Ricoeur, *Time and Narrative, Volume I.*

50. H. Bergson, *Introduction to Metaphysics.*

51. J. Derrida, *Of Grammatology.*

52. Derrida takes issue with Heidegger on the subject of the primacy of anticipation, arguing that events take us from behind, unexpectedly. Hence it is the back that is of more interest than the front. See J. Derrida, *The Postcard: From Socrates to Freud and Beyond.*

53. For Heidegger the "not yet" is best exemplified in the anticipation of death, a situation in which being ceases. We do not need to investigate Heidegger's concepts of death here. Refer also to E. Bloch, *The Principle of Hope*, which focuses on the "not yet" and Marxism.

54. M. Heidegger, *Being and Time*, 374.

55. The ancient philosophical term "a priori" captures something of this sense. A priori means "from the previous": a term is a priori if it is given or assumed, but there is no implication that there is a chronology to this givenness.

56. H. L. Dreyfus, *Being-in-the-World*, 200.

57. The way that "men from mars" served as surrogate communist infiltrators in science fiction movies is legendary.

Chapter 6

1. M. Pearce, "From urb to bit," 7.

2. Ibid.

3. M. Stefik, *Internet Dreams: Archetypes, Myths, and Metaphors*, xxiii.

4. K. A. Franck, "When I enter virtual reality, what body will I leave behind?" 23. Critics such as Robins regard such reflections as at "the fag-end of a Romantic sensibility." See K. Robins, "Cyberspace and the world we live in," 139.

5. S. Turkle, *Life on the Screen: Identity in the Age of the Internet*, 15.

6. Ibid.

7. M. Poster, "Postmodern virtualities," 80.

8. Ibid.

9. Ibid.

10. Turkle, *Life on the Screen*, 63.

11. Ibid., 20.

12. Ibid., 18.

13. Ibid., 20.

14. Ibid., 144.

15. Ibid., 22.

16. Ibid., 45.

17. Ibid., 139.

18. Ibid., 263.

19. D. J. Haraway, *Simians, Cyborgs, and Women: The Reinvention of Nature*, 152. Jencks concurs: "The distinctions between being born and being made are disappearing, those between first nature, culture and second nature are dissolving" (C. Jencks, *The Architecture of the Jumping Universe: A Polemic: How Complexity Science Is Changing Architecture and Culture,* 165). See D. Brande, "The business of cyberpunk: Symbolic economy and ideology in William Gibson," for further discussion of the nature of the cyborg as expounded by D. Haraway, W. Gibson and S. Zizek.

20. Haraway, *Simians, Cyborgs, and Women*, 152.

21. Ibid., 153.

22. Ibid.

23. Ibid.

24. See M. E. Clynes and N. S. Kline, "Cyborgs and space."

25. See V. Sobchack, "Beating the meat/surviving the text, or how to get out of this century alive," for a first person account of prosthetics opposed to Haraway's argument.

26. D. J. Haraway, *Simians, Cyborgs, and Women*, 178.

27. Ibid., 150.

28. Ibid., 151.

29. See A. Breton, *Manifestoes of Surrealism* and *What Is Surrealism? Selected Writings*.

30. M. Sarup, *Jacques Lacan*, 19.

31. M. Esslin, *The Theatre of the Absurd*, 235.

32. According to Esslin, nonsense connotes magic, and magic incantations often consisted of sequences of syllables that have rhyme or rhythm but seem to have lost any sense or meaning (M. Esslin, *The Theatre of the Absurd*, 244), a theme not lost on computer devotees who, as we considered in relation to romantic medieval-

ism (chapter 1), recognize and play on the "mysterious" properties of symbol sequences. For Esslin, surrealism also inherited the nihilism of Nietzsche, which presents "in anxiety or with derision, an individual human being's intuition of the ultimate realities as he experiences them; the fruits of one man's descent into the depths of his personality, his dreams, fantasies, and nightmares" (M. Esslin, *The Theatre of the Absurd*, 293).

33. J.-F. Lyotard, "The sublime and the avant-garde."

34. I. Kant, *Observations on the Feeling of the Beautiful and Sublime.*

35. E. Burke, *A Philosophical Enquiry into the Origin of Our Ideas of the Sublime and Beautiful.*

36. J.-F. Lyotard, "The sublime and the avant-garde," 204.

37. Ibid., 206. For Lyotard, the instantaneity and transience of information, and the dependence of the capitalist system on it, speak of the sublime, in that information speaks of lack and privation.

38. Dalí composed a storyboard of surrealist drawings for a sequence in *Fantasia*, but this was abandoned. See F. Thomas and O. Johnston, *Disney Animation: The Illusion of Life.*

39. Dalí designed the surrealist dream sequence in *Spellbound* (1951).

40. Overtly in the influence of Yoko Ono on John Lennon. Lennon was also a fan of Lewis Carroll.

41. By groups such as Orb, Future Sound of London, Node, and proponents of the multitude of remix styles, continuing the trajectory of the avant-garde musical assemblages of Steve Reich and others.

42. L. McCaffery, "An interview with William Gibson," 272.

43. M. C. Boyer, *Cybercities: Visual Perception in the Age of Electronic Communication.*

44. Max Ernst reported in Breton, *Manifestoes of Surrealism*, 275.

45. "Collage is mechanical whereas morphing is alchemical" (M. Novak, "Transmitting architecture: transTerraFirma/TidsvagNoll v2.0," 46).

46. Breton, *Manifestoes of Surrealism*, 174.

47. Pierre Reverdy, quoted in A. Breton, *Manifestoes of Surrealism,* 20.

48. J. Chénieux-Gendron, *Surrealism*, 65.

49. Ibid.

50. Ibid., 70.

51. McCaffery, "An interview with William Gibson," 269. According to Gibson: "Once I've hit on an image, a lot of what I do involves the controlled use of collage; I look around for ways to relate the image to the rest of the book" (281).

52. J. González, "Envisioning cyborg bodies: Notes from current research."

53. See Boyer, *Cybercities*. Punt also makes the connection between the flaneur and the net surfer. See M. Punt, "Accidental machines: the impact of popular participation in computer technology, 79.

54. See S. Buck-Morss, *The Dialectics of Seeing: Walter Benjamin and the Arcades Project.* For Benjamin, "no face is surrealistic in the same degree as the true face of a city" (W. Benjamin, "Surrealism," 182).

55. A. Breton, *Nadja*, 52.

56. Breton says that of all intellectual movements, "Surrealism is the only one to have armed itself against any inclination toward idealist fantasy" (A. Breton, *Manifestoes of Surrealism*, 272).

57. S. Alexandrian, *Surrealist Art*, 141.

58. See chapter 1 and E. Hooper-Greenhill, *Museums and the Shaping of Knowledge.*

59. McCaffery, "An interview with William Gibson," 275.

60. As is Virilio's "vision machine." See P. Virilio, *The Vision Machine.*

61. D. J. Haraway, "Cyborgs and symbionts: Living together in the new world order," xiv.

62. W. Gibson, *Neuromancer*, 51.

63. McCaffery, "An interview with William Gibson," 264.

64. I. Csicsery-Ronay, "Cyberpunk and neuromanticism," 190.

65. Franck, "When I enter virtual reality, what body will I leave behind?" 20.

66. Stefik, *Internet Dreams*, 390.

67. Alexandrian, *Surrealist Art*, 49.

68. Breton, *Manifestoes of Surrealism*, 14.

69. Sarup, *Jacques Lacan*, 19.

70. Breton, *Manifestoes of Surrealism*, 36.

71. Sarup, *Jacques Lacan*, 21.

72. Breton, *Manifestoes of Surrealism*, 277.

73. Ibid., 26.

74. Ibid., 277.

75. Ibid., 277.

76. Novak, "Transmitting architecture: transTerraFirma/TidsvagNoll v2.0."

77. J. H. Frazer, "The architectural relevance of cyberspace."

78. Sophisticated photorealistic imagery is often a collage of scanned photographs. A series of texture maps are scanned from physical materials and pictures of skies, or generated by algorithms, and then "wrapped" around geometrically defined objects. Parts of the image are therefore taken from photographs, distorted according to the rules of perspective, such that it is difficult to discern what is computer generated and what is not.

79. Frazer, "The architectural relevance of cyberspace," 76.

80. Alexandrian, *Surrealist Art*, 50.

81. G. F. Orenstein, *The Theater of the Marvelous: Surrealism and the Contemporary Stage*, 4.

82. Breton, *Manifestoes of Surrealism*, 26–27.

83. Ibid., 40.

84. Alexandrian, *Surrealist Art*, 29.

85. Sarup, *Jacques Lacan*, 17.

86. Breton, *Manifestoes of Surrealism*, 140.

87. Ibid., 158.

88. Ibid.

89. Ibid., 177.

90. Lyotard, "The sublime and the avant-garde," 211.

91. A. Breton, *Communicating Vessels*, 55.

92. Ibid.

93. Ibid., 56.

94. M. Heim, *The Metaphysics of Virtual Reality*, 83–108.

95. J. Chénieux-Gendron, *Surrealism*, 52.

96. Ibid., 51–52.

97. Breton, *Manifestoes of Surrealism*, 177.

98. Chénieux-Gendron, *Surrealism*, 54.

99. Breton, *Manifestoes of Surrealism*, 23.

100. Breton, *Entretiens*, with André Parinaud, 81, quoted in Chénieux-Gendron, *Surrealism*, 49. For Breton: "This is a condition that is held to be much more a part of the workings of Eastern thought than of Western, and presupposes on the part of the Western mind a more sustained tension and effort" (Ibid.).

101. Chénieux-Gendron, *Surrealism*, 50.

102. S. Dalí, "Le Surréalisme ASDLR," quoted in J. Chénieux-Gendron, *Surrealism*, 85.

103. Chénieux-Gendron, *Surrealism*, 62.

104. A. R. Stone, *The War of Desire and Technology at the Close of the Mechanical Age*.

105. S. Freud, "Delusions and dreams in Jensen's *Gradiva*."

106. Sarup, *Jacques Lacan*, 18.

107. Ibid., 18–19.

108. M. Caws, Introduction to *Communicating Vessels*, xiii.

109. Ibid.

110. Breton, *Communicating Vessels*, 152.

111. Chénieux-Gendron, *Surrealism*, 180.

112. S. Freud and C. Jung, *The Freud/Jung Letters*, 47.

113. . . . as implied by Stone. See A. R. Stone, *The War of Desire and Technology at the Close of the Mechanical Age*, 94.

114. Sophocles, *The Thebian Plays*.

115. E. Fromm, *The Forgotten Language: An Introduction to the Understanding of Dreams, Fairy Tales and Myths*, 173.

116. S. Freud, "Civilization and its discontents," 324.

117. Ibid., 326.

118. Ibid., 327. Freud also extends the language of individual psychosis to society as a whole: "under the influence of cultural urges, some civilizations, or some epochs of civilization—possibly the whole of mankind—have become 'neurotic'?" (338).

119. Ibid., 338.

120. S. Freud, "The dissolution of the Oedipus complex," 318.

121. For little girls, the phallus becomes an object of envy: "When she makes a comparison with a playfellow of the other sex, she perceives that she has 'come off badly' and she feels this is a wrong done to her and as a ground for inferiority" (S. Freud, "The dissolution of the Oedipus complex," 320).

122. Freud and Jung, *The Freud/Jung Letters*, 164.

123. C. G. Jung, *Four Archetypes: Mother, Rebirth, Spirit, Trickster*, 42; C. G. Jung, *Psychological Types*, 187.

124. E. Fromm, *The Forgotten Language*, 174–181.

125. A. C. Clarke, *2001: A Space Odyssey*.

126. For Boyer, the priestess is the Virgin Mother with her machine double, or dark side. See M. C. Boyer, *Cybercities, Visual Perception in the Age of Electronic Communication*, 110.

127. See J. Abbott, "the 'monster' reconsidered: *Blade Runner*'s replicant as romantic hero," for a similar account.

128. S. Freud, "Infantile sexuality," 116–117.

129. S. Freud, "On transformations of instinct as exemplified in anal eroticism," 299.

130. Ibid.

131. Ibid., 300.

132. L. Irigaray, *This Sex Which Is Not One*.

133. L. Armitt, *Theorising the Fantastic*.

134. Wolstenholme provides several Freudian interpretations of the story as presented in the original book by Frank Baum. As in many fairy tales, Dorothy is a young princess, "heir to more than what her immediate circumstances imply" (xxxi) and what her parents can offer her. There is the "obsessive return of mother figures" (xxxix), Aunt Em and the wicked witches whom Dorothy has to keep killing. The Wizard is an obvious father figure and an impotent fraud at that. The rite of passage theme is about the problem of keeping young girls down on the farm after they have seen the big city. See S. Wolstenholme, Introduction to *The Wonderful Wizard of Oz*.

135. By the end of *Alien Resurrection*, Ripley is herself both human and alien, consorting with a humanoid robot, with whom she is in sisterly embrace, fondly contemplating the spectacle of the wholeness of the earth from a space ship.

136. S. Freud, "The ego and the id," 376.

137. Ibid., 361. Rapaport identifies three successive developments in the concept of reality as used by Freud: (i) a defense against a real event so as to prevent its recurrence, (ii) the focus of the drive object, and (iii) the reality principle versus the pleasure principle arbitrated by the ego. See D. Rapaport, "The structure of psychoanalytic theory: A systematizing attempt," 97–101; Ricoeur, *Freud and Philosophy: An Essay in Interpretation*, 351, for two other categories developed by his successors.

138. S. Freud, "Why war?" 361.

139. Ibid.

140. Ibid.

141. Ricoeur summarizes the empiricist criticisms leveled against Freud in *Freud and Philosophy*.

142. O. L. Zangwill, "Freud," in *The Oxford Companion to the Mind*, 269.

143. J. Strachey, "Sigmund Freud: A sketch of his life and ideas," 22.

144. Ricoeur, *Freud and Philosophy*, 375.

145. M. Foucault, "What is an author?" 116.

146. The superego has a moralizing function and is the source of guilt. In popular terms, it is the conscience. In the early development of the child, the sexually oriented impulses of the Oedipus complex are in part desexualized and sublimated by being incorporated in the superego. They are in part inhibited in their aim and changed into impulses of affection. According to Freud: "The whole process has, on the one hand, preserved the genital organ—has averted the danger of its loss—and, on the other, has paralyzed it—has removed its function" (S. Freud, "The dissolution of the Oedipus complex," 319). The sexual impulses are then latent until puberty. For girls: ". . . the girl accepts castration as an accomplished fact, whereas the boy fears the possibility of its occurrence" (S. Freud, "The dissolution of the Oedipus complex," 321). Her Oedipus complex culminates in a desire to receive a baby from her father "as a gift—to bear him a child" (Ibid.). The complex is then given up by the girl as the desire is never fulfilled.

The third component of the psyche is the id, which pertains to uncoordinated instinct. So the psyche exhibits a topology in which "consciousness is the *surface* of the mental apparatus; that is, we have ascribed it as a function to a system which is spatially the first one reached from the external world" (Freud, "The ego and the id," 357). Physical stimuli reach the ego, which is the organized, realistic part of the psyche, through this "surface." The id pertains to uncoordinated instinctual trends. Whereas the ego deals in perceptions, the id pertains to instinct. The ego deals in reason and common sense, while the id deals in passions (Freud, "The ego and the id," 364). Pleasure reigns unrestricted in the id, were it not for the attempts by the ego to bring the influence of the external world to bear on it. The way the ego must guide the id is much as a rider controls a horse, which often requires letting the horse go where it wants to go if rider and horse are to stay together. The superego has a critical and moralizing function: "It is easy to show that the ego ideal [superego] answers to everything that is expected of the higher nature of man. As a substitute for a longing for the father, it contains the germ from which all religions have evolved" (Freud, "The ego and the id," 376). It is the "heir of the Oedipus complex": "By setting up this ego ideal [superego], the ego has mastered the Oedipus complex and at the same time placed itself in subjection to the id" (Freud, "The ego and the id," 376).

147. Turkle, *Life on the Screen: Identity in the Age of the Internet*, 263.

148. E. Fromm, *The Fear of Freedom*, 18.

149. Freud, "Infantile sexuality," 90.

150. S. Freud, "Repression," 147.

151. Ibid., 154.

152. Ibid., 155.

153. Ibid., 157.

154. Ibid., 157.

155. S. Freud, "The unconscious," 167. In fact, Freud formulated the psychic structure as comprising the conscious, the preconscious, and the unconscious. See Freud, "The ego and the id," 353.

156. Ricoeur, *Freud and Philosophy*, 372.

157. Jung introduces the concept of the *collective* unconscious into the debate, which comprises the "deposits of thousands of years of experience of the struggle for existence and for adaptation" (C. Jung, *Psychological Types*, 221).

158. H. Dreyfus, *What Computers Can't Do. The Limits of Artificial Intelligence.*

159. "The ego is first and foremost a bodily ego; it is not merely a surface entity, but is itself the projection of a surface" (Freud, "The ego and the id," 364).

160. Derrida tells us that the child was probably Freud's nephew in J. Derrida, "Coming into one's own."

161. S. Freud, "Beyond the pleasure principle," 284.

162. Ibid., 285.

163. Ibid., 286.

164. Ibid.

165. S. Freud, "A disturbance of memory on the Acropolis."

166. Ibid., 449.

167. Ibid., 452.

168. Ibid., 455.

169. Ibid., 456.

170. As we saw in chapter 5, the uncanny is also one of Heidegger's themes, which he relates to anxiety.

171. In fact, the worlds are uninhabitable. There are no outside public spaces, the transportation systems only carry one person at a time, the paths and bridges only take people in single file, and in spite of the excess of plumbing, there appear to be no lavatories.

172. Sarup, *Jacques Lacan*, 21.

173. S. Freud, "The 'uncanny,' " 353.

174. Ibid., 352.

175. The Oedipal theme is even more pronounced in *Riven*. There are two generations of fathers, each wanting to destroy the other. The grandfather is the most censorious, evil, and rule bound, and in one ending seems to shoot the losing player in the groin!

176. Freud, "The 'uncanny,' " 357.

177. Ibid., 358.

178. Ibid., 361.

179. Ibid., 371.

180. Ibid.

181. Ibid., 367.

182. Ibid., 372.

183. Robins, "Cyberspace and the world we live in," 139.

184. Freud, "Beyond the pleasure principle," 311.

185. Ibid., 278.

186. Ibid., 329.

187. Chénieux-Gendron, *Surrealism*, 122.

188. Ibid., 123.

189. Sarup, *Jacques Lacan*, 22.

190. Lévi-Strauss, *Structural Anthropology*, 210.

191. Ibid., 216.

192. Ibid., 217.

Chapter 7

1. See M. Jay, Downcast Eyes: The Denigration of Vision in Twentieth Century Thought, 211–262; M. Foucault, "Of other spaces"; and M. Foucault, *The Order of Things: An Archaeology of the Human Sciences*, on the subject of the mirror. In the latter, Foucault describes the painting *Las Meninas*, by Velázquez, in which: "No gaze is stable, or rather, in the neutral flow of the gaze piercing at a right angle through the canvas, subject and object, the spectator and the model, reverse their roles to infinity" (5).

2. S. Chaplin, "Cyberspace: Lingering on the threshold (architecture, post-modernism and difference)," 32.

3. Ibid., 33.

4. P. Tabor, "I am a videocam: The glamour of surveillance," 17.

5. R. Caillois, "Mimicry and legendary psychaesthenia."

6. M. Sarup, *Jacques Lacan*, 25.

7. J. Lacan, *The Four Fundamental Concepts of Psychoanalysis*, 82.

8. Ibid.

9. See, for example, S. Žižek, *Looking Awry: An Introduction to Jacques Lacan through Popular Culture, Enjoy Your Symptom! Jacques Lacan in Hollywood and Out*, and J. Forrester, *The Seductions of Psychoanalysis: Freud, Lacan and Derrida*.

10. M. Bowie, *Lacan*, 67. See R. Tallis, "The shrink from hell," and A. Sokal and J. Bricmont, *Intellectual Impostures: Postmodern Philosophers' Abuse of Science*, for less charitable views of Lacan and examples of the rage he has provoked in certain circles.

11. Sarup, *Jacques Lacan*, 113.

12. Lacan, *The Four Fundamental Concepts of Psychoanalysis*, 169.

13. According to Sarup: "Certain of Dalí's double or multiple images might be illustrations of Lacan's views on the mirror phase, and the narcissistic construction and function of the ego" (M. Sarup, *Jacques Lacan*, 23).

14. I am indebted to Elizabeth Cowley for pointing this out to me.

15. E. Casey and J. M. Woody, "Hegel, Heidegger, Lacan: The dialectic of desire," 88.

16. This holistic phase is also the cornerstone of the psychological theories of Piaget and Fromm. See E. Fromm, *The Fear of Freedom*, 19–23.

17. Casey and Woody summarize the position as follows: "They include the intrinsic incommensurabilities between the repressing and the repressed elements of the self, the signifier and the signified, language and speech, self and other, ego and Other" (E. Casey and J. M. Woody, "Hegel, Heidegger, Lacan," 88).

18. E. Grosz, *Jacques Lacan: A Feminist Introduction*, 35.

19. Lacan, *The Four Fundamental Concepts of Psychoanalysis*, 188.

20. See S. Žižek, *The Sublime Object of Ideology*, 161–164, 169–173, and D. Evans, *An Introductory Dictionary of Lacanian Psychoanalysis*, 159.

21. Quoted in Bowie, *Lacan*, 95. A direct quote from Lacan, *Ecrits: A Selection*, 388.

22. Lacan, *The Four Fundamental Concepts of Psychoanalysis*, 53.

23. Zizek, *The Sublime Object of Ideology*, 163.

24. Ibid., 164.

25. F. L. Baum, *The Wonderful Wizard of Oz*, 199.

26. Žižek, *The Sublime Object of Ideology*, 163. Žižek does not use the example of cyberspace.

27. Ibid., 171.

28. This is apparently compounded by the difficulty of translating from French to English. The English translator of *Écrits: A Selection* provides definitions of

Lacanian terms but admits that Lacan was unhelpful in the English translation of certain words. In the case of "*objet petit* a," Lacan prefers the reader to develop an appreciation of the concept in the course of its use: "Furthermore, Lacan insists that '*objet petit* a' should remain untranslated, thus acquiring, as it were, the status of an algebraic sign" (J. Lacan, *Écrits*, xi).

29. Evans, *An Introductory Dictionary of Lacanian Psychoanalysis*, 160.

30. Bowie, *Lacan*, 83. See also A. Wilden, "Lacan and the discourse of the other."

31. J. Lacan, *The Four Fundamental Concepts of Psychoanalysis*, 188.

32. Ibid.

33. "If I have said that the unconscious is the discourse of the Other (with a capital O), it is in order to indicate the beyond in which the recognition of desire is bound up with the desire for recognition" (J. Lacan, *Écrits*, 172).

34. Ibid., 196.

35. Lacan, *The Four Fundamental Concepts of Psychoanalysis*, 167.

36. Lacan implicates the phallus in the rift of our being and generalizes the concept of the phallus as a metalinguistic phenomenon. He explains the phallus as the supreme signifier, "the signifier of the signifier." He develops Saussure's pseudo-algebraic formulation S/s, which indicates the relationship of signifier (S) to signified (s), with the signifier (S) in the dominant position. For Lacan, the bar separating the two is the phallus. According to Casey and Woody, "The phallus signifies this bar in its simultaneously repressing and revealing role . . . Herein lies the origin of the split subject, barred from the urgent finalism of the desiring self" (E. Casey and J. M. Woody, "Hegel, Heidegger, Lacan," 109). The phallus is that which maintains the separation that is our being in language.

37. Lacan, *The Four Fundamental Concepts of Psychoanalysis*, 53–54.

38. Bowie, *Lacan*, 95.

39. Lacan, *The Four Fundamental Concepts of Psychoanalysis*, 154.

40. "The discrepancy between a disparate, incomplete subject and its imaged unity only anticipates a more profound splitting of the subject due to 'the agency of the letter,' to the subject's entrance into the symbolic order" (E. Casey and J. M. Woody, "Hegel, Heidegger, Lacan," 103).

41. Casey and Woody, "Hegel, Heidegger, Lacan," 103. According to Casey and Woody, the word "I," which replaces the image in the mirror, does not provide a unity. For Lacan, the word "I" is a fluid term and does not primarily signify the self. It is not dependent on individuality.

42. Grosz, *Jacques Lacan*, 59.

43. Ibid., 65.

44. Lacan, *Écrits*, 287.

45. Bowie, *Lacan*, 137.

46. See C. Alexander, *Notes on the Synthesis of Form.*

47. So technology and desire do not only meet as the site for people to have "computer sex," as Stone seems to suggest in *The War of Desire and Technology at the Close of the Mechanical Age;* rather, desire is at the heart of the technological imperative, the conflict inherent in the will to control.

48. Lacan, *Écrits*, 199. According to Gross: "In introjecting the name-of-the-father, the child (or rather, the *boy*) is positioned with reference to the father's name. He is now bound to the law, in so far as he is implicated in the symbolic 'debt,' given a name, and an authorized speaking position. The paternal metaphor is not a simple incantation but the formula by which the subject, through the construction of the unconscious, becomes an 'I,' and can speak in its own name" (Grosz, *Jacques Lacan*, 71).

49. Casey and Woody, "Hegel, Heidegger, Lacan," 108.

50. See S. D. O'Leary, "Cyberspace as sacred space: Communicating religion on computer networks."

51. S. Wolstenholme, Introduction to *The Wonderful Wizard of Oz*, xxxiv. By another reading, Dorothy is already connected with the earth on the farm in Kansas. The story is about Dorothy's return to the simple, honest country life.

52. Baum, *The Wonderful Wizard of Oz.*

53. Wolstenholme, Introduction to *The Wonderful Wizard of Oz*, xxxvi.

54. A. C. Clarke, *2001: A Space Odyssey*, 234.

55. Ibid.

56. Quoted in H. Marcuse, *Eros and Civilisation: A Philosophical Inquiry into Freud*, 258. See Fromm, *The Fear of Freedom.* According to Fromm, man "has no choice but to unite himself with the world in the spontaneity of love and productive work" (18). See also E. Fromm, *The Sane Society.*

57. Marcuse, *Eros and Civilisation*, 256.

58. Ibid., 67.

59. Ibid.

60. G. Deleuze and F. Guattari, *Anti-Oedipus: Capitalism and Schizophrenia.*

61. Deleuze and Guattari approve of Lacan's indications that "Oedipus is imaginary, nothing but an image, a myth; that this or these images are produced by an oedipalizing structure; that this structure acts only insofar as it reproduces the element of castration, which itself is not imaginary but symbolic" (G. Deleuze and F. Guattari, *Anti-Oedipus*, 310).

62. K. Marx and F. Engels, "Manifesto of the Communist Party," 66.

63. Deleuze and Guattari, *Anti-Oedipus*, 312.

64. Ibid., 275.

65. Ibid., 276.

66. ". . . first as a stimulus of departure, then as an aggregate of destination, and finally as an intermediary or an interception of communication" (G. Deleuze and F. Guattari, *Anti-Oedipus*, 276).

67. Ibid., 311.

68. Ibid.

69. M. Foucault, "Preface" xiii.

70. Compare with Kroker and Cook's characterization of television as "the consumption machine of late capitalism," in A. Kroker and D. Cook, "Television and the triumph of culture (from *The Postmodern Scene*)," 235.

71. D. Porush, "Hacking the brainstem: Stephenson's *Snow Crash*," 108.

72. Deleuze and Guattari, *Anti-Oedipus*, 10.

73. Ibid., 9.

74. Ibid, 36.

75. Ibid. They add, "Every machine has a sort of code built into it, stored up inside it. This code is inseparable not only from the way in which it is recorded and transmitted to each of the different regions of the body, but also from the way in which the relations of each of the regions with all the others are recorded" (38).

76. Ibid., 42. See also Markley on the subject of the desiring machine: "Cyberspace must destabilize liberal humanist conceptions of identity in favor of postmodern, fragmented, and performative subjectivities to market itself as a means to transform this always unstable and always incomplete identity into a thoroughly efficient desiring machine. The dream of cyberspace is the dream of infinite

production" (R. Markley, "Boundaries: mathematics, alienation, and the metaphysics of cyberspace," 74).

77. P. Virilio, *The Vision Machine*, 70.

78. See G. Ulmer, "The object of post-criticism." Derrida's style of analysis has also been identified with the tradition of Dada by Caputo. See J. Caputo, *Radical Hermeneutics: Repetition, Deconstruction and the Hermeneutic Project*, 212. The surrealist connection has also been made by E. D. Ermarth, *Sequel to History: Postmodernism and the Crisis of Representational Time.*

79. Derrida says that it is more difficult to see the world otherwise than metaphysically. It is not something we should necessarily resist, but we should recognize that any assertion is "under erasure." We can be aware of our metaphysical assumptions, and unsettle them, even as we assert them. The writing of Deleuze and Guattari as well as Haraway could be construed as such writing, though theirs is not the only way.

80. But then a social contructivist view of the real is also metaphysical in affirming the centrality of human culture. In fact, Derrida's project is most effective when confronting discourses that claim they are against metaphysics in some way.

81. Caputo, *Radical Hermeneutics*, 135.

82. J. Derrida, "Freud and the scene of writing."

83. J. Derrida, "Coming into one's own."

84. Ibid., 115.

85. Derrida, "Freud and the scene of writing," 202.

86. Ibid.

87. Ibid.

88. Caputo, *Radical Hermeneutics*, 139.

89. See W. Benjamin, "The work of art in the age of mechanical reproduction."

90. Caputo, *Radical Hermeneutics*, 146.

91. But Derrida is able to show how even Lacan's discourse betrays a commitment to what it is he seems to be resisting. See B. Johnson, "The frame of reference: Poe, Lacan, Derrida"; E. A. Poe, "The purloined letter"; and J. Lacan, Seminar on "The purloined letter."

92. A. Giddens, *The Transformation of Intimacy: Sexuality, Love and Eroticism in Modern Societies*, 114.

93. P. Ricoeur, *Freud and Philosophy: An Essay on Interpretation*, 30.

94. Freud, Marcuse, Deleuze and Guattari, and others pick up on the concept of Cartesian doubt, but whereas Descartes doubted things, they have introduced doubt into the very process of reflection on that doubt, on consciousness. According to Ricoeur, psychoanalysis is a process where "to seek meaning is no longer to spell out the consciousness of meaning, but to *decipher its expressions*" (P. Ricoeur, *Freud and Philosophy*, 33).

95. P. Virilio, *Open Sky*, 20.

96. See H.-G. Gadamer, *Truth and Method*.

97. Giddens, *The Transformation of Intimacy*, 117. According to Rorty, Freud "suggested that we praise ourselves by weaving idiosyncratic narratives—case histories, as it were—of our success in self-creation, our ability to break free from an idiosyncratic past" (R. Rorty, *Contingency, Irony, and Solidarity*, 33).

98. Johnson, "The frame of reference," 163.

99. Ibid.

100. P. Ricoeur, *Freud and Philosophy*, 519.

101. These rereadings of the Oedipus myth seem to take it further and further from the issue of sex and the libido. But there is sufficient in the interpretive version to cover sex, and there is enough in Lacan, Marcuse, Irigaray, and Deleuze and Guattari to support the case that the sex drive is but a particular manifestation of Eros, that the quest for understanding is already erotic.

102. Johnson captures the nature of this interpretive turn and what it can do for an understanding of psychoanalysis. She argues that psychoanalysis is the original moment, "the primal scene" it is seeking: "It is the *first* occurrence of what has been repeating itself in the patient without ever having occurred" (Johnson, "The frame of reference," 167–168). She adds that psychoanalysis is not "the *interpretation* of repetition," but the "repetition of a *trauma of interpretation*—called 'castration' or 'parental coitus' or 'the Oedipus complex' or even 'sexuality.' It is the traumatic deferred interpretation not *of* an event, but *as* an event that never took place as such. The 'primal scene' is not a scene but an *interpretative infelicity* whose result was to situate the interpreter in an intolerable position. And psychoanalysis is the reconstruction of that interpretative infelicity not as *its* interpretation, but as its first and last act. Psychoanalysis has content only insofar as it repeats the dis-content of what never took place" (Ibid.).

103. S. Gallagher, *Hermeneutics and Education*.

104. P. Ricoeur, *Freud and Philosophy*, 413. Clearly not all schools of psychoanalytic practice endorse Freud's language, which promotes further resistance, and we can say the same of the application of Freud to culture and technology studies.

105. C. M. Boyer, *Cybercities: Visual Perception in the Age of Electronic Communication*.

106. I. Calvino, *Invisible Cities*.

107. For a technical summary of VR capabilities and problems, see J. Wann and M. Mon-Williams, "What does virtual reality NEED?: Human factors issues in the design of three-dimensional computer environments."

108. For a summary of issues pertaining to "telepresence," see E. Paulos and J. Canny, "Ubiquitous tele-embodiment: Applications and implications."

109. For applications of VR and "telepresence" in medicine and surgery, see M. Solaiyappan, T. Poston, P. A. Heng, E. R. McVeigh, M. A. Guttman and E. A. Zerhouni, "Interactive visualization for rapid noninvasive cardiac assessment"; and F. L. Kitson, T. Malzbender, and V. Bhaskaran, "Opportunities for visual computing in healthcare."

110. S. Fish, *Doing What Comes Naturally: Change, Rhetoric, and the Practice of Theory in Literary and Legal Studies*, 141.

111. Ibid., 150.

112. Ibid.

113. Ricoeur, *Freud and Philosophy*, 407.

114. Giddens, *The Transformation of Intimacy*.

115. There is no going back to this situation, and Giddens argues that attempts to reassert tradition in the modern world constitute a kind of fundamentalism, the unreflexive assertion of traditional practices in a highly reflexive, nontraditional society. We are already reflexive about almost every aspect of our lives, and to deny this reflexivity can only be accomplished under some kind of tyranny.

116. Giddens, *The Transformation of Intimacy*, 30.

117. For Giddens, the value of Freud is that he constructed a particular language game and presented particular connections as problematic: "The importance of Freud was not that he gave the modern preoccupation with sex its most cogent formulation. Rather, Freud disclosed the connections between sexuality and self identity when they were still entirely obscure and at the same time showed those connections to be problematic" (A. Giddens, *The Transformation of Intimacy*, 30). Freud's language game provides a language for enabling people to construct self-

narratives: "It provides a setting, and a rich fund of theoretical and conceptual resources, for the creation of a reflexively ordered narrative of self" (31).

118. Ibid., 74. He adds: "Moreover . . . such choices are not just 'external' or marginal aspects of the individual's attitudes, but define who the individual 'is.' In other words, life-style choices are constitutive of the reflexive narrative of self" (74-75). According to E. Goffman: "The self, then, as a performed character, is not an organic thing that has a specific location, whose fundamental fate is to be born, to mature, and to die; it is a dramatic effect arising diffusely from a scene that is presented, and the characteristic issue, the crucial concern, is whether it will be credited or discredited" (E. Goffman, *The Presentation of the Self in Everyday Life*, 245).

119. Giddens, *The Transformation of Intimacy*, 147. Giddens translates contemporary symptoms of neuroses, such as domestic violence, that men are "unable to express feelings," and the breakup of relationships, away from the Oedipus myth to the problem of constructing narratives of the self: "Instead, we should say that many men are unable to construct a narrative of self that allows them to come to terms with an increasingly democratized and reordered sphere of personal life" (117).

120. Ricoeur, *Freud and Philosophy*, 382.

121. Ibid.

122. Ibid., 482.

123. Ibid., 383.

124. See R. Coyne, *Designing Information Technology in the Postmodern Age: From Method to Metaphor*, 264–270.

125. Foucault, *Discipline and Punish.*

126. Giddens extends the body theme to consider the body as "a visible carrier of self-identity," which is increasingly integrated into the "life-style decisions which an individual makes" (A. Giddens, *The Transformation of Intimacy*, 31). Giddens uses the example of diet, which is a matter of individual choice, and a bewildering choice, made in the context of a bewildering array of guides, manuals, cookbooks, tv shows, etc. We can now add the prospect of choice as to bodily form presented by IT "manuals."

127. It also accords with Milton's identification of the difference between humans and animals: humans can stand upright and look up to God (J. Milton, *Paradise Lost*, 393).

128. M. Weiser, "The computer for the 21st century."

129. The movie *Lawnmower Man* provides one example. For Vitruvius, the man was lying, inscribed in a circle and square, with hands and feet touching the circumference of a circle and the perimeter of a square.

130. Ricoeur, *Freud and Philosophy*, 415.

131. A. Kay, "User interface: A personal view."

132. H.-G. Gadamer, *Truth and Method*, 98.

133. It is on this point that Heidegger has come under a great deal of criticism, for directing attention away from the particulars of pollution, the housing problem, Nazi extermination camps, in favor of the primordial conditions of technological enframing, dwelling, and thinking. See R. Bernstein, *The New Constellation: The Ethical-Political Horizon of Modernity/Postmodernity.*

134. M. Heidegger, "The origin of the work of art," 48–49. Heidegger adds: "But the relation between world and earth does not wither away into the empty unity of opposites unconcerned with one another. The world, in resting upon the earth, strives to surmount it. As self-opening it cannot endure anything closed. The earth, however, as sheltering and concealing, tends always to draw the world into itself and keep it there" (49). See G. Hill, "The architecture of circularity: Design, Heidegger and the earth," for an interesting elaboration (and application) of Heidegger's theme of earth.

135. Gadamer, *Truth and Method*, 230.

136. Ibid., 231.

137. Ibid., 232.

138. Ibid., 236.

139. D. Haraway, *Simians, Cyborgs, and Women: The Reinvention of Nature*, 150.

140. Gadamer, *Truth and Method*, 245. On the subject of identity, and anticipating (or echoing) Lacan, Gadamer adds: "The focus of subjectivity is a distorting mirror. The self-awareness of the individual is only a flickering in the closed circuits of historical life" (245). This is a view echoed by Maffesoli: "Each social actor is less acting than acted upon. Each person is diffracted into infinity . . . Social life is then a stage upon which, for an instant, crystallization takes place" (M. Maffesoli, *The Time of the Tribes: The Decline of Individualism in Mass Society*, 145).

141. Note that Gadamer does not discuss care or desire. Having elevated the process of interpretation to the ontological, perhaps Gadamer regards care as

subsumed within the concept of expectation, fore-projection. According to Gadamer's application of Heidegger, we already come to an interpretive situation with understanding: "A person who is trying to understand a text is always performing an act of projecting" (H.-G. Gadamer, *Truth and Method*, 236). We read a text "with particular expectations in regard to a certain meaning" (236). This expectation undergoes revision during the encounter with the text, in the manner of the hermeneutical play.

142. Ibid., 276.

143. Ricoeur, *Freud and Philosophy*, 387.

144. See S. Gallagher, *Hermeneutics and Education*; and J. Caputo, *Radical Hermeneutics*.

145. This is according to White in H. White, *Tropics of Discourse: Essays in Cultural Criticism*.

Chapter 8

1. See P. Ricoeur, *Time and Narrative*, *Volume I*, 39, on the subject of narrative form. For Ermarth and A. Gibson, contemporary postmodern narrative is rather characterized as "a temporal instance of collage," an indeterminate play that seems to defy the old categories of structure, form, theme, representation, and subject. See E. D. Ermarth, *Sequel to History: Postmodernism and the Crisis of Representational Time*; and A. Gibson, *Towards a Postmodern Theory of Narrative*.

2. G. W. F. Hegel, *Hegel's Science of Logic*, 664–704.

3. See D. I. A. Cohen, *Introduction to Computer Theory*, 5.

4. This is the Ames room illusion.

5. In spite of the best efforts of the design methods movement, the language of combinatorics is entirely inadequate to the design process, which, in terms of the surrealist project, is best accounted for through the workings of metaphor and difference, or in the language of hermeneutics in terms of interpretation, judgment, and play. See A. Snodgrass and R. Coyne, "Is designing hermeneutical?"; and A. Snodgrass and R. D. Coyne, "Models, metaphors and the hermeneutics of designing."

6. See B. Lawson, *Design in Mind*, for an elaboration of this view.

7. See A. Clark, *Being There: Putting Brain, Body, and World Together Again*, 75.

8. P. Prusinkiewicz, "Visual models of morphogenesis."

9. Clark, *Being There*, 215.

10. For an example of this approach to design and its problems, see A. Bijl, *Computer Discipline and Design Practice: Shaping Our Future*; and A. Bijl, *Ourselves and Computers: Difference in Minds and Machines*.

11. See Clark, *Being There*.

12. W. Ockham, *Philosophical Writings*, 97.

13. S. Hawking, *Black Holes and Baby Universes and Other Essays*, 33.

14. F. von Schlegel, "On the limits of the beautiful," 414.

15. S. Žižek, *Tarrying with the Negative: Kant, Hegel, and the Critique of Ideology*, 44.

16. C. Shannon and W. Weaver, *The Mathematical Theory of Communication*.

17. Gregory captures the tenor of this view in a discussion of the importance of negotiation skills: "When two people are getting to know each other they are consciously or unconsciously transacting. Each is 'performing' for the benefit of the other, and each is employing cue perception and skills of interpretation to understand the other's meaning. At the same time each conveys to the other the impression the other has created, and each is motivated to conform to the mirror-image received from the other, or to correct distortions" (R. L. Gregory, *The Oxford Companion to the Mind*, 508).

18. M. E. Clynes and N. S. Kline, "Cyborgs and space."

19. Ibid., 31.

20. Ibid., 29.

21. Ibid., 31.

22. Ibid., 33.

23. For other cyborg commentators, these technologies lead to an "increasing awareness of a sense of self." See A. R. Stone, *The War of Desire and Technology at the Close of the Mechanical Age*, 20.

24. M. E. Clynes, "Cyborg II: Sentic space travel," 41.

25. Stone, *The War of Desire and Technology at the Close of the Mechanical Age*, 17.

26. There are counternarratives that dissociate themselves from the pathological fantasizing of cyborg sensibility. On occasion, Clynes retreats to a straightforward empiricist account of the cyborg. Becoming a cyborg does not alter one's "ability

to experience emotions, no more than riding a bicycle does," and "it hasn't altered their essential identity" (C. H. Gray, "An interview with Manfred Clynes," 49).

27. D. Haraway, "Cyborgs and symbionts: Living together in the new world order," xvi.

28. S. Turkle, *Life on the Screen: Identity in the Age of the Internet*, 185.

29. Ibid.

30. Ibid., 263.

31. Ibid., 259.

32. Ibid., 241.

33. Ibid., 185.

34. Ibid., 268.

35. Ibid.

36. C. H. Gray, S. Mentor, and H. J. Figueroa-Sarriera "Cyborgology: Constructing the knowledge of cybernetic organisms," 7.

37. M. C. Boyer, *Cybercities: Visual perception in the Age of Electronic Communication*, 111.

38. Žižek, *Tarrying with the Negative*, 40.

39. Ibid.

40. Ibid., 41.

41. See R. Coyne, S. McLaughlin, and S. Newton, "Information technology and praxis: a survey of computers in design practice"; and R. D. Coyne, F. Sudweeks, and D. Haynes, "Who needs the Internet? Computer-mediated communication in design firms." See also M. Pantzar, "Domestication of everyday life technology: Dynamic views on the social histories of artifacts," for an account of the various theories that situate technologies in practice contexts.

42. That such disclosures come about through fictions should cause us no disquiet. In Lacan's terms, as interpreted by Žižek, in making love the lover's attention is constructed through fantasy, which is a precondition for its recognition as real: " 'Reality' is always framed by a fantasy, i.e., for something real to be experienced as part of 'reality,' it must fit the preordained coordinates of our fantasy-space (a sexual act must fit the coordinates of our imagined fantasy-scripts, a brain must fit the functioning of a computer, etc.)" (S. Žižek, *Tarrying with the Negative*, 43, 44).

43. C. H. Gray, S. Mentor, and H. J. Figueroa-Sarriera, "Cyborgology: Constructing the knowledge of cybernetic organisms," 2.

44. D. Haraway, *Simians, Cyborgs, and Women: The Reinvention of Nature*, 150.

45. G. Deleuze, *Difference and Repetition*, 1.

46. Ibid., 3.

References

Abbott, Joe. 1993. The "monster" reconsidered: *Blade Runner*'s replicant as romantic hero, *Extrapolation* 34 (4): 340–350.

Adorno, Theodor W. 1973. *The Jargon of Authenticity*, trans. Knut Tarnowski. London: Routledge and Kegan Paul.

Adorno, Theodor W., and Max Horkheimer. 1979. *Dialectic of Enlightenment*, trans. J. Cumming. London: Verso. First published in German in 1944.

Alexander, Christopher. 1964. *Notes on the Synthesis of Form*. Cambridge: Harvard University Press.

Alexandrian, Sarane. 1970. *Surrealist Art*, trans. Gordon Clough. London: Thames and Hudson.

Allen, Reginald E. 1966. *Greek Philosophy: Thales to Aristotle*. New York: Free Press.

Aquinas, Thomas. 1993. *Selected Philosophical Writings*, ed. Timothy McDermott. Oxford: Oxford University Press.

Arendt, Hannah. 1958. *The Human Condition*. Chicago: University of Chicago Press.

Armitt, Lucie. 1996. *Theorising the Fantastic*. London: Arnold.

Augustine. [413–426] 1984. *City of God*, trans. Henry Bettenson. London: Penguin.

———. [397–400] 1991. *Confessions*, trans. Henry Chadwick. Oxford: Oxford University Press.

Austin, John. 1966. *Hou to Do Things with Words*. Cambridge, Mass.: Harvard University Press.

Ayer, A. J. 1990. *Language, Truth and Logic*. London: Penguin.

Baer, Marc. 1992. The memory of the Middle Ages: From history of culture to cultural history. In *Medievalism in England*, ed. Leslie J. Workman, 290–309. Cambridge: D. S. Brewer.

Bakhtin, Mikhail. 1984. *Rabelais and His World*, trans. Hélène Iswolsky. Bloomington: Indiana University Press.

Barnsley, M. 1988. *Fractals Everywhere*. Boston, Mass.: Academic Press.

Barr, Avron, and Edward A. Feigenbaum, eds. 1981. *The Handbook of Artificial Intelligence: Volume I*. Los Altos, Calif.: William Kaufmann.

Barrett, Edward, ed. 1994. *Sociomedia: Multimedia, Hypermedia, and the Social Construction of Knowledge*. Cambridge, Mass.: MIT Press.

Barthes, Roland. 1973. *Mythologies*, trans. Annette Lavers. London: Paladin.

Baudrillard, Jean. 1988. Simulacra and simulations. In *Jean Baudrillard: Selected Writings*, 166–184, Stanford, Calif.: Stanford University Press.

Baum, Frank L. 1997. *The Wonderful Wizard of Oz*. Oxford: Oxford University Press. First published in 1900.

Bellamy, Edward. 1967. *Looking Backward 2000–1887,* ed John L. Thomas. Cambridge, Mass.: Harvard University Press. First published in 1888.

Benedikt, Michael. 1991a. *Cyberspace: First Steps,* Cambridge, Mass.: MIT Press.

———. 1991b. Cyberspace: Some proposals. In *Cyberspace: First Steps,* ed. Michael Benedikt, 119–224. Cambridge, Mass.: MIT Press.

Benjamin, Walter. 1969. The work of art in the age of mechanical reproduction. In *Illumination*, 217–251. New York: Schocken Books.

———. 1978. Surrealism. In *Reflections: Essays, Aphorisms, Autobiographical Writings*, trans. Edmund Jephcott, 177–192, New York: Schocken.

Bergson, Henri. 1919. *Time and Free Will*, trans. F. L. Pogson. London: George Allen and Unwin. First published in French in 1889.

———. 1980. *Introduction to Metaphysics*. Indianapolis: Bobbs-Merrill.

Berkeley, George. 1871. Siris: A chain of philosophical reflexions and inquiries, etc. In *The Works of George Berkeley, D.D. Volume II*, ed. Alexander Campbell Fraser, 365–508. Oxford: Clarendon Press.

———. 1962. *The Principles of Human Knowledge: With Other Writings*. London: Collins. First published in 1710.

Bernstein, Richard J. 1983. *Beyond Objectivism and Relativism*. Oxford: Basil Blackwell.

———. 1991. *The New Constellation: The Ethical-Political Horizon of Modernity/Postmodernity*. Cambridge: Polity.

Bijl, Aart. 1989. *Computer Discipline and Design Practice: Shaping Our Future*. Edinburgh: Edinburgh University Press.

———. 1995. *Ourselves and Computers: Difference in Minds and Machines*. Basingstoke, Hampshire: Macmillan.

Bloch, Ernst. 1986. *The Principle of Hope*, trans. Neville Plaice, Stephen Plaice, and Paul Knight. Oxford: Basil Blackwell.

Bohm, David, and F. David Peat. 1989. *Science, Order and Creativity*. London: Routledge and Kegan Paul.

Bohr, Niels. 1985. Light and life. In *Niels Bohr: A Centenary Volume*, ed. A. French and P. Kennedy, 311–324. Cambridge, Mass.: Harvard University Press.

———. 1985. The Bohr-Einstein dialogue. In *Niels Bohr: A Centenary Volume*, ed. A. French and P. Kennedy, 121–140. Cambridge, Mass.: Harvard University Press.

Bonnel, Roland. 1994. Medieval nostalgia in France, 1750–1789: The Gothic imaginary at the end of the old regime. In *Medievalism in Europe*, ed. Leslie J. Workman, 139–163. Cambridge: D. S. Brewer.

Borgman, Albert. 1984. *Technology and the Character of Contemporary Life: A Philosophical Inquiry*. Chicago, Ill.: University of Chicago Press.

Bowie, Malcolm. 1991. *Lacan*. London: Fontana.

Boyer, M. Christine. 1996. *Cybercities: Visual Perception in the Age of Electronic Communication*. New York: Princeton Architectural Press.

Brande, David. 1996. The business of cyberpunk: Symbolic economy and ideology in William Gibson. In *Virtual Realities and Their Discontents*, ed. Robert Markley, 79–106. Baltimore: Johns Hopkins University Press.

Breton, André. 1960. *Nadja*, trans. Richard Howard. New York: Grove Press. First published in French in 1928.

———. 1972. *Manifestoes of Surrealism*, trans. Richard Seaver and Helen R. Lane. Ann Arbor: University of Michigan Press. First published in 1929 and in 1953, respectively.

———. 1978. *What Is Surrealism? Selected Writings*, ed. Franklin Rosemont. London: Pluto Press.

———. 1990. *Communicating Vessels*, trans. Mary Ann Caws and Geoffrey T. Harris. Lincoln: University of Nebraska Press.

Buck-Morss, Susan. 1990. *The Dialectics of Seeing: Walter Benjamin and the Arcades Project*. Cambridge: MIT Press.

Burke, Edmund. 1958. *A Philosophical Enquiry into the Origin of Our Ideas of the Sublime and Beautiful*. London: Routledge and Kegan Paul. First published in 1757.

Bush, Vannevar. 1945. As we may think, *The Atlantic Monthly* (July). Reprinted at http://www.isg.sfu.ca/~duchier/misc/vbush/vbush-all.shtml

Caillois, Roger 1984. Mimicry and legendary psychaesthenia, *October* 31 (Winter, 1984): 17–32.

Calvino, Italo 1972. *Invisible Cities*, trans. William Weaver. New York: Harcourt Brace Jovanovich.

Caputo, John. 1987. *Radical Hermeneutics: Repetition, Deconstruction and the Hermeneutic Project*. Bloomington: Indiana University Press.

Carey, James W. 1989. *Communication as Culture: Essays on Media and Society.* London: Routledge and Kegan Paul.

Carnap, Rudolf. 1967. *The Logical Structure of the World: Pseudoproblems in Philosophy*, trans. Rolf A. George. London: Routledge and Kegan Paul.

Carroll, Lewis. 1962. *Alice's Adventures in Wonderland and Through the Looking Glass*, London: Puffin. First published in 1865 and 1872.

Casey, Edward S., and J. Melvin Woody. 1983. Hegel, Heidegger, Lacan: The dialectic of desire. In *Interpreting Lacan*, ed. Joseph H. Smith and William Kerrigan, 75–112. New Haven, Conn.: Yale University Press.

Castells, Manuel. 1989. *The Informational City: Information Technology, Economic Restructuring, and the Urban-Regional Process*. Oxford: Basil Blackwell.

Caws, Mary Ann. 1990. Introduction to *Communicating Vessels*, André Breton, trans. Mary Ann Caws and Geoffrey T. Harris, ix–xxiv. Lincoln: University of Nebraska Press.

Chaplin, Sarah. 1995. Cyberspace: lingering on the threshold (architecture, post-modernism and difference). *In Architectural Design Profile No. 118: Architects in Cyberspace*, 32–35. London: Academy Edition.

Chénieux-Gendron, Jacqueline. 1990. *Surrealism*, trans. Vivian Folkenflik. New York: Columbia University Press.

Chesher, Chris. 1994. Computers and the power of invocation. *Working paper.* Sydney: University of Macquarrie.

Clark, Andy. 1997. *Being There: Putting Brain, Body, and World Together Again.* Cambridge: MIT Press.

Clarke, Arthur C. 1990. *2001: A Space Odyssey*. London: Legend.

Clynes, Manfred E. [1970] 1995. Cyborg II: Sentic space travel. In *The Cyborg Handbook*, ed. Chris H. Gray, 35–42. New York: Routledge.

Clynes, Manfred E., and Nathan S. Kline. 1995. Cyborgs and space. In *The Cyborg Handbook*, ed. Chris H. Gray, 29–33. New York: Routledge. From an article published in 1960.

Cohen, Daniel I. A. 1986. *Introduction to Computer Theory.* New York: John Wiley.

Commager, H. S. 1978. *The Empire of Reason: How Europe Imagined and America Realized the Enlightenment.* London: Weidenfeld and Nicholson.

Comte, August. 1974. *The Essential Comte*, ed. Stanislav Andreski, trans. Margaret Clarke. London: Croom Helm. From *Cours de Philosophie Positive,* first published in 1830–1842.

Coomaraswamy, A. K. 1977a. Imitation, expression, and participation. In *Coomaraswamy: 1: Selected Papers: Traditional Art and Symbolism,* ed. Roger Lipsey, 276–285. Princeton, N.J.: Princeton University Press.

———. 1977b. Literary symbolism. In *Coomaraswamy: 1: Selected Papers: Traditional Art and Symbolism*, ed. Roger Lipsey, 323–330. Princeton, N.J.: Princeton University Press.

———. 1977c. The symbolism of the dome. In *Coomaraswamy: 1: Selected Papers: Traditional Art and Symbolism*, ed. Roger Lipsey, 415-464. Princeton, N.J.: Princeton University Press.

Coyne, Richard D. 1988. *Logic Models of Design*. London: Pitman.

———. 1995. *Designing Information Technology in the Postmodern Age: From Method to Metaphor*. Cambridge, Mass.: MIT Press.

———. 1997. Language, space and information. In *Intelligent Environments: Spatial Aspects of the Information Revolution*, ed. Peter Droege, 495–516. Amsterdam: Elsevier.

Coyne, Richard D., M. A. Rosenman, A. D. Radford, M. Balachandran, and J. S. Gero. 1990. *Knowledge-Based Design Systems*. Reading, Mass.: Addison Wesley.

Coyne, Richard D., Fay Sudweeks, and David Haynes. 1996. Who needs the Internet? Computer-mediated communication in design firms, *Environment and Planning B: Planning and Design* 23: 749–770.

Coyne, Richard D., Sally McLaughlin, and Sidney Newton. 1996. Information technology and praxis: A survey of computers in design practice, *Environment and Planning B: Planning and Design* 23: 515–551.

Crawford, Alan. 1997. Ideas and objects: The Arts and Crafts Movement in Britain, *Design Issues* 13 (1): 15–26.

Csicsery-Ronay, Istvan. 1991. Cyberpunk and neuromanticism. In *Storming the Reality Studio: A Casebook of Cyberpunk and Postmodern Science Fiction*, ed. Larry McCaffery, 182–193. Durham, N.C.: Duke University Press.

Cunningham, A. and Jardine, N., eds. 1990. *Romanticism and the Sciences.* Cambridge: Cambridge University Press.

Dante Alighieri. 1985. *The Divine Comedy, Vol. II: Purgatory*, trans. Mark Musa. New York: Penguin. Commenced about 1308.

Dator, James A. 1979. The futures of culture or cultures of the future. In *Perspectives on Cross-Cultural Psychology*, ed. Anthony J. Marsella, Roland G. Tharp, and Thomas J. Ciborowski, 369–388, New York: Academic Press.

Davies, Paul. 1985. *Superforce: The Search for the Grand Unified Theory of Nature*. London: Unwin.

———. Introduction to *Physics and Philosophy: The Revolution in Modern Science*, Werner Heisenberg. London: Penguin.

Dawkins, Richard. 1986. *The Blind Watchmaker*. London: Penguin.

Deleuze, Gilles. 1991. *Bergsonism*, trans. Hugh Tomlinson and Barbara Habberjam. New York: Zone Books.

———. 1993. *The Deleuze Reader*, ed. Constantin V. Boundas. New York: Columbia University Press.

———. 1994. *Difference and Repetition*, trans. Paul Patton. London: Athlone Press.

Deleuze, Gilles, and Félix Guattari, 1984. *Anti-Oedipus: Capitalism and Schizophrenia*, trans. Robert Hurley, Mark Seem, and Helen R. Lane. London: Athlone Press.

Dennett, Daniel C. 1981. Where am I? In *The Mind's I: Fantasies and Reflections on Self and Soul*, ed. Douglas R. Hofstadter and Daniel C. Dennett, 217–231. Harmondsworth, Middlesex: Penguin.

———. 1991. *Consciousness Explained*. London: Penguin.

Derrida, Jacques. 1976. *Of Grammatology*, trans. Gayatri Chakravorty Spivak. Baltimore: Johns Hopkins University Press.

———. 1978a. Coming into one's own. In *Psychoanalysis and the Question of the Text*, ed. Geoffrey H. Hartman, 114–148. Baltimore: Johns Hopkins University Press.

———. 1978b. Freud and the scene of writing. In *Writing and Difference*, trans. Alan Bass, 196-231, Chicago, Ill: Chicago University Press.

———. 1979. *The Postcard: From Socrates to Freud and Beyond*, trans. Alan Bass. Chicago: University of Chicago Press.

———. 1988. *Limited Inc.*, trans. Samuel Weber, ed. G. Graff. Evanston, Ill.: Northwestern University Press.

Dery, Mark. 1996. *Escape Velocity: Cyberculture at the End of the Century*. London: Hodder and Stoughton.

Descartes, René. 1964. Space and matter. In *Problems of Space and Time*. ed. J. J. C. Smart, 73–80. New York: Macmillan. From Descartes's *Principles of Philosophy*, first published in 1644.

———. 1968. *Discourse on Method and the Meditations*, trans. F. E. Sutcliffe. Harmondsworth, Middlesex: Penguin.

Dibbell, Jullian. 1996. A rape in cyberspace. In *Internet Dreams: Archetypes, Myths, and Metaphors*, ed. Mark Stefik, 293–315. Cambridge, Mass.: MIT Press.

Domenico, Pietropaolo. 1993. Eco on medievalism. In *Medievalism in Europe*, ed. Leslie J. Workman. Cambridge: D. S. Brewer, 127–138.

Dovey, Kimberly. 1993. Putting geometry in its place: Towards a phenomenology of the design process. In *Dwelling, Seeing and Designing: Toward a Phenomenological Ecology*, ed. David Seamon, 247–269, New York: SUNY Press.

Dreyfus, Hubert L. 1972. *What Computers Can't Do: The Limits of Artificial Intelligence.* New York: Harper and Row.

———. 1990. *Being-in-the-World: A Commentary on Heidegger's* Being and Time, *Division I*. Cambridge, Mass.: MIT Press.

Dunlop, C., and R. Kling, eds. 1991. *Computerization and Controversy: Value Conflicts and Social Choices*, Boston, Mass.: Academic Press.

Dyer, Richard. 1993. Entertainment and utopia. In *The Cultural Studies Reader*, ed. Simon During, 271–283. London: Routledge.

Eckhart, Meister. 1994. *Meister Eckhart: Selected Treatises and Sermons*, trans. James M. Clark and John Skinner. London: Harper Collins. First published in Latin and German in 1302–1329.

Eco, Umberto. 1986. *Travels in Hyperreality*, trans. William Weaver. San Diego, Calif.: Harcourt Brace Jovanovich.

Einstein, Albert. 1993. *Relativity: The Special and General Theory*, trans. Robert W. Lawson. London: Routledge. First published in 1916.

Eisenman, Peter, and R. Krauss. 1987. *Peter Eisenman: House of Cards.* New York: Oxford University Press.

Eliade, Mercea. 1965. *The Two and the One*, trans. J. M. Cohen. London: Harvill Press.

Ermarth, Elizabeth Deeds. 1992. *Sequel to History: Postmodernism and the Crisis of Representational Time.* Princeton: Princeton University Press.

Esslin, Martin. 1961. *The Theatre of the Absurd.* London: Eyre and Spottiswood.

Evans, Dylan. 1996. *An Introductory Dictionary of Lacanian Psychoanalysis.* London: Routledge.

Feigenbaum, Edward A., and Pamela McCorduck. 1983. *Fifth Generation: Artificial Intelligence and Japan's Computer Challenge to the World.* Reading, Mass.: Addison-Wesley.

Feyerabend, Paul K. 1981. *Realism, Rationalism and Scientific Method: Philosophical Papers Volume 1.* Cambridge: Cambridge University Press.

Fine, Arthur. 1986. *The Shaky Game: Einstein, Realism and the Quantum Theory.* Chicago: University of Chicago Press.

Fish, Stanley. 1989. *Doing What Comes Naturally: Change, Rhetoric, and the Practice of Theory in Literary and Legal Studies.* Durham, N. C.: Duke University Press.

Forester, T., ed. 1989. *Computers in the Human Context: Information, Technology, Productivity and People.* Oxford: Blackwell.

Forrester, John. 1990. *The Seductions of Psychoanalysis: Freud, Lacan and Derrida.* Cambridge: Cambridge University Press.

Foucault, Michel. 1970. *The Order of Things: An Archaeology of the Human Sciences.* New York: Vintage.

———. 1977. *Discipline and Punish: The Birth of the Prison.* London: Penguin.

———. 1984a. Preface to *Anti-Oedipus: Capitalism and Schizophrenia*, Gilles Deleuze and Félix Guattari, trans. Robert Hurley, Mark Seam, and Helen R. Lane, xi–xiv. London: Athlone Press.

———. 1984b. What is an author? In *The Foucault Reader*, ed. Paul Rabinow, 101–120. London: Penguin.

———. 1986. Of other spaces, *Diacritics* 16 (1): 23–27.

Fourier, Charles. 1996. *The Theory of the Four Movements*, ed. Gareth Stedman Jones and Ian Patterson. Cambridge: Cambridge University Press. First published in French in 1808.

Fox, C. J. 1983. *Information and Misinformation: An Investigation of the Notions of Information, Misinformation, Informing, and Misinforming.* Westport, Conn.: Greenwood Press.

Franck, Karen, A. 1995. When I enter virtual reality, what body will I leave behind? In *Architectural Design Profile No. 118: Architects in Cyberspace*, 20–23. London: Academy Edition.

Frazer, John H. 1995. The architectural relevance of cyberspace. In *Architectural Design Profile No. 118: Architects in Cyberspace*, 76–77. London: Academy Edition.

Freud, Sigmund. 1990a. Delusions and dreams in Jensen's *Gradiva*. In *The Penguin Freud Library, Volume 14: Art and Literature*, ed. Albert Dickson, 27–118. Harmondsworth, Middlesex: Penguin. First published in German in 1907.

———. 1990b. The 'uncanny.' In *The Penguin Freud Library, Volume 14: Art and Literature*, ed. Albert Dickson, 335–376. Harmondsworth, Middlesex: Penguin. First published in German in 1919.

———. 1991a. A disturbance of memory on the Acropolis. In *The Penguin Freud Library, Volume 11: On Metapsychology*, ed. Angela Richards, 443–456, Harmondsworth, Middlesex: Penguin. First published in German in 1936.

———. 1991b. Beyond the pleasure principle. In *The Penguin Freud Library, Volume 11: On Metapsychology*, ed. Angela Richards, 269–338. Harmondsworth, Middlesex: Penguin. First published in German in 1920.

———. 1991c. Civilization and its discontents, in *The Penguin Freud Library, Volume 12: Civilization, Society and Religion*, ed. Albert Dickson, 251–340. Harmondsworth, Middlesex: Penguin. First published in German in 1930.

———. 1991d. Infantile sexuality, in *The Penguin Freud Library, Volume 7: On Sexuality*, ed. Angela Richards, 88–126. Harmondsworth, Middlesex: Penguin. First published in German in 1905.

———. 1991e. On transformations of instinct as exemplified in anal eroticism. In *The Penguin Freud Library, Volume 7: On Sexuality*, ed. Angela Richards, 295–302. Harmondsworth, Middlesex: Penguin. First published in German in 1917.

———. 1991f. Repression, in *The Penguin Freud Library, Volume 11: On Metapsychology*, ed. Angela Richards, 145–158, Harmondsworth, Middlesex: Penguin. First published in German in 1915.

———. 1991g. The dissolution of the Oedipus complex. In *The Penguin Freud Library, Volume 7: On Sexuality*, ed. Angela Richards, 315–322. Harmondsworth. Middlesex: Penguin. First published in German in 1924.

———. 1991h. The ego and the id. In *The Penguin Freud Library, Volume 11: On Metapsychology*, ed. Angela Richards, 339–407. Harmondsworth, Middlesex: Penguin. First published in German in 1923.

———. 1991i. The unconscious. In *The Penguin Freud Library, Volume 11: On Metapsychology*, ed. Angela Richards, 167–222, Harmondsworth, Middlesex: Penguin. First published in German in 1915.

———. 1991j. Why war? In *The Penguin Freud Library, Volume 12: Civilization, Society and Religion*, ed. Albert Dickson, 343–362. Harmondsworth, Middlesex: Penguin. First published in German in 1932.

Freud, Sigmund, and Carl Jung. 1974. *The Freud/Jung Letters*, ed. William McGuire. London: Penguin.

Fromm, Erich. 1942. *The Fear of Freedom*. London: Routledge and Kegan Paul.

———. 1952. *The Forgotten Language: An Introduction to the Understanding of Dreams, Fairy Tales and Myths*. London: Victor Gollancz.

———. 1956. *The Sane Society*. London: Routledge and Kegan Paul.

Furst, Lillian R. 1969. *Romanticism in Perspective: A Comparative Study of Aspects of the Romantic Movements in England, France and Germany*. London: Macmillan.

Gadamer, Hans-Georg. 1975. *Truth and Method*. London: Sheed Ward.

Gallagher, Shaun. 1991. *Hermeneutics and Education*. Albany, N. Y.: SUNY Press.

Gates, Bill. 1996. *The Road Ahead*. London: Penguin.

Gell-Mann, Murray. 1994. *The Quark and the Jaguar: Adventures in the Simple and the Complex*. London: Abacus.

Gere, Charles. 1996. *The Computer as an Irrational Cabinet*. Unpublished Ph.D. thesis. London: Middlesex University.

Gibson, Andrew. 1996. *Towards a Postmodern Theory of Narrative*. Edinburgh: Edinburgh University Press.

Gibson, William. 1993. *Neuromancer*. London: Harper Collins.

Giddens, Anthony. 1984. *The Constitution of Society: Outline of the Theory of Structuration*. Cambridge: Polity.

———. 1992. *The Transformation of Intimacy: Sexuality, Love and Eroticism in Modern Societies*, Cambridge: Polity Press.

———. 1994. *Beyond Left and Right: The Future of Radical Politics*. Cambridge: Polity Press.

Gleick, James. 1987. *Chaos: Making of a New Science*. London: Cardinal.

Gödel, Kurt. 1962. *On Formally Unprovable Propositions*. New York: Basic Books.

Goffman, Erving. 1959. *The Presentation of Self in Everyday Life*. London: Penguin.

González, Jennifer. 1995. Envisioning cyborg bodies: Notes from current research. In *The Cyborg Handbook*, ed. Chris H. Gray, 267–279. New York: Routledge and Kegan Paul.

Goodman, Nelson. 1968. *Languages of Art*. Indianapolis: Bobbs-Merrill.

Gotlieb, C. C., and A. Borodin. 1973. *Social Issues in Computing*. London: Academic Press.

Gray, Chris Hables. 1995. An interview with Manfred Clynes. In *The Cyborg Handbook*, ed. Chris H. Gray, 43–53. New York: Routledge.

Gray, Chris Hables, Steven Mentor, and Heidi J. Figueroa-Sarriera. 1995. Cyborgology: Constructing the knowledge of cybernetic organisms. In *The Cyborg Handbook*, ed. Chris H. Gray, 1–14. New York: Routledge.

Gregory, Bruce. 1988. *Inventing Reality: Physics as Language*. New York: Wiley.

Grosz, Elizabeth. 1990. *Jacques Lacan: A Feminist Introduction*. London: Routledge.

Grusin, Richard. 1996. "What is an electronic author? Theory and the technological fallacy." In *Virtual Realities and Their Discontents*, ed. Robert Markley, 39-53. Baltimore: Johns Hopkins University Press.

Habermas, Jürgen. 1987a. *The Philosophical Discourse of Modernity: Twelve Lectures*, trans. F. G. Lawrence. Cambridge: Polity Press.

———. 1987b. *Theory of Communicative Action*, trans. T. McCarthy. Cambridge: Polity.

Hafner, Katie, and John Markoff. 1994. *Cyberpunk: Outlaws and Hackers on the Computer Frontier*. London: Corgi.

Hale, Constance, ed. 1996. *Wired Style: Principles of English Usage in the Digital Age*. San Francisco, Calif.: Hardwired.

Haraway, Donna J. 1991. *Simians, Cyborgs, and Women: The Reinvention of Nature*. London: FAb.

———. 1995. Cyborgs and symbionts: Living together in the new world order. In *The Cyborg Handbook*, ed. Chris H. Gray, xi–xx. New York: Routledge.

Hawkes, Terence. 1977. *Structuralism and Semiotics*. London: Methuen.

Hawking, Stephen W. 1988. *A Brief History of Time: From the Big Bang to Black Holes*. New York: Bantam.

———. 1993. *Black Holes and Baby Universes and Other Essays.* London: Bantam.

Hayle, N. Katherine. 1999. *How We Became Posthuman: Virtual Bodies in Cybernetics, Literature, and Informatics.* Chicago: University of Chicago Press.

Hegel, Georg W. F. 1969. *Hegel's Science of Logic*, trans. A. V. Miller. Atlantic Highlands, N.J.: Humanities Press International. First published in German in 1812.

Heidegger, Martin. 1962. *Being and Time*, trans. J. Macquarrie and E. Robinson. London: SCM Press.

———. 1971. The origin of the work of art. In *Poetry, Language, Thought*, Martin Heidegger, trans. A. Hofstadter, 15-87. New York: Harper and Rowe.

———. 1977. *The Question Concerning Technology and Other Essays*, trans. W. Lovitt. New York: Harper and Row.

Heim, Michael. 1991. The metaphysics of virtual reality. In *Virtual Reality: Theory, Practice and Promise*, ed. S. K. Helsel and J. P. Roth, 27–34. Westport, Conn.: Meckler.

———. 1993. *The Metaphysics of Virtual Reality.* New York: Oxford University Press.

———. 1998. *Virtual Realism.* New York: Oxford University Press.

Heisenberg, Werner. 1989. *Physics and Philosophy: The Revolution in Modern Science.* London: Penguin. First published in 1958.

Hesse, Mary. 1970. *Models and Analogies in Science.* Notre Dame, Ind.: University of Notre Dame Press.

———. 1987. Unfamiliar noises—tropical talk: The myth of the literal. *Journal of the Aristotelian Society* (July): 297–311.

Hickman, L. A. 1992. *John Dewey's Pragmatic Technology.* Bloomington, Ind.: Indiana University Press.

Hill, Glen. 1997. The architecture of circularity: Design, Heidegger and the earth. Unpublished Ph.D. Thesis. Sydney: University of Sydney.

Hiltz, Starr Roxanne and Murray Turoff. 1994. *The Network Nation: Human Communication via Computer.* Cambridge, Mass.: MIT Press. First published in 1978.

Hönnighausen, Lothar. 1988. *The Symbolist Tradition in English Literature: A Study of Pre-Raphaelitism and* Fin de Siécle, trans. Gisela Hönnighausen. Cambridge: Cambridge University Press.

Hooper-Greenhill, Eileen. 1992. *Museums and the Shaping of Knowledge.* London: Routledge.

Hospers, John. 1967. *An Introduction to Philosophical Analysis.* London: Routledge and Kegan Paul.

Houlgate, Stephen. 1991. *Freedom, Truth and History: An Introduction to Hegel's Philosophy.* London: Routledge.

Inayatulla, Sohail. 1990. "Deconstructing and reconstructing the future: Predictive, cultural and critical epistemologies," *Futures* (March): 115–141.

Irigaray, Luce. 1985. *This Sex Which Is Not One*, trans. Catherine Porter and Carolyn Burke. Ithaca, N.Y.: Cornell University Press.

Jakobson, Roman, and Morris Halle. 1956. *Fundamentals of Language*. The Hague: Mouton.

Jameson, Frederic. 1972. *The Prison-House of Language: A Critical Account of Structuralism and Russian Formalism*. Princeton, N.J.: Princeton University Press.

Jay, Martin. 1994. *Downcast Eyes: The Denigration of Vision in Twentieth Century Thought*. Berkeley: University of California Press.

Jencks, Charles. 1995. *The Architecture of the Jumping Universe: A Polemic: How Complexity Science Is Changing Architecture and Culture*. London: Academy Editions.

Joergensen, Joergen. 1970. The development of logical empiricism. In *Foundations of the Unity of Science: Toward an International Encyclopedia of Unified Science*, Vol. 2, Nos. 1–9, ed. Otto Neurath, Rudolf Carnap, and Charles Morris, 845–936. Chicago: University of Chicago Press.

Johnson, Barbara. 1978. The frame of reference: Poe, Lacan, Derrida. In *Psychoanalysis and the Question of the Text*, ed. Geoffrey H. Hartman, 149–171. Baltimore: Johns Hopkins University Press.

Johnson, Mark. 1987. *The Body in the Mind: The Bodily Basis of Meaning, Imagination, and Reason*. Chicago: University of Chicago Press.

Jung, Carl G. 1986. *Four Archetypes: Mother, Rebirth, Spirit, Trickster*, London: Ark. (First published in German in 1934–56.)

———. 1991. *Psychological Types*, trans. H. G. Baynes. London: Routledge. First published in 1921 in German.

Kant, Immanuel. 1960. *Observations on the Feeling of the Beautiful and Sublime*, trans. John T. Goldthwait. Berkeley: University of California Press. First published in 1764.

———. 1970. *Immanuel Kant's Critique of Pure Reason*, trans. Norman Kemp Smith. London: Macmillan. First published in 1781.

Kaplan, S., and R. Kaplan, 1982. *Cognition and Environment: Functioning in an Uncertain World*. New York: Praeger.

Kay, Alan. 1990. User interface: A personal view. In *The Art of Human-Computer Interface Design*, ed. B. Laurel, 191–207. Reading, Mass.: Addison-Wesley.

Kendrick, Michael. 1996. Cyberspace and the technological real. In *Virtual Realities and Their Discontents*, ed. Robert Markley, 143–160. Baltimore: Johns Hopkins University Press.

Kern, Stephen. 1983. *The Culture of Time and Space: 1880–1918*. Cambridge, Mass.: Harvard University Press.

Kitson, Frederick Lee, Tom Malzbender, and Vasudeve Bhaskaran. 1997. Opportunities for visual computing in healthcare. *IEEE Multimedia* 4 (2): 46–57.

Kling R., and S. Iacono. 1988. The mobilization of support for computerization: The role of computerization movements, *Social Problems* 35: 3.

Körner, Stephan. 1955. *Kant.* London: Penguin.

Kroker, A. 1984. Processed world: Technology and culture in the thought of Marshall McLuhan, *Philosophy of the Social Sciences* 14: 433–459.

Kroker, Arthur, and David Cook. 1991. Television and the triumph of culture (from *The Postmodern Scene*). In *Storming the Reality Studio: A Casebook of Cyberpunk and Postmodern Science Fiction*, ed. Larry McCaffery, 229–238. Durham, N.C.: Duke University Press.

Kuhn, Thomas. 1970. *The Structure of Scientific Revolutions.* Chicago: University of Chicago Press.

Lacan, Jacques. 1973. Seminar on "The purloined letter," *Yale French Studies* 48, Special issue: French Freud: 38–72.

———. 1977. *Écrits: A Selection*, trans. Alan Sheridan. London: Routledge and Keegan Paul. From work originally published in French in 1966.

———. 1979. *The Four Fundamental Concepts of Psychoanalysis*, trans. Alan Sheridan. London: Penguin.

Lacey, A. R. 1976. *A Dictionary of Philosophy.* London: Routledge and Kegan Paul.

Lakoff, George. 1987. *Women, Fire, and Dangerous Things: What Categories Reveal about the Mind*, Chicago: University of Chicago Press.

Lakoff, George, and Mark Johnson. 1980. *Metaphors We Live By.* Chicago: University of Chicago Press.

Lansdown, John. 1987. *Teach Yourself Computer Graphics.* London: Hodder and Stoughton.

Lansdown, John, and Simon Schofield. 1995. Expressive rendering: A review of nonphotorealistic techniques, *IEEE Computer Graphics & Applications* 15 (3): 29–37.

Latour, Bruno, and Steve Woolgar. 1986. *Laboratory Life: The Construction of Scientific Facts.* Princeton, N.J.: Princeton University Press.

Lawson, Bryan. 1994. *Design in Mind.* Oxford: Butterworth Architecture.

Leary, Timothy. 1991. The cyberpunk: the individual as reality pilot. In *Storming the Reality Studio: A Casebook of Cyberpunk and Postmodern Science Fiction*, ed. Larry McCaffery, 245–258. Durham, N. C.: Duke University Press.

Lefebvre, Henri. 1991. *The Production of Space*, trans. D. Nicholson-Smith. Oxford: Blackwell. First published in French in 1974.

Leibniz, Gottfried. 1964. The relational theory of space and time. In *Problems of Space and Time*, ed. J. J. C. Smart, 89–98. New York: Macmillan. From *The*

Leibniz-Clark Correspondence, ed. G. Alexander. Manchester: Manchester University Press, 1956.

Lemonick, Michael D. 1995. Future tech is now. *Time Australia* 28 (July 17): 44–73.

Lévi-Strauss, Claud. 1977. *Structural Anthropology.* trans. C. Jacobson and B. G. Schoepf. London: Penguin.

Levitas, Ruth. 1990. *The Concept of Utopia.* New York: Philip Allan.

Lévy, Pierre. 1998. *Becoming Virtual: Reality in the Digital Age*, trans. Robert Bononno. New York: Plenum.

Locke, John. 1960. *An Essay Concerning Human Understanding*, ed. A.D. Woozley. London: Collins. First published in 1698.

Long, R. J. 1987. *New Office Automation Technology: Human and Managerial Implications.* London: Croom Helm.

Lovejoy, Arthur O. 1960. *The Great Chain of Being: A Study of the History of an Idea.* New York: Harper and Row.

Luckhurst, Roger. 1996. (Touching On) Tele-Technology. In *Applying: To Derrida*, ed. John Brannigan, Ruth Robbins, and Julian Wolfreys, 171–183. Houndmills, Basingstoke, Hampshire: Macmillan.

Lynch, Kevin. 1960. *The Image of the City.* Cambridge, Mass.: MIT Press.

Lyotard, Jean-François. 1986. *The Postmodern Condition: A Report on Knowledge.* Manchester: Manchester University Press.

———. 1989. The sublime and the avant-garde. In *The Lyotard Reader*, trans. Andrew Benjamin, 196–211. Oxford: Basil Blackwell.

Madge, Pauline. 1997. Ecological design: A new critique, *Design Issues* 13 (2): 44–54.

Maffesoli, Michel. 1996. *The Time of the Tribes: The Decline of Individualism in Mass Society*, trans. Don Smith. London: Sage.

Malory, Thomas. 1969. *Le Morte d'Arthur*, ed. Janet Cowen. Harmondsworth, Middlesex: Penguin. First published in 1485.

Mandelbrot, Benoit. 1983. *The Fractal Geometry of Nature.* New York: Freeman.

March, Lionel, and Philip Steadman. 1971. *The Geometry of Environment: An Introduction to Spatial Organisation in Design.* London: RIBA.

Marcuse, Herbert. 1987. *Eros and Civilisation: A Philosophical Inquiry into Freud.* London: Routledge and Kegan Paul.

———. 1988. *One-Dimensional Man: Studies in the Ideology of Advanced Industrial Society.* London: Routledge.

Marías, Julián. 1967. *History of Philosophy*, trans. Stanley Applebaum and Clarence C. Stowbridge. New York: Dover.

Markley, Robert. 1996. Boundaries: Mathematics, alienation, and the metaphysics of cyberspace. In *Virtual Realities and Their Discontents*, ed. Robert Markley, 55–77. Baltimore: Johns Hopkins University Press.

Marx, Karl, and Friedrich Engels. 1959. Manifesto of the Communist Party. In *Marx and Engels: Basic Writings on Politics and Philosophy*, ed. Lewis S. Feuer, 43–82. London: Fontana. First published in German in 1848.

McCaffery, Larry. 1991. An interview with William Gibson. In *Storming the Reality Studio: A Casebook of Cyberpunk and Postmodern Science Fiction*, ed. Larry McCaffery, 263–285. Durham, N.C.: Duke University Press.

McLuhan, Marshall. 1962. *The Gutenberg Galaxy: The Making of Typographic Man.* Toronto: University of Toronto Press.

———. 1964. *Understanding Media: The Extensions of Man.* London: Routledge and Kegan Paul.

Meadows, Donella H., Nennis L. Meadows, Jørgen Randers, and William W. Behrens. 1972. *The Limits of Growth, a Report for the Club of Rome's Project on the Predicament of Mankind.* London: Potomac.

Merleau-Ponty, Maurice. 1968. *The Visible and the Invisible.* Evanston, Ill.: Northwestern University Press.

Meyrowitz, Joshua. 1985. *No Sense of Place: The Impact of Electronic Media on Social Behavior.* New York: Oxford University Press.

Mill, John Stuart. 1866. *August Comte and Positivism.* London: Trübner.

Milton, John. 1975. *Paradise Lost*, ed. Scott Elledge. New York: Norton. First published in 1667.

Milton, Kay, 1996. *Environmentalism and Cultural Theory: Exploring the Role of Anthropology in Environmental Discourse.* London: Routledge and Kegan Paul.

Minsky, Marvin. 1985. *The Society of Mind.* New York: Simon and Schuster.

Mitchell, William J. 1990. *The Logic of Architecture: Design, Computation, and Cognition.* Cambridge, Mass.: MIT Press.

———. 1995. *City of Bits: Space, Place and the Infobahn.* Cambridge, Mass.: MIT Press.

Moore, G. E. 1970. The refutation of idealism. In *Berkeley: Principles of Human Knowledge*, ed. Colin Murray Turbayne. 57-84. Indianapolis: Bobbs-Merrill. First published in 1903.

Moravec, Hans. 1988. *Mind Children: The Future of Robot and Human Intelligence.* Cambridge, Mass.: Harvard University Press.

More, Thomas. 1965. *Utopia*, trans. Paul Turner. Harmondsworth, Middlesex: Penguin.

Morris, William. 1970. *News From Nowhere*, ed. James Redmond. London: Routledge and Kegan Paul.

———. 1973. The society of the future. In *Political Writings of William Morris*, ed. A. L. Morton, 188–104. Berlin: Seven Seas Books.

Nagel, Thomas. 1986. *The View From Nowhere*. Oxford: Oxford University Press.

Negroponte, Nicholas. 1995. *Being Digital*. London: Hodder and Stoughton.

Newton, Isaac. 1964. Absolute space and time. In *Problems of Space and Time*, ed. J. J. C. Smart, 81–88. New York: Macmillan. From Newton's *Mathematical Principles of Natural Philosophy*, first published in 1687.

Nietzsche, Friedrich. 1961. *Thus Spoke Zarathustra*, trans. R. J. Hollingdale. London: Penguin. First published in German in 1892.

Norberg-Schulz, Christian. 1980. *Genius Loci: Towards a Phenomenology of Architecture*. London: Academy Edition.

Novak, Marcos. 1991. Liquid architectures in cyberspace. In *Cyberspace: First Steps*, ed. Michael Benedikt, 225–254. Cambridge, Mass.: MIT Press.

———. 1995. Transmitting architecture: transTerraFirma/TidsvagNoll v2.0, *Architectural Design Profile No. 118: Architects in Cyberspace*, 43–47. London: Academy Edition.

Ockham, William. 1957. *Philosophical Writings*, trans. Philotheus Boehner. Edinburgh: Nelson. (Writings from the fourteenth century.)

Ogden, C. K., and I. A. Richards. 1985. *The Meaning of Meaning: A Study of the Influence of Language upon Thought and the Science of Symbolism*. London: Ark. First published in 1923.

O'Leary, Stephen D. 1996. Cyberspace as sacred space: Communicating religion on computer networks, *Journal of the American Academy of Religion* 64 (4): 781–808.

Ong, Walter J. 1982. *Orality and Literacy: The Technologizing of the Word*. London: Routledge.

———. 1972. *Ramus: Method, and the Decay of Dialogue from the Art of Discourse to the Art of Reason*. New York: Octagon.

Orenstein, Gloria Feman. 1975. *The Theater of the Marvelous: Surrealism and the Contemporary Stage*. New York: New York University Press.

Orwell, George. 1984. *1984*. Oxford: Clarendon Press.

Otway, H. J., and M. Peltu, eds. 1983. *New Office Technology: Human and Organisational Aspects*. London: Frances Pinter.

Owen, Robert. 1969. *Report to the County of Lanark and A New View of Society*, ed. V. A. C. Gatrell. Harmondsworth, Middlesex: Penguin. First published 1813–1821.

Pantzar, Mika. 1997. Domestication of everyday life technology: Dynamic views on the social histories of artifacts, *Design Issues* 13 (3): 52–65.

Paulos, Eric, and John Canny. 1997. Ubiquitous tele-embodiment: Applications and implications. *International Journal of Human-Computer Studies* 46: 861–877.

Pearce, Martin. 1995. From urb to bit. In *Architectural Design Profile No. 118: Architects in Cyberspace* 7. London: Academy Edition.

Penrose, Roger. 1989. *The Emperor's New Mind: Concerning Computers, Minds, and the Laws of Physics*. London: Vintage.

Pepper, David. 1989. *The Roots of Modern Environmentalism*. London: Routledge.

Peterson, Aage. 1985. The philosophy of Niels Bohr. In *Niels Bohr: A Centenary Volume*, ed. A. French and P. Kennedy, 299–310. Cambridge, Mass.: Harvard University Press.

Piaget, Jean. 1970. *Structuralism*, trans. Chaninah Maschler. New York: Basic Books.

Plant, Sadie. 1998. *Zeros and Ones: Digital Women and the New Technoculture*. London: Fourth Estate.

Plato. 1941. *The Republic of Plato*, trans. Francis MacDonald Cornford. London: Oxford.

———. 1965. *Timaeus and Critias*, trans. Desmond Lee. London: Penguin.

Plattel, Martin G. 1972. *Utopian and Critical Thinking*. Pittsburgh, Penn.: Duquesne University Press.

Plotinus. 1948. *The Essence of Plotinus: Extracts from the Six Enneads and Porphyry's Life of Plotinus*, ed. G. H. Turnbull, trans. Stephen Mackenna. New York: Oxford University Press.

Poe, Edgar Allan. 1951. The purloined letter. In *Tales of Mystery and Imagination*, 454–471. London: Dent.

Popper, Karl. 1972. *Objective Knowledge: An Evolutionary Approach*. London: Oxford University Press.

———. 1991. *The Poverty of Historicism*. London: Routledge.

Porter, Tom. 1997. *The Architect's Eye: Visualisation and Depiction of Space in Architecture*. London: E & FN Spon.

Porush, David. 1996. Hacking the brainstem: Stephenson's *Snow Crash*. In *Virtual Realities and Their Discontents*, ed. Robert Markley, 107–141. Baltimore: Johns Hopkins University Press.

Poster, Mark. 1992. *The Mode of Information*. Cambridge: Polity Press.

———. 1995. Postmodern virtualities. In *Cyberspace, Cyberbodies, Cyberpunk: Cultures of Technological Embodiment*, ed. Mike Featherstone and Roger Burrows, 79–95. London: Sage.

Prusinkiewicz, Przemyslaw. 1995. Visual models of morphogenesis. In *Artificial Life: An Overview*, ed. Christopher G. Langton, 61–74. Cambridge, Mass.: MIT Press.

Punt, Michael. 1998. Accidental machines: The impact of popular participation in computer technology, *Design Issues*, 14 (1): 54–80.

Rapaport, David. 1958. The structure of psychoanalytic theory: A systematizing attempt. In *Psychology: A Study of a Science 3*, ed. S. Koch, 55–183. New York: McGraw Hill.

Reddy, Michael. 1979. The conduit metaphor—A case of frame conflict in our language about language. In *Metaphor and Thought*, ed. Andrew Ortony, 284–324. Cambridge: Cambridge University Press.

Reedy, W. Jay. 1994. Ideology and utopia in the medievalism of Louis de Bonald. In *Medievalism in Europe*, ed. Leslie J. Workman, 164–182. Boston: D. S. Brewer.

Reichenbach, Hans. 1964. Non-Euclidean spaces. In *Problems of Space and Time*, ed. J. J. C. Smart, 214–224. New York: Macmillan. From Reichenbach's. *The Philosophy of Space and Time*, trans. Maria Reichenbach and John Freud, first published in 1958.

Relph, Edward. 1985. Geographical experiences and being-in-the-world: The phenomenological origins of geography. In *Dwelling, Place and Environment: Towards a Phenomenology of Person and World*, ed. David Seamon and Robert Mugerauer, 15–31. Dordrecht, The Netherlands: Nijhoff.

Rheingold, Howard. 1991. *Virtual Reality*. London: Secker and Warburg.

———. 1993. *The Virtual Community: Homesteading on the Electronic Frontier*. Reading, Mass.: Addison Wesley.

Richens, Paul, and Simon Schofield. 1995. Interactive computer rendering, *ARQ: Architectural Research Quarterly* 1: 82–91.

Ricoeur, Paul. 1970. *Freud and Philosophy: An Essay on Interpretation*, trans. Denis Savage. New Haven, Conn.: Yale University Press.

———. 1977. *The Rule of Metaphor*, trans. Robert Czerny with Kathleen McLaughlin and John Costello. London: Routledge and Kegan Paul.

———. 1999. *Time and Narrative, Volume I*, trans. Kathleen McLaughlin and David Pellauer. Chicago: University of Chicago Press.

Robins, Kevin. 1995. Cyberspace and the world we live in. In *Cyberspace, Cyberbodies, Cyberpunk: Cultures of Technological Embodiment*, ed. Mike Featherstone and Roger Burrows, 135–155, London: Sage.

Rorty, Richard. 1980. *Philosophy and the Mirror of Nature*. Oxford: Basil Blackwell.

———. 1987. Unfamiliar noises: Hesse and Davidson on Metaphor, *Journal of the Aristotelian Society* (July): 283–296.

———. 1989. *Contingency, Irony and Solidarity*. Cambridge: Cambridge University Press.

Rousseau, Jean Jacques. 1968. *The Social Contract*, trans. Maurice Cranston, Harmondsworth: Penguin.

———. 1974. *Emile*, trans. Barbara Foxley, London: Dent.

Rushkoff, Douglas. 1995. *Cyberia: Life in the Trenches of Hyperspace*. London: Harper Collins.

Russell, Bertrand. 1980. *The Problems of Philosophy*. Oxford: Oxford University Press. First published in 1912.

Ryle, Gilbert. 1929. Review of Martin Heidegger's *Sein und Zeit, Mind* 38: 355–370.

Ryle, Gilbert. 1963. *The Concept of Mind.* Harmondsworth, Middlesex: Penguin.

Sacks, Warren. 1997. Artificial human nature, *Design Issues* 13 (1): 55–64.

Saint-Simon, Henri de. 1964. *Social Organization, The Science of Man and Other Writings*, trans. Felix Markham. New York: Harper and Row. First published in French in 1803–1825.

Sarup, Madan. 1992. *Jacques Lacan.* New York: Harvester Wheatsheaf.

Saussure, Ferdinand de. 1983. *Course in General Linguistics,* trans. Roy Harris. London: Duckworth. Originally published as *Cours de Linguistique Générale* in Paris in 1916.

Scarce, Ric. 1990. *Eco-Warriors: Understanding the Radical Environmental Movement.* Chicago: Noble Press.

Schank, Roger. 1990. What is AI anyway? In *The Foundations of Artificial Intelligence: A Source Book*, ed. D. Partridge and Y. Wilks, 3–13. Cambridge: Cambridge University Press.

Schilpp, P. A., ed. 1949. *Albert Einstein: Philosopher Scientist.* La Salle, Ill.: Open Court.

Schlegel, Frederick von. 1860. On the limits of the beautiful. In *The Aesthetic and Miscellaneous Works of Frederick Von Schlegel*, trans. E. J. Millington, 413–424. London: Henry G. Bohn. First published in German in 1794.

Schmitt, Ronald. 1993. Mythology and technology: The novels of William Gibson, *Extrapolation*, 34 (1): 64–78.

Schneiderman, Ben. 1994. Education by engagement and construction: a strategic education initiative for a multimedia renewal of American education. In *Sociomedia: Multimedia, Hypermedia, and the Social Construction of Knowledge*, ed. Edward Barrett, 13–26. Cambridge, Mass.: MIT Press.

Scholem, Gershom G. 1955. *Major Trends in Jewish Mysticism.* London: Thames and Hudson.

Searle, John, R. 1987. Minds, brains and programs. In *Artificial Intelligence: The Case Against*, ed. Rainer Born, 18–40. Beckenham, Kent: Croom Helm.

Shannon, Claud, and William Weaver. 1971. *The Mathematical Theory of Communication.* Urbana: University of Illinois Press.

Simon, Herbert. 1969. *The Sciences of the Artificial.* Cambridge, Mass.: MIT Press.

Snodgrass, Adrian B. 1990. *Architecture, Time and Eternity: Studies in the Stellar and Temporal Symbolism of Traditional Buildings, Volumes I and II.* New Delhi, India: Aditya Prakashan.

———. 1996. Can design assessment be objective?, *Architectural Theory Review* 1 (1): 30–47.

Snodgrass, Adrian B., and Richard D. Coyne. 1992. Models, metaphors and the hermeneutics of designing. *Design Issues* 9 (1): 56–74.

———. 1997. Is designing hermeneutical?, *Architectural Theory Review*, 2 (1): 65–97.

Sobchack, Vivian. 1995. Beating the meat/surviving the text, or how to get out of this century alive. In *Cyberspace, Cyberbodies, Cyberpunk: Cultures of Technological Embodiment*, ed. Mike Featherstone and Roger Burrows, 205–214. London: Sage.

Sokal, Alan, and Jean Bricmont. 1998. *Intellectual Impostures: Postmodern Philosophers' Abuse of Science*, 495–516. Amsterdam: Elsevier.

Solaiyappan, Meiyappan, Tim Poston, Pheng Ann Heng, Elliot R. McVeigh, Michael A. Guttman, and Elias A. Zerhouni. 1996. Interactive visualization for rapid noninvasive cardiac assessment, *IEEE Computer* 29 (1): 55–62.

Sophocles. [429–420 BC] 1947. King Oedipus, in *The Thebian Plays*, trans. E. F. Watling, 23–68. Harmondsworth, Middlesex: Penguin.

Spinosa, Charles, Fernando Flores, and Hubert L. Dreyfus. 1997. *Disclosing New Worlds: Entrepreneurship, Democratic Action, and the Cultivation of Solidarity*. Cambridge, Mass.: MIT Press.

Stallabrass, Julian. 1995. Empowering technology: The exploration of cyberspace, *New Left Review* 211 (May/June): 3–32.

Stallybrass, Peter, and Allon White. 1993. Bourgeois hysteria and the carnivalesque. In *The Cultural Studies Reader*, ed. Simon During, 368–381. London: Routledge.

Stefik, Mark (ed.) 1996. *Internet Dreams: Archetypes, Myths, and Metaphors*. Cam bridge, Mass.: MIT Press.

Steiner, Rudolph. 1994. *Theosophy: An Introduction to the Spiritual Processes in Human Life and in the Cosmos*, trans. Catherine E. Creeger. Hudson, NY: Anthroposophic Press. First published in 1910.

Stokes, Michael C. 1971. *One and Many in Presocratic Philosophy*. Cambridge, Mass.: Harvard University Press.

Stone, Allucquére Rosanne. 1995. *The War of Desire and Technology at the Close of the Mechanical Age*. Cambridge, Mass.: MIT Press.

Strachey, James. 1991. Sigmund Freud: A sketch of his life and ideas. In *The Penguin Freud Library, Volume 11: On Metapsychology*, ed. Angela Richards, 13–25. Harmondswoth, Middlesex: Penguin.

Straw, Will. 1993. Characterizing rock music culture: The case of heavy metal. In *The Cultural Studies Reader*, ed. Simon During, 285–292. London: Routledge and Kegan Paul.

Sullivan-Trainor, Michael. 1994. *Detour: The Truth About the Information Superhighway*. San Mateo, Calif.: IDG Books Worldwide.

Tabor, Philip. 1995. I am a videocam: the glamour of surveillance. In *Architectural Design Profile No. 118: Architects in Cyberspace*, 15–19. London: Academy Edition.

Tallis, Raymond. 1997. The shrink from hell: Book review of *Jacques Lacan* by Elizabeth Roudinesco, *The Times Higher Education Supplement* (October 31): 20.

Thomas, Frank, and Ollie Johnston. 1981. *Disney Animation: The Illusion of Life*. New York: Abbeville Press.

Tofler, Alvin. 1980. *The Third Wave*. London: Pan Books.

Tolkien, John Ronald Reuel. 1974. *Lord of the Rings*. London: Unwin Books.

Tomas, David. 1991. Old rituals for new space: Rites de passage and William Gibson's cultural model of cyberspace. In *Cyberspace: First Steps*, ed. Michael Benedikt, 31–47. Cambridge: MIT Press.

Toulmin, Stephen, and June Goodfield. 1968. *The Architecture of Matter*. Harmondsworth, Middlesex: Penguin.

Tucker, P. E. 1962. Chivalry in the Morte. In *Essays on Malory*, ed. J. A. W. Bennett, 64–103. Oxford: Clarendon Press.

Turing, Allan M. 1995. Computing machinery and intelligence. In *Computers and Thought*, ed. Edward A. Feigenbaum and Julian Feldman, 11–35. Cambridge, Mass.: MIT Press.

Turkle, Sherry. 1995. *Life on the Screen: Identity in the Age of the Internet*. London: Weidenfeld and Nicolson.

Turnbull, G. H. 1948. Appendix to *The Essence of Plotinus: Extracts from the Six Enneads and Porphyry's Life of Plotinus*, ed. G. H. Turnbull, trans. Stephen Mackenna, 223–276. New York: Oxford University Press.

Ulmer, Gregory L. 1987. The object of post-criticism, In *Postmodern Culture*, ed. Hal Foster, 83–110. London: Pluto.

Ulmer, Gregory. 1989. *Teletheory: Grammatology in the Age of Video*. London: Routledge.

Virilio, Paul. 1994. *The Vision Machine*. Bloomington: Indiana University Press.

———. 1995. "Red alert in cyberspace," *Radical Philosophy* 74: 2–4.

———. 1997. *Open Sky*, trans. Julie Rose. London: Verso.

Voller, Jack G. 1993. Cyberspace and the sublime, *Extrapolation* 34 (1): 18–29.

Wainwright, J., and A. Francis. 1984. *Office Automation, Organisation and the Nature of Work*. Aldershot, Hampshire: Gower.

Wallis, R. T. 1972. *Neoplatonism*. London: Duckworth.

Wann, John, and Mark Mon-Williams. 1996. What does virtual reality NEED?: Human factors issues in the design of three-dimensional computer environments, *International Journal of Human-Computer Studies* 44: 829–847.

Wasson, R. 1972. Marshall McLuhan and the politics of modernism, *Massachusetts Review* 13 (4): 567–580.

Weber, Max. 1992. *The Protestant Ethic and the Spirit of Capitalism*, trans. Talcott Parsons. London: Routledge and Kegan Paul.

Weinsheimer, J. C. 1985. *Gadamer's Hermeneutics: A Reading of Truth and Method.* New Haven, Conn.: Yale University Press.

Weiser, Mark. 1991. The computer for the 21st century. *Scientific American* 265 (3): 66–75.

Wells, H. G. 1994. *World Brain: H. G. Wells on the Future of World Education.* London: Adamantine Press. First published in 1938.

White, Hayden. 1978. *Tropics of Discourse: Essays in Cultural Criticism.* Baltimore: Johns Hopkins University Press.

Whitehead, Alfred North. 1926. *Science and the Modern World.* Cambridge: Cambridge University Press.

Whyte, Iain Boyd. 1993. The expressionist sublime. In *Expressionist Utopias: Paradise, Metropolis, Architectural Fantasy*, ed. Timothy O. Benson, 118–137. Los Angeles, Calif.: Los Angeles County Museum of Art.

Wiener, Norbert. 1950. *The Human Use of Human Beings: Cybernetics and Society.* Boston: Houghton Mifflin.

Wilden, Anthony. 1968. Lacan and the discourse of the other. In *The Language of the Self: The Function of Language in Psychoanalysis*, Jacques Lacan, trans. Anthony Wilden, 159–311. Baltimore: Johns Hopkins University Press.

Winch, Peter. 1988. *The Idea of a Social Science: and Its Relation to Philosophy.* London: Routledge. First published in 1958.

Winograd, Terry, and Fernando Flores. 1986. *Understanding Computers and Cognition: A New Foundation for Design.* Reading, Mass.: Addison Wesley.

Winterbourne, Anthony. 1988. *The Ideal and the Real: An Outline of Kant's Theory of Space, Time and Mathematical Construction.* Dordrecht: Kluwer Academic.

Wittgenstein, Ludwig. 1953. *Philosophical Investigations.* Oxford: Basil Blackwell.

Wolstenholme, Susan. 1997. Introduction to *The Wonderful Wizard of Oz.* In *The Wonderful Wizard of Oz*, Frank L. Baum ix–xliii. Oxford: Oxford University Press.

Yates, Frances A. 1966. *The Art of Memory.* London: Routledge and Kegan Paul.

Zangwill, O. L. 1987. Freud. In *The Oxford Companion to the Mind*, ed. Richard L. Gregory, 268–270. Oxford: Oxford University Press.

Žižek, Slavoj. 1989. *The Sublime Object of Ideology.* London: Verso.

———. 1991. *Looking Awry: An Introduction to Jacques Lacan through Popular Culture.* Cambridge, Mass: MIT Press.

———. 1992. *Enjoy Your Symptom! Jacques Lacan in Hollywood and Out.* New York: Routledge and Kegan Paul.

———. 1993. *Tarrying with the Negative: Kant, Hegel, and the Critique of Ideology.* Durham, N.C.: Duke University Press.

Index